An introduction to plant cell development

AN INTRODUCTION TO

PLANT CELL DEVELOPMENT

JEREMY BURGESS

Department of Cell Biology, John Innes Institute, Norwich

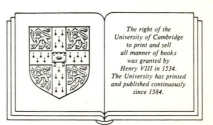

The right of the
University of Cambridge
to print and sell
all manner of books
was granted by
Henry VIII in 1534.
The University has printed
and published continuously
since 1584.

CAMBRIDGE UNIVERSITY PRESS

Cambridge

London New York New Rochelle

Melbourne Sydney

Published by the Press Syndicate of the University of Cambridge
The Pitt Building, Trumpington Street, Cambridge CB2 1RP
32 East 57th Street, New York, NY 10022, USA
10 Stamford Road, Oakleigh, Melbourne 3166, Australia

First published 1985

Printed in Great Britain at the University Press, Cambridge

Library of Congress catalogue card number: 84–19974

British Library Cataloguing in Publication Data

Burgess, Jeremy
An introduction to plant cell development

1. Plant cells and tissues
I. Title
581.87   QK725

ISBN 0 521 30273 0 hard covers
ISBN 0 521 31611 1 paperback

118938

CONTENTS

In every act of looking there is an expectation of meaning. This expectation should be distinguished from a desire for an explanation. The one who looks may explain afterwards; but, prior to any explanation, there is the expectation of what appearances themselves are about to reveal.
(John Berger, 1982)

PREFACE

The subject of this volume is the plant cell and the way in which as an individual it contributes to the survival of the whole plant. It is written from the standpoint of cell structure, with deliberate emphasis where possible on spatial considerations rather than on the fine details of biochemistry or molecular biology. I have adopted this stance because that is where my own interest in the subject lies. I think it is justified in view of the comparative rarity in undergraduate teaching for students to have direct experience of structural studies. I also feel that biology is entering a phase in which the importance of spatial events within cells is re-emerging in the light of the striking advances made in the areas of protein and nucleic acid biochemistry.

This volume is in two parts. The first four chapters describe the idealised undifferentiated cell and the ways in which the details of its structure are modulated to give rise to specialised functions. The second part of the volume is concerned with the ways in which cells interact and organise themselves into tissues. Consideration is given to possible mechanisms underlying such organisation. By the end of the volume, I hope the reader will emerge with a deeper understanding of the complexity of the cellular environment and, perhaps, with the wish to contribute personally to the further elucidation of that complexity.

Historically, science has been a demystifying influence on the human mind. It is tempting to conclude, faced with the rapid advance of knowledge in the areas of genetics and cell chemistry, that the final mystery of biology is imminently to be solved. The reader of this book will quickly realise that this is a view which is not held by its author. I have chosen deliberately to highlight outstanding areas of doubt and controversy. My intention by so doing has been to stimulate the reader to a greater curiosity. My success or failure in this aim will no doubt reflect the reader's temperament as well as my own.

I should like to thank those of my colleagues who have advised me on the scope and content of the book, and who provided illustrations. Specifically I thank Professor A. Bajer, Jim Dunwell, Basil Galatis, Professor B. Gunning, David Hanke, Graham Hus-

sey, Mike Jones, Clive Lloyd, Professor E. Schnepf and Martin Willison. My thanks go to Keith Roberts for allowing me access to some of his excellent drawings, and to Roger Turner for the use of scanning electron microscope facilities at the Food Research Institute. I thank Pat Phillips for her rapid and accurate typing of the manuscript, Peter Silver of Cambridge University Press for his encouragement, and Valerie Neal for her sympathetic sub-editing. As is customary, and quite correct, I conclude by saying that all remaining errors, omissions and personal emphases in the text are entirely my own responsibility.

July 1984   Jeremy Burgess

1

The plant cell

Introduction

The meaning of development

All plant bodies can be regarded as being comprised of fundamental compartments called cells. Many lower plants consist of only one cell, or of simple arrangements of relatively small numbers of cells. The cell population of a higher plant may amount to tens of millions. Each cell is a compartment capable of a limited series of functions which contribute to the survival and wellbeing of the plant as a whole. Prolonged growth involves the production of new cells by division of existing ones – the progeny of these divisions then proceed to grow and take on specialised functions corresponding to their position in the whole plant. A study of plant cell development seeks to describe and perhaps explain how this process of specialisation takes place.

The term 'development' encompasses three types of process. First, new cells are produced by division. In the higher plant, this occurs most commonly, although not exclusively, in regions called meristems. Next, there is a phase of growth or cell enlargement. Finally, the cells differentiate into their mature and specialised states. It is important to grasp from the outset that these three phases of development are not necessarily separated either in space or in time. Divisions may occur in cells which are actively enlarging, and in certain circumstances even in cells which would ordinarily be considered as mature and fully differentiated. Differentiation is usually regarded as the attainment of some final state of stability by a cell, coupled to a specific function (such as transport in the vascular system or photosynthesis in the mesophyll). However, the situation is not always final. For example, the aerial epidermis of a plant consists of cells of a special type which can be correctly regarded as differentiated. Nonetheless, some of these cells differentiate further to produce the stomata, whilst others may produce glandular hairs. The term 'differentiated' can therefore have a comparative meaning as well as an absolute one. In a simple linear differentiation sequence, such as the formation of a xylem element in a root, the final cell is clearly differentiated in an absolute sense; it is incapable of further change. Equally, however, all the cells in the lineage which ends with that xylem element are differentiated from each other, since they are at different stages on the pathway.

The importance of differentiation

The significance of differentiation can be expressed quite simply. Without this ability, living things would consist of single cells, or shapeless blobs of cells, or perhaps they might grow as multicellular organisms of simple geometric shape. Moreover, such organisms would be ill-equipped to withstand environmental variation or stress. They would be non-motile, living out a precarious existence in places where major changes in temperature and humidity never occurred. The enormous range of diversity in the shape, size and behaviour of living creatures could not have evolved without specialisation of their constituent cells. This is the importance of cell differentiation.

In plants, its gross manifestations are easy enough to see. The towering trunk of a redwood tree is clearly different from its root system, and it takes only an intelligent guess to realise that these differences are likely to be reflected in individual cells. However, even the humblest plants are capable of a degree of specialisation in their behaviour when this is to their advantage. Desiccation of a unicellular alga may elicit specialised wall formation, for example. The process of sexual reproduction involves considerable departures from the normal activities associated with vegetative growth. The general course of evolution amongst organisms is one of increasing ability to produce specialised cell types.

The way in which plant cells develop and differ-entiate into functioning populations is the subject of this book. The study can be made at several levels. It is axiomatic, for example, that cell development depends upon the regulated expression of the genetic material. There is considerable evidence to suggest that many highly differentiated cell types within a plant body retain all the genetic information required to specify the structure of an entire plant – they are totipotent. This fact implies that, within any given cell of the plant body, only part of the genetic information is being expressed at one time. At this level, development is concerned with the orderly and sequential expression of parts of the genetic material. This type of study emphasises strongly the temporal aspects of development.

Differentiation can also be expressed in terms of biochemical activities. The cells of the root cap, for example, display a different set of biochemical functions to those found in, say, photosynthetic mesophyll. Biochemical specialisation may be limited to quite small populations of cells: for example, pigment production might only occur in the petals; very specialised proteins are found in the pollen grain wall which contribute to incompatibility reactions ensuring cross pollination in many species. The biochemistry of the cell is mediated through enzymes whose structure is, of course, specified by genes. Thus biochemical differentiation is an extension of the ideas concerned with regulation of gene expression. It is concerned with changes in time but also, in many cases, chemical reactions may take place in localised areas of the cell, so that a spatial component is introduced.

The third level of study, and the one which forms the backbone of this book, is that of structure. Modern techniques of light and electron microscopy allow a very detailed description of the changes in cell structure which are part of development and differentiation. The structural approach to the study of development has one great advantage over other methods. It can define the changes which occur during development at the level of the individual cell, something which is simply not possible with current methods of biochemistry or physiology. The study of structure enforces an appreciation of the complexity of the cell and its behaviour, and also of the importance of interactions between cells. Structural studies are primarily concerned with changes in form and space; as such they represent a stimulating antidote to regarding the cell as merely a vehicle for the expression of genes, or as a bag of enzymes. Cells can be regarded as compartments within the organism. Each compartment contributes something to its neighbours and, in turn, is partly reliant upon them. Cells are themselves compartmentalised in various ways. The study of structure attempts to understand how this diversity is integrated in the whole plant.

The undifferentiated cell

Differentiation and development allow the plant to generate specialised cells whose activities contribute to its survival. The very words 'differentiation' and 'specialised' imply a process of change from some undifferentiated and unspecialised starting point. Is it possible to recognise such a starting point within the plant?

In the case of a plant which reproduces sexually, all its cells can be traced back to the original single cell from which they arose, the fertilised zygote. Compared to the zygote, all other cells in the plant might be said to be differentiated, in that their progeny will normally only produce a part of the plant, whereas the progeny of the zygote produce the whole plant. However, this is not a particularly useful or experimentally convenient concept. Multicellular plants in general retain throughout their life a series of centres of production of new cells which allow continued growth. Such centres are known as meristems. They occur in root tips, shoot tips, as an annular layer in the bark of stems in higher plants, and at the growing points of tissues of many lower plants. Compared to the zygote, these meristematic regions contain cells which are already specialised. The root tip meristem, for example, will under normal growth conditions only produce cells which are characteristic of roots. Nevertheless, the meristems are a persistent feature throughout the life of the plant and the cells which comprise them do not show any drift in structure or behaviour with age. Their speciality, if it can be so called, is to divide and donate progeny cells which subsequently embark on the changes associated with development. This pattern of continuing division at the meristem both defines its function and ensures its unspecialised character. The cells of different meristems will give rise to different mature structures; those of the root tip will produce the structures of the root, whereas those at the shoot tip will initially produce the stem and the leaves, but may

also eventually produce thorns, buds and flowers. It is precisely a definition of the meristematic cell that the future development of its progeny cannot be predicted from a study of the meristematic cell itself. For this reason, the standard of the 'undifferentiated' cell, the starting point for developmental change, will be taken as the meristematic cell.

The structure of the meristematic cell

The plant cell consists of a protoplast surrounded by a restraining wall. The protoplast is membrane-bound and contains a variety of subcellular structures (the organelles) which exist within a fluid environment (the cytoplasm). The development process involves changes to both the wall and the organelles of the cell, and these aspects will be examined in detail in later chapters. The meristematic cell is unspecialised but it contains in some form all the major elements which make up the structure of differentiated cells. For the purposes of examining these structural elements, we shall take as our model cells from the root tip of the maize plant, *Zea mays*. These represent as good an example as it is possible to find of a generalised plant cell (Fig. 1.1).

The cell wall

The cell wall of a meristematic cell is thin and, in the root tip, it gives the cell a roughly isodiametric outline. This is not always true – cells of the cambium, for example, are very long and thin. During growth, the cell wall expands so that the volume of the cell is increased. In mature parts of a linear organ such as a root, the preferred direction of this expansion follows the long axis of the organ. In the meristem however, expansion is more equal over the entire cell surface and continued cell divisions maintain the cell volume within quite narrow limits. The cell wall is an important indicator of the state of differentiation of a cell, as we shall see in Chapters 3 and 4.

In meristematic cells the wall retains its so called primary structure, without any of the specialised thickening or lignification which is found in many types of mature cells. Each cell of the meristem is surrounded by others with which it shares its wall. The shapes of the cells are therefore interdependent since they are in intimate contact with each other. The only exceptions to this are the external cells which abut the environment, and those parts of the cell wall of internal cells which face air spaces between cells. Although the structure of a meristem is not usually discernible as a geometric pattern, nonetheless the progeny of cells in different parts of the meristem will develop along different pathways. The importance of the wall in this process is to define the position of a particular cell, and also to mediate the exchange of materials between the neighbouring cells.

During growth, material is contributed to the primary wall from both sides, so that it retains a more or less symmetrical structure across the middle lamella. The middle lamella is chemically different from the rest of the wall since it represents a remnant of the first-formed structure after division, the cell plate (p. 34).

The cell wall defines the shape of the cell. The wall also confers integrity and rigidity to the tissue, and serves as a physical barrier to the movement of

Fig. 1.1. A section through two meristematic cells in the root tip of the maize plant (*Zea mays*). The main features of the meristematic state – a large nucleus, small vacuoles, proplastids and thin cell walls – are clearly demonstrated. Bar = 5 μm.

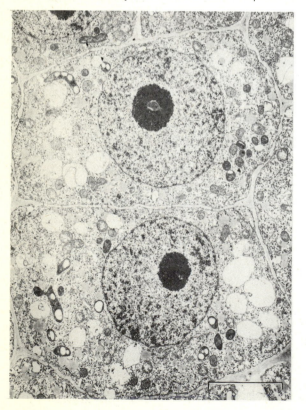

particulate material, invading micro-organisms, and so on. The cell wall may be regarded as the external medium in which the living protoplasts are made to grow. It represents a continuous and integrated extracellular space. This concept of the 'apoplasm' has important implications for development in terms of the movement of hormones and small effector molecules. Such substances may, in principle, move through the plant body without having to enter any particular cell.

The cell wall has been extensively studied for both its chemical and physical properties, its mode of synthesis and its structure. The form of the cell wall is central to many developmental processes, and for this reason it will be considered in detail in Chapter 3.

Plasmodesmata

The cell wall of our generalised meristematic cell is crossed by pores which are called plasmodesmata (Fig. 1.2). The diameter of the pores ordinarily varies from 30–60 nm, and is usually non-uniform. The pores are lined with the plasma membrane which forms the outer boundary of each protoplast. They therefore represent a series of direct connections between the cytoplasm of adjacent cells. Just as there is a unified extracellular space, the apoplasm, so too is there a continuous intracellular space, the symplasm. However, plasmodesmata are certainly not just simple tubes allowing free passage of materials between neighbouring cells.

Plasmodesmata are formed at the end of the mitotic cycle when fragments of endoplasmic reticulum become trapped between fusing vesicles which are contributing material to the cell plate (p. 34). The pore in the wall is therefore lined with the plasma membrane but blocked by a membraneous tubule derived from the endoplasmic reticulum. This structure is called the desmotubule and is up to 20 nm in diameter. Its presence means that the pore is closed to all the particulate contents of the cytoplasm. It also means that, in theory at least, communication between cells may take place by a series of different routes involving the plasmodesmata (Fig. 1.3). Transfer of materials might take place along the surface of the plasma membrane, which is continuous through the wall, between neighbouring cells. The annulus of cytoplasm in the plasmodesma might serve as a second channel for such transfer. The dimensions of this annulus are such that it is likely to act as a filter at the molecular level, offering comparatively little resistance to small organic molecules and solute ions, but much higher resistance to polymeric molecules such as proteins. The membrane of the desmotubule and its lumen are other conceivable pathways of communication. This last possibility is of particular interest, since, as we shall see (p. 10), the endoplasmic reticulum is in continuity with the nuclear envelope, and hence in principle a direct and enclosed route could exist for the transfer of materials between the nucleoplasm of adjacent cells mediated via the desmotubule. It has to be said that this type of transfer has not been demonstrated experimentally, and indeed some electron microscope pictures suggest that the desmotubule may be blocked. Nevertheless, the existence of plasmodesmata between the cells in a tissue has undoubted significance in the coordination of their activities.

Fig. 1.2. A section through a primary wall in the root tip of maize. Two plasmodesmata are visible, with their associated endoplasmic reticulum. The plasma membrane appears as two parallel lines. A single microtubule is also visible next to the plasma membrane. Bar = 0.1 μm.

Occurrence and significance of plasmodesmata

Attempts to compute or measure the degree of intercellular communication which is mediated by plasmodesmata often come to rest on assumptions about their structure which are as yet not entirely clear. In terms of the area of the wall which is perforated by plasmodesmata, it might be thought that their significance was slight – typically they occupy only 1% or so of the area of the wall. It has been calculated however that, at least in cells with thickening walls, the area of the plasma membrane which is enclosed within plasmodesmata might represent as much as half the total area of plasma membrane surrounding each cell. In these terms, the importance of plasmodesmata could be great.

On an evolutionary scale, plasmodesmata are always found in complex three dimensional tissues, but they may be absent in filamentous or two dimensional lower plants where each cell is in contact with the external medium. Where they do occur in simple systems, it is possible to demonstrate their importance experimentally. In growing filaments of fern gametophytes, for example, each cell is connected via plasmodesmata to its neighbour in the filament. Divi-

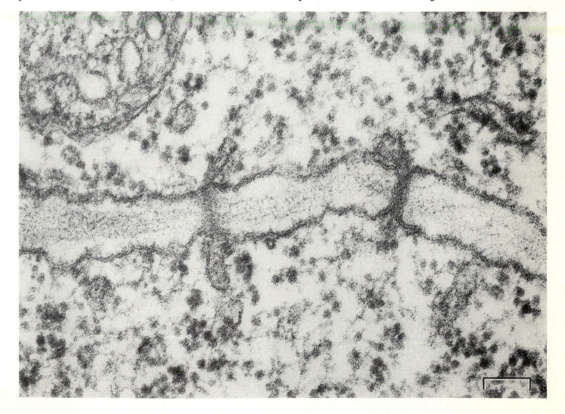

sions of the cell at the tip of the filament occur so as to increase its length. Eventually, two dimensional growth is initiated by a division in one of the cells of the filament at 90° to the axis of the filament. This event may be highly predictable (p. 189). However, if a filament is subjected to a temporary plasmolysis procedure, the intercellular connections will be broken. Following this, any or all of the cells in the filament may proceed to divide at right angles to the filament axis. This simple experiment points strongly to a role for the plasmodesmata in the control of normal development processes.

Plasmodesmata are not fixed in their structure throughout the life of the cell. They may disappear selectively as the cell grows, giving rise to fields of plasmodesmata in localised regions of mature walls. There is some evidence from chimaeras which shows that plasmodesmata may be formed at times other than during cytokinesis following mitosis. The structure of individual plasmodesmata can be modified. For example, virus particles are in general too large to pass through the normal plasmodesma, but they are sometimes seen inside the pore in the wall of infected plants. This implies that they are capable of causing a change in the structure of the plasmodesmata which results in effective enlargement of the pore size. In normal development, plasmodesmata in a modified form play an essential role in the long-distance transport of materials in the phloem. A progressive and organised series of changes in the structure of the plasmodesmata within the end walls of developing

Fig. 1.3. Diagrammatic representation of a plasmodesma. The plasma membrane is continuous through the pore in the wall. The desmotubule is connected on both sides to the endoplasmic reticulum.

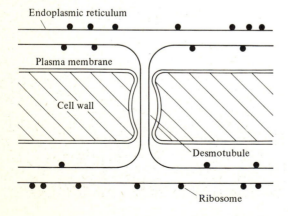

sieve elements results in the formation of the sieve plate. This will be described in Chapter 4 (p. 97). It is interesting to notice that this long-distance transport system is intracellular; the vascular system in animals is a modification of the extracellular space.

The plasma membrane

The structure which encompasses the cell within its restraining wall, and which mediates the exchange of materials between the intracellular and extracellular spaces is the plasma membrane. In electron micrographs it appears as a three-layered structure, consisting of two dark lines separated by a space (Figs. 1.2 and 1.25). The lines are 2.5–3.5 nm thick, and the space between them is about 3.5 nm wide, so that the whole membrane is about 10 nm across. The outermost line (the one nearest the wall) is sometimes more densely stained than the inner line, suggesting that the whole structure is not simply symmetrical.

The plasma membrane follows the contours of the wall surface exactly, since it is under pressure. If the cell wall were to be removed, the membrane-bound protoplast within would expand by osmosis until it burst. An osmotic pressure equivalent to about 0.3 M of an inert solute such as mannitol is required to restrain the intrinsic expansion of the protoplast. The plasma membrane serves as a highly selective barrier between the contents of the cell and its external environment.

The structure of the membrane is that of a lipid bilayer in which 'float' the proteins which carry out many of the activities of the membrane. The technique of freeze–fracturing enables the surface of the two leaflets which comprise the membrane to be examined. It reveals the presence of many particles embedded in an amorphous matrix material (Fig. 1.4). The particles are thought to represent the proteins, whilst the matrix is thought to be the lipid phase (Fig. 1.5). The present model of the membrane regards its structure as a 'fluid mosaic'; it is argued that the proteins are free to move within the surface of the lipid layers, and this carries the important implication that the membrane may not be uniform in composition or activity over its entire surface. Frequently local changes in plasma membrane properties appear as the result of developmental change. Such local differences are not detectable by the examination of thin sections, although clustering of particles within local regions of the membrane may be detected by freeze–fracture techniques. The fluid mosaic model of membrane structure is not limited in its application to the plasma membrane; it is a useful general description of membrane structure.

Chemical properties of the plasma membrane

The presence of the cell wall has been a considerable barrier to progress in research on the plant plasma membrane. In general, up to 50% of the membrane appears to be lipid in nature, comprising phospholipids, glycolipids, sterols and neutral lipids. 35–40% of the membrane is protein. The remainder is thought to be made up of carbohydrate materials in the form of glycolipids, glycoproteins and possibly a certain amount of nascent wall material. A detailed description of the enzymic properties of the plasma membrane is not yet possible. However, amongst the biochemical properties of the membrane which may have importance in development may be mentioned

the ability to bind a specific auxin transport inhibitor, N-1-naphthylphthalic acid. This implies that the membrane has a function related to hormone action (p. 202). In some cells, the plasma membrane is also thought to be the site of photoreceptor molecules; these are important in photomorphogenesis (p. 218).

As the boundary of the protoplast, the plasma membrane is obviously important for the control of growth processes. It is no doubt responsible for the maintenance of the intracellular environment in a state appropriate for the activities of the enzymes and organelles within. In order to achieve this, the plasma membrane must be able selectively to transport ions and small solutes across its width. The plasma membrane also has important functions in relation to the synthesis of the cell wall. The cellulose component of the wall is probably synthesised by enzymes which are an integral part of the plasma membrane (p. 77). This may occur at different rates in different localities on the cell surface. The other components of the wall are

Fig. 1.4. A pair of metal replicas which show the internal surfaces of the fractured plasma membrane in cultured cells of the white ash (*Fraxinus americana*). The left-hand replica is from the cytoplasmic side, and the right-hand replica is from the external (wall) side. Both surfaces show the presence of particles. The large circular areas in the pictures represent plasmodesmata. Bar = 0.1 μm. Pictures courtesy of Dr Martin Willison.

exported from the cell as the contents of membrane-bound vesicles. These vesicles must fuse with the plasma membrane in order to expel their contents. The existence of this type of secretion process means that new membrane material is continuously being added to the plasma membrane during cell growth. In order to maintain a balance in the amount of material present in the membrane, recycling mechanisms must exist to return excess materials back to the cytoplasm. The complexity of the membrane systems within the cytoplasm will be described below (p. 15). The plasma membrane thus emerges as a dynamic structure, non-uniform in its properties in space, and undergoing constant renewal and recycling in time.

The nucleus

Within the cytoplasm of the meristematic cell the most prominent structure is the nucleus (Fig. 1.1). It may occupy as much as 75% of the volume of the cell at this stage, although in maturer cells this proportion decreases markedly, due mainly to the great expansion in the volume of vacuoles. There is usually only one nucleus in a cell, and certainly this is always the case in cells of meristems. Certain developmental pathways may result in the elimination of the nucleus altogether, exemplified by the formation of xylem elements (p. 103). In other situations several, even hundreds of nuclei may co-exist in an uncompartmentalised cytoplasm. This is true of the entire

Fig. 1.5. Diagrammatic representation of a unit membrane such as the plasma membrane. Proteins (solid black) traverse the membrane and 'float' in a sea of lipid molecules which have their hydrophobic chains directed towards the central space.

Fig. 1.6. Nuclear pores in the root tip of maize (*Zea mays*). *a*, Appearance of the pores when the membrane is sectioned transversely. The detailed structure of the pore is obscure; bar = 0.2 μm. *b*, Appearance of the pores when the membrane is cut across its surface. The pores appear to be circular, composed of subunits, and may have a central dot. The pores exclude the chromatin from the nuclear envelope; bar = 0.2 μm.

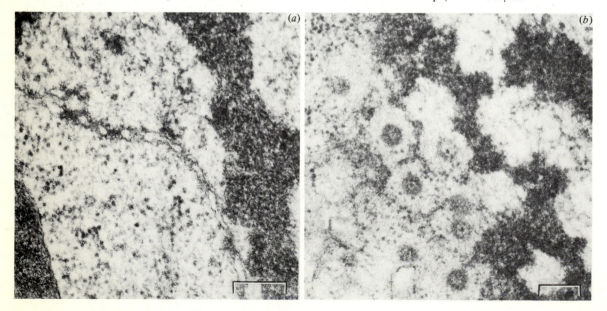

vegetative stage of a coenocytic alga such as *Caulerpa* (p. 187); the same situation may occur transiently in higher plants during embryo or endosperm development. However, it is convenient and generally true to regard a cell as a compartment with only one nucleus.

The nucleus is the site of most of the genetic material of the cell. Its activity in the encoding of protein structure necessarily involves a considerable amount of interchange of materials between the internal nucleoplasm and the cytoplasm outside the nucleus. Protein synthesis takes place in the cytoplasm, and is mediated by RNA molecules of various types which are produced within the nucleus. The DNA of the nucleus is associated with protein, and this protein is made in the cytoplasm. Thus efficient exchange of materials between nucleus and cytoplasm is essential for the continued function of both compartments. In developmental terms, the nucleus directs the type of proteins which are made, to correspond to the state of the cell. However, the cytoplasm is not a passive partner in this arrangement, since it can influence the activities of the nucleus. These aspects of development will be discussed in more detail in Chapter 8.

The nuclear envelope

The nucleus is bounded by a double membrane called the nuclear envelope. Each of the leaflets of this double membrane is a three-layered structure. The thickness of each leaflet is about 7.5 nm, i.e. it is slightly thinner than the plasma membrane. The leaflets are separated by a space, known as the perinuclear space, which is in the range 10–40 nm in width. The whole envelope is crossed in many places by pores. Nuclear pores consist of an opening caused by local fusion of the inner and outer leaflets of the envelope. In cross section they are often difficult to see, but in the surface view they are clearly more visible (Fig. 1.6).

The structure of the pore is complex and its details are to some extent disputed. The margin of the pore is octagonal in outline, and is associated on both sides with a circular annulus. The lumen of the pore is not simply an open hole, but may be filled with fibrous material, or occasionally blocked by an electron dense granule or particle. These features are described diagrammatically in Fig. 1.7. The area of the nuclear envelope which is crossed by pores may be considerable – up to 20%. A nucleus with a diameter of 6 μm might have as many as 10 000 pores. The giant vegetative nucleus of *Acetabularia* has over 2 000 000 pores. Very little is understood of the distribution of nuclear pores, although isolated observations suggest that it may be highly ordered. In germinating spores of *Equisetum*, for example, the pores have been shown to occur in rows. In water-stressed protoplasts of the moss *Physcomitrella*, the pores cluster at one pole of the nucleus. It is very likely that normal thin-sectioning techniques used by electron microscopists seriously underestimate the degree of ordering of the distribution of nuclear pores.

The presence of pores in the envelope surrounding an organelle which is known to exchange materials with the cytoplasm naturally raises the question of their role in this exchange. Although the pores appear to be reasonably 'open', electrical measurements have shown that the nuclear envelope has a resistance some three orders of magnitude higher

Fig. 1.7. Diagrammatic representation of a region of the nuclear envelope. The pores have eight-fold symmetry and may have a centrally located particle. The outer nuclear envelope carries ribosomes (open circles).

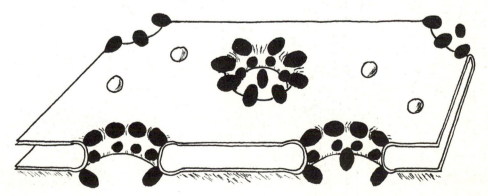

than would be expected if the pores allowed free passage of ions. More direct measurements of effective pore size have been attempted using markers such as colloidal gold or ferritin. These experiments, carried out with *Amoeba*, suggest that particles greater than about 10 nm in diameter are unable to pass through the nuclear pore, whilst those below about 9 nm can pass through relatively easily. In the natural state this would mean that cytoplasmic particles, including ribosomes, would be unable to penetrate the nuclear pore.

The outer leaflet of the nuclear envelope frequently carries ribosomes on its surface, and may sometimes be seen in direct continuity with the endoplasmic reticulum (Fig. 1.8). This raises the possibility that the perinuclear space of adjacent cells might be in some degree interconnected via the plasmodesmata (Fig. 1.9). The inner nuclear envelope does

not bear ribosomes, and there is evidence to suggest that it may form sites of attachment for the chromatin within the nucleus.

The nucleolus

The most obvious structure within the nucleus itself is the nucleolus. This is a dark-staining region which can often be resolved into granular and fibrillar parts (Fig. 1.10). The nucleolus may also contain areas with very light-staining properties, nucleolar vacuoles, and this is regarded as a sign of a highly active nucleolus. Undifferentiated cells such as those of meristems generally have larger nucleoli than do mature or dormant cells. A nucleus may have one nucleolus, or it may have several; numbers up to four are quite common (Fig. 1.11).

The major function of the nucleolus is the manufacture of the RNA for ribosome production in the cytoplasm. The cytoplasmic ribosome is a particle containing both RNA and protein, and its assembly involves a series of steps. The RNA components of the ribosome are synthesised in the nucleolus, then combined with proteins which have been synthesised in the cytoplasm and transported into the nucleus.

Fig. 1.8. Section through the nuclear envelope of the nucleus in a cell of the moss *Physcomitrella patens*. The endoplasmic reticulum is clearly connected to the outer nuclear envelope. Bar = 0.5 μm.

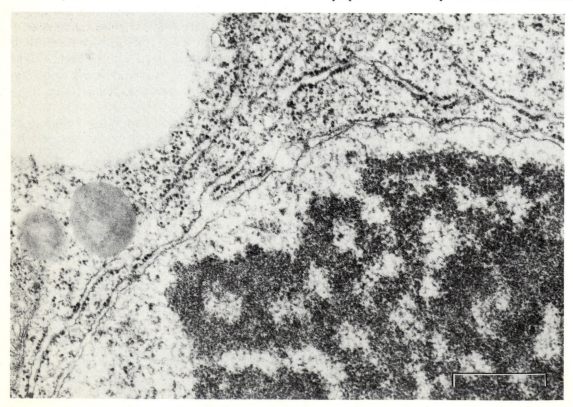

The RNA–protein complexes so formed are exported from the nucleus via the nuclear pores, and only when they reach the cytoplasm does final assembly of the complete ribosomal particle take place. This system uses the nuclear pores as a route for the movement of subribosomal particles and, indeed, it has been claimed that the granule which is occasionally seen to block the nuclear pore in electron micrographs represents such a ribosomal precursor. Experiments with radioactive precursors of RNA have shown that it is the fibrillar portion of the nucleolus which is the initially synthesised RNA. Label passes from this region to the granular part of the nucleolus. Quiescent nucleoli have almost no granular component and, taken together, these facts suggest that the granular component represents precursors of ribosomes awaiting export from the nucleus.

The ribosomal RNA is encoded by special regions of chromosomes called nucleolar organising regions. The nucleolus disappears at the start of mitosis, and reappears at telophase in the form of small nucleoli whose number corresponds to the number of chromosomes containing a nucleolar organising region. Subsequent growth and fusion of

these small initial nucleoli may produce just one large unit during the interphase period.

Chromatin and the organisation of the nucleoplasm

Aside from the nucleolus, the nucleoplasm at interphase is very variable in its appearance. Regions of dark-staining material occur apparently at random throughout the nucleoplasm (Figs. 1.1, 1.10, 1.11). The relationship between this image and the state of the DNA is only just becoming clear.

The DNA of a eukaryotic cell does not consist of a naked molecule, but is always found in combination with proteins. This complex is called chromatin. Chromatin is a word which was originally associated with light microscope studies of the nucleus, and the words 'heterochromatin' (dark-staining chromatin) and 'euchromatin' (light-staining chromatin) are likewise functional descriptions originally based on staining properties rather than biological activities.

The problem of organisation in the nucleoplasm is a very great one. A typical plant cell nucleus contains enough DNA to form a continuous fibril of between 1 and 10 metres in length (10^6–10^7 μm). This amount of DNA has to be fitted into an organelle which may be 10 μm in diameter or less, and yet retain its capacity for selective action and replication. Current opinion favours the view that the initial ordering of the DNA involves the formation of a 'beads on a string' structure, utilising the basic proteins of the nucleus, the histones. This initial folding of the DNA

Fig. 1.9. Diagram to illustrate a possible route of communication between adjacent cells. The nuclear envelope is part of the same membrane system as the endoplasmic reticulum, which is itself continuous through the plasmodesmata.

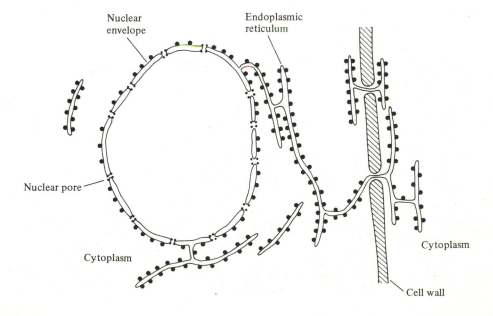

Nuclear envelope

Endoplasmic reticulum

Nuclear pore

Cytoplasm

Cytoplasm

Cell wall

produces a fibril of about 10 nm in diameter. Higher orders of folding of this fibril, again mediated by proteins, result in the necessary contraction ratio for the DNA to be fitted into the nucleus. The precise modelling of DNA structure has important implications for the control of gene expression, and discussion of this aspect of the problem will be deferred until Chapter 8.

The degree of condensation of the chromatin affects both its staining properties and its activity. Thus the heterochromatin of light microscopy is considered to be highly condensed and comparatively inactive in transcription, whereas the euchromatin is less condensed and more active. The physical organisation of the chromatin within the nucleus is poorly understood as yet. It appears likely that a major skeletal element may occur in the form of a lamina attached to the inside of the inner leaflet of the nuclear envelope. This nuclear lamina is a fibrous network composed of a small number of proteins. It is often not visible in micrographs, but its presence can be inferred by the fact that it is possible to strip the membrane from isolated nuclei without loss of the overall shape of the organelle. If the chromatin is then removed, a 'ghost' remains which consists of the fibrous lamina.

The lamina interacts with the inner leaflet and the nuclear pores on its outside surface, and with the chromatin on its inner surface. It is thought that DNA synthesis may take place at the points where chromosomes are attached to the lamina; the lamina also serves to clear a space around each nuclear pore so that it is not physically blocked by associated chromatin (Fig. 1.6). The nuclear space itself is also crossed by a network of protein material, the intranuclear matrix. All these observations point strongly to the fact that although the nucleus appears to be a random structure, it is in fact very far from being so. No doubt, future work will define the complexity of the nucleus in more detail.

The mitochondrion

All eukaryotic cells contain a nucleus, and they also all contain organelles of the next type to be considered – mitochondria. Mitochondria are essential for cellular respiration and the production of ATP from reduced organic molecules such as the sucrose produced by photosynthesis. They have a relatively constant structure in whichever cell they are found.

The mitochondrion consists of an outer membrane which encloses a space containing an inner membrane. The inner membrane is folded into deep

Fig. 1.10. Section through a nucleolus in an interphase cell of the maize root tip. The nucleolus has a central vacuole and consists of aggregated granular and fibrillar material. Bar = 1 μm.

Fig. 1.11. Section through a nucleus in the root tip of maize. Four nucleoli are visible in this interphase nucleus. Bar = 5 μm.

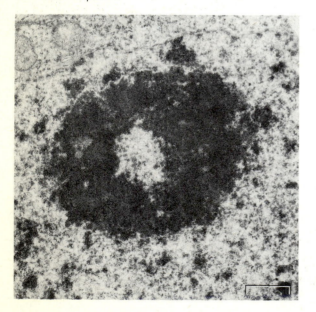

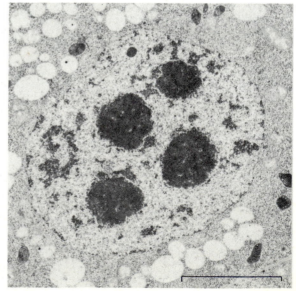

invaginations called cristae, and so defines two types of space within the mitochondrion – the matrix, or central region, and the inter-cisternal region. The membranes comprising the double envelope of the mitochondrion cannot normally be resolved into a tripartite structure in thin sections prepared for electron microscopy. In shape, plant mitochondria usually appear circular or elliptical in section, with dimensions of 1–3 μm in the short and long axis. In meristematic cells of the maize root, the cristae are in fact rather poorly developed (Fig. 1.12).

The inner and outer membranes of the mitochondrion are different chemically and function-ally. The outer membrane is characterised by a high lipid content, and is permeable to small solutes. The inner membrane is the site of the energy-producing chain of ubiquinone and the cytochromes, and it is likely that these molecules are present in highly organised arrangements in the membrane. The matrix of the mitochondrion contains the enzymes of the Krebs cycle, with the possible exception of suc-cinic dehydrogenase, which is membrane-bound. The formation of cristae from the inner membrane increases its surface area, and thus the rate at which energy production can occur. In cells which are undergoing rapid respiration, the presence of many cristae in the mitochondria is a common feature. The most striking examples of this modification are to be found in flight muscle; in plants, the *Arum* spadix is a tissue which shows the same effect.

Mitochondria contain DNA and ribosomes, and these are often both visible within the matrix (Fig.

Fig. 1.12. Section through part of the peripheral cytoplasm in a root tip cell of maize. Two mitochondria are visible, together with endoplasmic reticulum sectioned across its surface to show spiral patterns of ribosomes. Cortical microtubules are visible next to the cell wall. Bar = 0.5 μm.

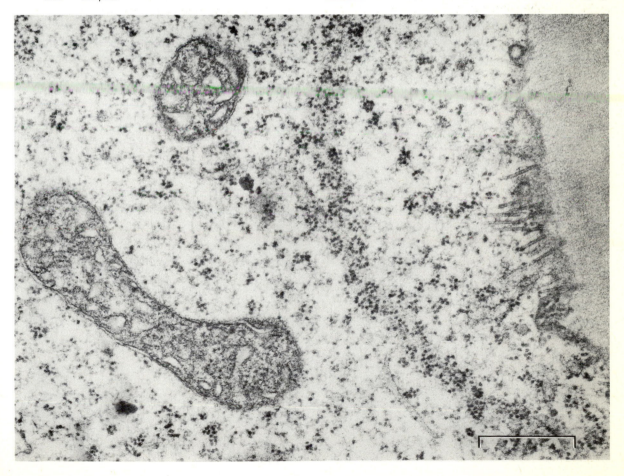

1.13). The mitochondrion is capable of making some of its own proteins. Mitochondria divide by a process of fission; it has long been speculated that they may have arisen on an evolutionary scale from invading prokaryotic organisms which were taken over by their host. This idea is made more attractive by the finding that mitochondrial ribosomes (which are smaller than cytoplasmic ribosomes) show the same pattern of sensitivity to inhibitors of protein synthesis as do ribosomes from bacteria. Thus, cycloheximide inhibits protein synthesis by cytoplasmic ribosomes, but not by isolated mitochondria or bacterial cells. Conversely, mitochondrial protein synthesis is inhibited by chloramphenicol, which has no effect on the 80S ribosomes of the cytoplasm. Despite their semiautonomous nature, it is clear that mitochondria are dependent upon cytoplasmic mechanisms of macromolecular synthesis for the maintenance of their function. Typically the molecular weight of the DNA in a mitochondrion is about 10^7 daltons. This is enough to code for 30 proteins with an average molecular weight of 20 000, and this is certainly inadequate for all the functions of the organelle.

Mitochondria are present within a cell as a consequence of their presence in the previous cell before it divided. This argument can be carried right back to their presence in the fertilised zygote. Earlier ideas that mitochondria might be formed *de novo* from the nuclear envelope have not been confirmed. The numbers of mitochondria within a cell vary considerably. The alga *Chlorella* for example has only one, although it is very branched and convoluted in shape. Higher plant cells usually have hundreds or thousands; the number depends on the cell type, and the state of the cell cycle. For example in the stem apex of willow it has been calculated that just before division cells contain about 120 mitochondria, so that the daughter cells immediately after division contain about 60 each. In the meristematic cells of the root cap of maize, about 200 mitochondria are present. This increases to 2000 in the mature central zone cells; this increase, however, matches the increase in cell volume, so that the effective 'concentration' of mitochondria stays roughly constant.

The overwhelming impression of the mitochondrion is that it is a remarkably constant and stable feature of cells. In cells of yeast it is possible to alter mitochondrial structure by placing the cells under anaerobic conditions. This sensitivity to the external environment is, however, not shown by mitochondria of higher plants (presumably because it is never required). It seems likely that respiratory activity within higher plant cells is regulated as much by pure numbers of mitochondria as it is by drastic developmental changes in the organelle. This stability and

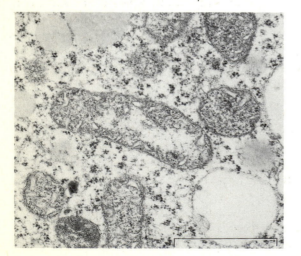

Fig. 1.13. A group of mitochondria within the root tip of maize. The central clear space of the mitochondrion in the middle of the picture is crossed by fine fibrils, corresponding to mitochondrial DNA. Bar = 1 μm.

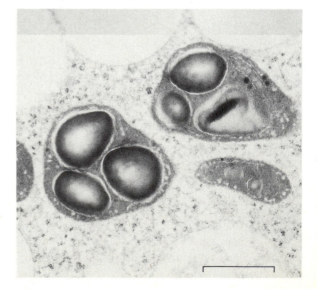

Fig. 1.14. A group of proplastids within the root tip of maize. Proplastids contain a few scattered membranes and may contain starch or protein deposits. Bar = 1 μm.

The proplastid

The proplastid is another cytoplasmic organelle bounded by a double envelope. Proplastids are the forerunners of a family of organelles, the plastids, which are characteristic of plants. The proplastid is the simplest member of the family and is the typical plastid found in our generalised undifferentiated cell. Specialised cells contain other types of plastid, and these will be considered in detail in Chapter 2.

The proplastid of the meristematic cell is a simple organelle, usually slightly larger than a mitochondrion, but with a much less well developed internal membrane system (Fig. 1.14). Proplastids contain DNA and ribosomes, and they may also accumulate small amounts of storage materials. These include starch, lipid droplets and phytoferritin, a protein which contains iron. The importance of the proplastid in the economy of the meristematic cell is uncertain, mainly due to the difficulty of isolating these organelles free from contamination.

Proplastids divide by a fission process and, like mitochondria, are partitioned between the daughter cells following mitosis, so that the plastids within any particular cell are derived from the cytoplasm of its parent cell. The plastids of the fertilised zygote are in many instances derived solely from the female parent plant. As a family of organelles, the plastids are particularly well studied and have considerable interest in terms of their developmental capacities. These aspects of plastid biology form the subject of Chapter 2.

The endomembrane system

The importance of membranes in the functioning of the plant cell should already be clear. Membranes delimit the protoplast itself, and they define the specialised compartments of biochemical activity which are known as the organelles. The organelles do not simply float in a formless and uniform cytoplasm however. The cytoplasm itself is filled with a variety of other membraneous elements which, because of their interrelationship, are known as the endomembrane system. The principal components of this system are the endoplasmic reticulum, the Golgi apparatus and the vacuoles.

constancy contrasts sharply with the next organelle type to be considered – the plastids.

The endoplasmic reticulum

The central pivot of the endomembrane system is the endoplasmic reticulum (ER). This is a network of paired membranes which permeates the cytoplasm as a series of flattened cisternae. The membranes of the ER therefore enclose a space which is separate from the cytoplasm. The ER is continuous with the nuclear envelope, and may pass through plasmodesmata in the form of the desmotubule. It therefore represents a potentially important pathway for the exchange of materials both between the nucleus and the cytoplasm, and between adjacent cells. It is best revealed by selective staining techniques (Fig. 1.15). In meristematic cells it appears to be more or less randomly disposed throughout the cytoplasm, with a tendency to form a peripheral layer under the plasma membrane.

The endoplasmic reticulum may bear ribosomes on both its cytoplasmic surfaces (rough ER), or on one only; it may also be devoid of ribosomes entirely (smooth ER). These different forms of the ER represent local modifications and should not be regarded as being due to two different membrane systems. Rough ER membranes are usually more or less parallel to one another, whereas smooth ER exists more commonly in the form of distended sheets or tubules. In the meristematic cell, rough ER predominates; in plants smooth ER is usually only found in quantity in specialised cells (p. 116). Rough ER is the site of much of the protein synthesis which takes place in the cell and, in favourable sections which cut across the surface of one of the membranes, it is possible to see that the ribosomes occur in spiral patterns (Fig. 1.12). These 'polysomes' are presumed to represent sites of active protein synthesis.

The importance of the endoplasmic reticulum in quantitative terms may be appreciated by the fact that it has been calculated that in liver cells each cubic centimetre of cell volume contains ER with a surface area of 1 m². The ER may increase in extent when protein synthesis needs to be greatly stimulated. The fate of the proteins synthesised on the ER is variable, and may be highly directed. For example, in developing cotyledons, storage protein is synthesised on the ER and transported into vacuoles. In protein-secreting glands, the product of the ER protein synthesis is removed from the cell altogether. In oil-secreting glands there is a marked increase in the amount of smooth ER, but not of rough ER. Finally, in nectaries (which are engaged in the secretion of sugars), the ER network is highly developed; the ER seems in this case not to be involved in a synthetic role at all, but rather with the selective transport of materials from the phloem (Chapter 4).

The ER is thus a complex network, probably more physically organised than is apparent from sections, which can engage in a number of different activities, both synthetic and transport related. The meristematic cell shows it in its least differentiated form.

The Golgi apparatus

The second part of the endomembrane system is the Golgi apparatus. The Golgi apparatus of a cell consists of a series of membraneous organelles called dictyosomes. Each dictyosome is a stack of flattened membrane sacs which are known as cisternae (Figs. 1.16, 1.17). A cisternal sac is a folded sheet of membrane, with no associated ribosomes on any surface. The diameter of each sac is usually of the order of 1–2 μm; at its edges it may show considerable branching into tubules, or activity in the production of vesicles. A dictyosome is a stack of these cisternal sacs, although it is possible for a dictyosome to consist of only one cisterna. In our generalised meristematic cells the dictyosomes usually consist of 4–6 cisternal sacs. In algal cells numbers as high as 20 or 30 are quite commonly found. There are usually several dictyosomes in each cell and, taken together, they comprise the Golgi apparatus of that cell. The number of dictyosomes in a Golgi apparatus varies from only one up to a quoted figure of 25 000 in a rapidly elongating pollen tube. The functions of the Golgi apparatus are very important for the membrane chemistry of the cell, and they also vary considerably during various types of developmental change.

The functional unit of the Golgi apparatus then

Fig. 1.15. Section through meristematic cells in the root tip of maize. Prolonged treatment of the material with osmic acid has resulted in preferential staining of the endoplasmic reticulum and nuclear envelope. The association of the endoplasmic reticulum with the nuclear envelope, and with the wall at positions of plasmodesmata, is shown clearly. Bar = 5 μm.

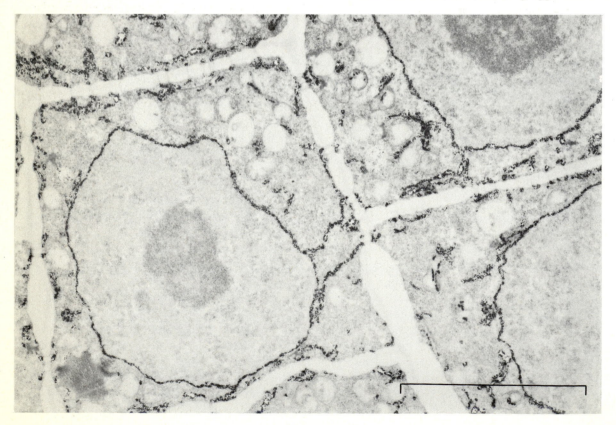

is the dictyosome, which consists of a stack of cisternae. The individual cisternae within a stack are separated by a distance of about 10 nm, and are held together by a bonding substance which can sometimes be seen as a series of fine fibrils. The cisternae are often curved, so that a dictyosome has a convex and a concave face. The membranes of each cisterna are not all the same, and it is obvious, even on morphological grounds alone, that the dictyosome is an important site of membrane modification.

The model of the dictyosome describes an organelle which has a temporal difference expressed across its structure (Fig. 1.17). At the forming face, the membranes comprising the initial cisterna are of low contrast and difficult to resolve into a three-layered structure. In this regard they resemble the membranes of the ER or the nuclear envelope. As the cisternal stack is crossed, the membranes become

more densely stained and more easily resolvable into three layers. They come to resemble the plasma membrane in appearance. This morphological description fits well with the known functions of the Golgi apparatus in cells. Dictyosomes are most prominent in cells which are growing rapidly or secreting an extracellular product. A well studied example of this is found in the outer cells of the root cap which secrete a mucopolysaccharide to lubricate the passage of the root tip through the soil. The dictyosomes in such cells are very highly developed, and the individual cisternae are filled with a stained material which resembles the excreted slime, and radioactive labelling experiments have confirmed this observation chemically (p. 108).

Such secretion products are packaged by the dictyosomes in the form of vesicles budded off from the maturing face of the cisternal stack. The membrane around the vesicle resembles the plasma membrane, and is able to fuse with it, expelling its contents to the extracellular space. Clearly, if membrane material is being continually lost to the dictyosome by the formation of vesicles, there must exist some mechanism for its replacement. This new membrane

Fig. 1.16. A dictyosome in a meristematic cell of the root tip of maize. The difference in membrane staining and thickness across the stack of cisterna is visible. The edges of each cisterna are inflated into membrane sacs. Bar = 0.2 μm.

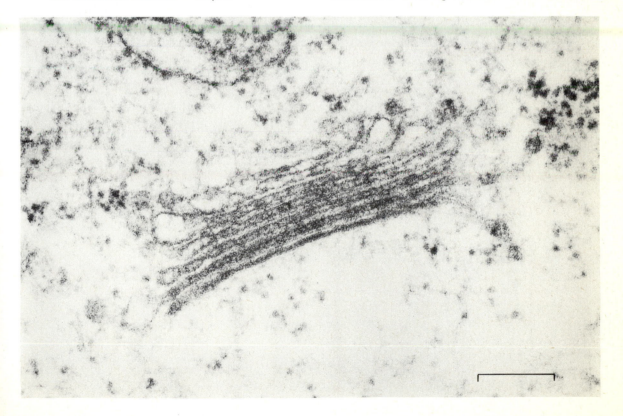

is added in the form of new cisternal stacks at the forming face of the dictyosome. The source of this new membrane is thought to be the ER or nuclear envelope, both because of the morphological resemblance between these membrane systems, and also because in some cells the dictyosomes are clearly associated with ER or the nucleus. In fact, this behaviour is not seen in meristematic cells of the maize root, but nonetheless, the concept of membrane flow through the dictyosome from the ER to the plasma membrane is generally applicable.

The Golgi apparatus in synthesis and secretion

Membrane modification is an important function of the Golgi apparatus, but it is not the only one which it has. Studies on isolated dictyosomes show that they contain a range of enzymes for the synthesis of wall components. In certain specialised areas, this activity can be visualised directly. The marine alga *Chrysochromulina chiton*, for example, has an extracellular matrix which is composed of sculptured scales. These scales are clearly recognisable by their shape. The single dictyosome of the cell of *Chrysochromulina* contains these scales in successively more mature form until, at the maturing face of the stack, the cisterna are distended and contain fully mature scales. The scales are produced entirely within the stack of

cisternae, showing its synthetic capabilities. Even more remarkable is the finding that in cells which have an extracellular matrix composed of two different types of scales, these can be synthesised in adjacent cisternae of a single dictyosome. This suggests that it is the cisterna which is the synthesising 'unit' in the Golgi apparatus, and not the dictyosome. It also suggests that individual cisternae move through the stack as integral units, and do not pass on their product to the next cisterna for further processing. However, this does not seem to be universal behaviour; in other cases it appears that the product of one cisterna may be transferred to the next, possibly by the intermediate formation of tubular or vesicular outgrowths of the edges of the cisternal membrane. The number of cisternae within a dictyosome can be varied experimentally; treatments which slow down the export of the product from the maturing face will lead to an increase in the number of cisternae within the stack.

Secretion of material through the plasma membrane carries a further implication for the cell. Often such secretion occurs in cells which are not expanding – again, the outer cells of the root cap are an example of this circumstance. The delivery of the secreted product to the plasma membrane necessarily involves the delivery of membrane also, since the vesicle containing the product is membrane-bound. Clearly, if the surface area of the protoplast is not increasing because the cell is not growing in size, secretory processes will inevitably lead to an excess amount of membrane being present at the cell surface. This implies that a mechanism must exist for the return of

Fig. 1.17. Diagrammatic representation of the role of the dictyosome in membrane transformation. The forming face of the dictyosome resembles the endoplasmic reticulum, whereas the maturing face resembles the plasma membrane.

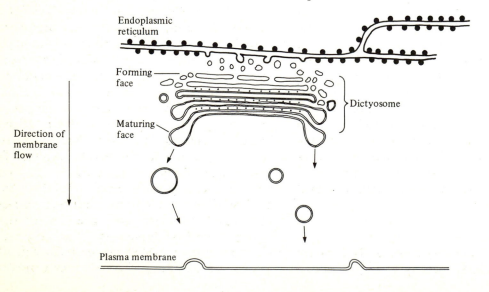

excess membrane material back to the cell interior. Calculations show how pressing this problem can be. In the water glands of *Monarda*, for example, it has been calculated that the 850 dictyosomes in the cell, which each consist of 8 cisternae, can deliver an area of membrane in the form of vesicles which equals the area of the entire cell surface in just 20 s. In the outer cells of the root cap, the dictyosome vesicles deliver an area of membrane equal to the area of the cell surface in about 1 hr.

Several methods can be envisaged for recycling the huge excesses of membrane which these figures represent. The process might take place at a molecular level, with membrane components being withdrawn and returned to cytoplasmic pools. Alternatively, whole fragments of membrane might be returned in the form of empty vesicles. A possible candidate for this type of process is the 'coated' vesicle which is commonly found at or near the surface of cells which are undergoing active secretion or rapid tip growth. The characteristic coat on these vesicles is composed of a protein, clathrin, and this might serve as a targeting device for the return of the vesicle to a specific site within the cell. It would be particularly elegant for this description if the returned membrane were incorporated into the forming face of the dictyosomes. However, evidence for this has not been found; in addition, of course, it has to be remembered that the forming face membrane is different from the plasma membrane.

Membranes have a common structure in terms of gross interactions between lipids and proteins, but it is becoming increasingly recognised that they are highly specific in the composition of their minor components, particularly glycoproteins. It is possible in this context that different constituents of the plasma membrane might be returned to different sites within the cell. Selective removal of components of the plasma membrane in this way would not only explain the observed behaviour of the maintenance of membrane integrity, but it could also be invoked in a speculative fashion to allow a drift in the properties of the plasma membrane with time and space. Since many of the changes associated with development are expressed in the extracellular space, such a drift could be highly significant.

The vacuole

In terms of secretion of materials from the cytoplasm, the plasma membrane is not the only

boundary. There is a space which is geometrically within the cell but which nevertheless can be regarded as 'outside' the cytoplasm. This space is the vacuole, and the membrane which separates it from the cytoplasm is called the tonoplast. Meristematic cells tend to have many small vacuoles (Fig. 1.1), and during growth these expand enormously and may coalesce into a smaller number of units, eventually forming one continuous space. The small size of the vacuoles within a meristematic cell is in fact the most diagnostic feature of such cells at the level of the light microscope – they appear to be 'full of cytoplasm'. By contrast, mature cells often appear to be 'empty'. Over 90% of the volume of a mature parenchyma cell may be occupied by its vacuole and, in these circumstances, the cytoplasm is often confined to a thin layer around the cell periphery.

The tonoplast is superficially similar to the plasma membrane; it has a width of about 10 nm and a well developed three-layered structure. That the membranes are different is shown by a variety of specific staining reactions. It is also reasonable that they are different on logical grounds – they show different chemical properties insofar as the plasma membrane abuts and contributes to the wall, whereas the tonoplast abuts a liquid-filled space, the vacuole. The tonoplast assumes the shape of the cell at maturity, but in meristematic cells, it encloses a more or less spherical space (Fig. 1.1).

The functions of vacuoles vary with the state of differentiation of the cell, but the most general is that of the regulation of water potential in the cytoplasm. Vacuoles contain dissolved solutes, usually in the range of concentration of 0.4–0.6 M. This means that the vacuole tends to increase in volume due to osmosis. The vacuole thus contributes in an important way to the turgor of the cell; that is, the force which keeps the whole protoplast tightly pressed against the enclosing wall. It is a function which can be shown experimentally by the procedure of plasmolysis. If a piece of tissue is placed into a solution containing a higher concentration of an inert solute than the concentration of solutes in the vacuole, then the whole protoplast shrinks away from the wall, due to osmotic removal of water from the vacuole. This effect can be produced by a 0.7 M solution of mannitol, for example. The degree of shrinkage is related to the degree of vacuolation in individual cells. Thus mature cells with large vacuoles will shrink much more than the cells of the meristem, where vacuoles represent a much smaller proportion of the volume of the cell.

The vacuole in normal growth therefore provides a regulatory mechanism for the cell in times of stress. By pumping solutes across the tonoplast, the water potential of the cytoplasm can be maintained within the fine limits necessary for its function, even under conditions of water stress. This chemical process can be translated into mechanical movement in specialised situations. For example, in guard cells of the stomata, the size of the pore which limits gas exchange and transpiration is controlled by changes in the turgidity of the guard cell vacuole. A rapid increase in the solute permeability of the tonoplast is also thought to be the underlying change which causes the collapse of the leaves of the sensitive plant, *Mimosa pudica*. As will be seen later (p. 120) the size and shape of the vacuole can restrain the position of the mitotic spindle in certain cases.

The vacuole is not just a passive sink for solutes, however. It contains a variety of enzymes which are capable of breaking down products of the plant's metabolism. This function is of particular importance to plants since they have no possibility of exporting waste products, or indeed materials which have outlived their usefulness in a developmental sequence. This lysosomal function of the vacuole is most clearly seen in the terminal stages of vascular development, and also in the recycling of materials during tissue senescence. In a less dramatic way it can be envisaged that the continuous turnover of high molecular weight materials which occurs throughout the plant cell's life may be at least in part mediated by enzymes within the vacuole. The vacuole may also be used as a storage space, particularly in seed development, where protein reserves are often held in the vacuoles in a highly dehydrated form. Vacuoles are also the site of some of the pigments found in the petals of flowers. They may contain other products of importance in the plant's relationship to its environment. For example, the vacuoles of solanaceous plants contain a protease inhibitor which probably serves to reduce the palatability of the tissues to animals. The storage and digestive functions of vacuoles come together during seed germination, when protein reserves are mobilised for the growth of the young seedling.

Origins of vacuoles

The origin of the vacuolar system has been difficult to determine, mainly due to the uncertainty of identifying a small precursor to a vacuole amongst a mass of other cytoplasmic vesicles and tubular elements. Vital staining of meristematic cells of root tips with neutral red shows that their lysosomal activity is contained within a network of tubes and small spherical bodies. The structures are close to the limits of resolution of the light microscope. A recent advance

Fig. 1.18. Diagram showing possible routes of origin of vacuoles. *a*, Dictyosome vesicles engulfing a small region of cytoplasm to give provacuoles; *b*, vesicles from the endoplasmic reticulum fusing with or being engulfed by provacuoles. Both origins are possible within one cell.

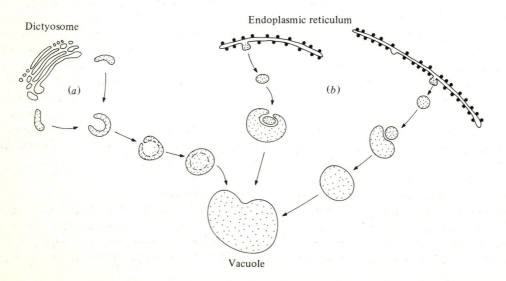

in the use of very high voltage electron microscopy which allows the examination of thick (1 μm) sections has led to the suggestion that vacuoles originate from cisternae of the dictyosomes. Tubular extensions at the margins of the cisternae have been shown to be the sites of accumulation of hydrolytic enzymes, and these tubular elements seem to associate together to form provacuoles. In one model, the spherical vacuole is formed by the enclosure of a region of cytoplasm within a network of fusing tubules, followed by its digestion (Fig. 1.18). If this is correct, it serves as another example of the complexity of membrane differentiation within cells. In this case it is necessary that one side of the membrane tubule originally derived from a dictyosome cisterna becomes permeable to hydrolytic enzymes within the tubule lumen, whilst the other side does not. Clearly if there were leakage of hydrolytic enzymes at random into the cytoplasm, the result would be disastrous for the cell. This model for the formation of vacuoles also implies a special function of the Golgi apparatus in this situation. Meristematic cells do not show rapid growth or secretory activity, and in such circumstances, a role for the Golgi apparatus in the formation of the vacuolar system could be a major activity.

Microbodies

Finally, there is one more type of recognisable structure which may be regarded as part of the endomembrane system. This is the microbody. Microbodies are in fact not commonly found in meristematic cells, but are rather a feature of leaf cells and also of storage tissues. Microbodies often contain crystalline inclusions, and are commonly found in association either with lipid droplets in storage tissues, or with chloroplasts in leaves. Isolation of microbodies from these situations shows that they are adapted to a specialised function concerned with lipid metabolism on the one hand and photosynthesis on the other.

The microbodies which are found in association with lipid droplets are called glyoxysomes. They contain enzymes which catalyse fatty-acid breakdown and which, in simple biochemical terms, give rise to hydrogen peroxide. This potentially toxic molecule is broken down within the glyoxysome by the enzyme peroxidase. Thus, in a single sequestered compartment, the energy stored in the form of lipid can be made available to the cell as a whole without the

formation of a toxic by-product. A further advantage of sequestration of the enzymes may be that the conditions for their action can be optimised with the microbody. The pH within the microbody is thought to be higher than it is in the surrounding cytoplasm.

The microbodies which are found in association with the chloroplasts are termed peroxisomes. They too function as a multienzyme system within a sequestered compartment, this time oxidising a product of photosynthesis (glycolic acid). The reaction product, glyoxylic acid, is returned to the chloroplast for reduction. The specificity of this association is remarkable, and it suggests that special permeability properties have been developed in the membranes of both the microbody and the chloroplast to allow the interchange of the low molecular weight organic acids involved in the reaction. Once again, the oxidation process results in the transient appearance of hydrogen peroxide which is detoxified *in situ* by the enzyme catalase. These are two specific examples of the importance of microbodies; no doubt they represent only a small part of the capabilities of this group of structures.

That microbodies may be considered part of the endomembrane system is based on their apparent origin from the endoplasmic reticulum. It seems well established that their limiting membrane is derived from this source, although the routes of synthesis and incorporation of the enzymes which they contain is largely unknown. They are not easily recognisable or associated with any particular type of organelle in the meristematic cell and, as such, represent a more 'differentiated' type of cytoplasmic structure.

The endomembrane system is thus a complicated network of different structures, each with a variety of functions which may only be found in certain cell types. The existence of the endomembrane system means that a great variety of different compartments exists within the cell, aside from the major organelles. This allows for the development of biochemical diversity, and it also vastly complicates the interpretation of very many chemical and physiological experiments with plants.

Microtubules

Microtubules, like microbodies, are named descriptively; they are small tubular elements which are widespread in plant cells and appear to have important functions to perform in matters related to directed motion. Unlike most of the organelles and membrane systems of the cell which perform a variety of biochemical functions by means of enzymes, microtubules are basically not sites of chemical action, but are mechanical elements. Motion of course requires energy for its execution and, in this sense, microtubules are chemically active, but they are not involved in synthetic or degradative pathways. Their implied mechanical function leads to an immediate caution. The presence of microtubules is most commonly detected by electron microscopy of fixed material; when it is known that motion or a mechanical process is occurring in their presence, it is frequently assumed that the microtubules are directly involved. This conclusion is not always justified, and it is important to realise that the coincidence of two events in time and space does not necessarily imply that they are causally related.

Substructure of microtubules

There is no such ambiguity about the structure or chemistry of microtubules. Microtubules are assembled from protein subunits. The subunit is called tubulin, and is a dimer of molecular weight 110 000. This protein polymerises under favourable conditions to yield the microtubule, which has a diameter of 24–5 nm, with a central core of diameter 12 nm. The substructure of the microtubule can sometimes be seen directly in section; it consists of 13 subunits in a circular array when sectioned transversely. The 13 subunits represent 13 filaments of the tubulin, so arranged as to lie in a helix around the wall of the microtubule. Tubulin is a dimer of two different proteins, called α and β tubulin, and these are arranged alternately along the length of each filament. This means that the microtubule has an intrinsic polarity, which has important consequences in terms of microtubule polymerisation and also possible mechanisms of action. Microtubules are often, though not always, surrounded by a narrow zone of empty space when viewed in section; this may be simply an artefact of preparation, or it may represent a zone of influence round the microtubule which could mediate in microtubule interactions with other cell components. In certain situations obvious bridges exist between adjacent microtubules. In plants these tend to be correlated with circumstances where rigidity of the microtubule array would appear to be needed. An example of this is in the generative cell of

the pollen grain, where the cell shape may be maintained not by a rigid wall, but by a skeletal arrangement of crosslinked microtubules.

The chemistry of the microtubule is important for an understanding of the way in which microtubules may appear and disappear at specific sites within the cell. Microtubules are polymers, and the equilibrium between the polymer (the microtubule itself) and the monomers (the tubulin protein dimer) is easily disturbed. Since microtubules are polarised, the association constants of the equilibrium are different at the two ends. In experiments with isolated tubulin, it is found that at high concentrations of the monomer, microtubule growth occurs faster at one end than at the other. As the concentration of the monomer is lowered, a situation arises where growth occurs only at one end; below this concentration net loss of monomer occurs and the microtubule disassociates.

Since the net growth or shrinkage of a microtubule is itself the sum of addition and loss of subunits from both ends simultaneously, but at different rates, an interesting consequence follows. At the steady-state concentration, net gain of one subunit at the growing end is compensated by net loss of one subunit at the shrinking end. However, these net changes are accomplished at different rates; for example, by the gain of eleven subunits and the loss of ten at the growing end, and the gain of three and loss of four at the shrinking end. This means that a particular subunit within a microtubule, even when the microtubule itself is not growing, will move along its length, eventually being lost at the slow end. This is known as 'treadmilling'. It serves as an illustration of the great complexity of the control of the presence or absence of microtubules within the cytoplasm.

Physiologically, the equilibrium is sensitive to the levels of calcium and magnesium ions. Experimentally, it can be altered by various treatments. Disassociation of microtubules into subunits is favoured by low temperatures, high hydrostatic pressures, and the presence of various drugs of which colchicine is the most commonly used. Heavy water favours the presence of the microtubule. These experimental treatments are often used to test whether a particular function is dependent upon the presence of microtubules. For this purpose, colchicine is particularly widely applied, as will be seen later. The danger of such experiments is, of course, that an agent which is supposed to act solely on microtubules may, in fact, have several unrelated effects in the same cell.

The consequence of these equilibrium effects in physiological terms is that the cell can precisely control the appearance and disappearance of microtubules at different sites and different times. Microtubules will appear when the equilibrium is shifted in favour of the polymer; this may be as a result of a local change in the concentration of subunits, or of calcium or magnesium ions, or as a result of the capping of existing microtubules to prevent turnover at one end. Microtubules frequently arise from recognisable areas of cytoplasm called microtubule

organising centres (MTOCs). This rather clumsy descriptive name is applied to centres of amorphous material from which microtubules regularly arise. MTOCs occur at the spindle poles, and in the phragmoplast (p. 34), but their precise nature is at present poorly understood.

The distribution of microtubules

In the non-dividing meristematic cell, the microtubules are found in the peripheral cytoplasm, lying close to the plasma membrane. They are found preferentially at right angles to the major growth axis. They appear to encircle the cell, although recent evidence suggests that an individual microtubule probably does not form a complete hoop. Microtubules in these cortical arrays are usually more or less parallel to one another (Fig. 1.12, 1.19).

Microtubules have widely different distributions in other cells, and also in cells which are dividing. These are matters which will be considered

Fig. 1.19. Glancing section through a primary wall in the root tip of maize. Cortical microtubules are visible within the peripheral cytoplasm. They are approximately parallel to one another.
Bar = 0.5 μm.

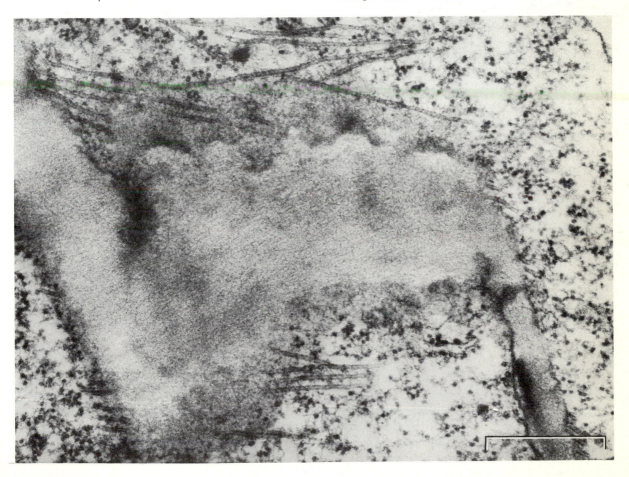

later. In the interphase meristematic cell it is quite uncertain what the function of microtubules might be. It has been suggested that they in some way influence the direction of cellulose fibres formed within the wall. This will be discussed in detail in Chapter 3. Other suggested functions are the direction of the movement of cellular organelles, or vesicles from the dictyosomes. Both these ideas spring from a coincidence of the presence of microtubules in positions where such directed movements may be assumed to be occurring. The best example of the effect is seen in the development of the cell plate at cytokinesis (p. 34).

Microfilaments

In considering movements within plant cells, one further structural feature should be mentioned. These are the microfilaments – fine fibrous elements resembling actin filaments – which are found in some cells of higher and lower plants.

The most generally accepted function of these microfilaments is in the generation of the motion of cytoplasmic streaming. Cytoplasmic streaming is seen at its most spectacular in giant algal cells, such as those of *Nitella*. Here bundles of microfilaments are found at the interface between the moving and stationary parts of the cytoplasm. It is clear that such filament bundles play a direct role in the generation of the force for motion of cytoplasmic particles, via the action of actin.

Considerable speculative interest attaches to the appearance of filaments within the cytoplasm of plant cells, since in many ways they appear to correspond to the complex of fibrous and filamentous proteins which comprise the 'cytoskeleton' of animal cells. However, it should be emphasised that the chemistry of the microfilament system, where it occurs in plant cells, is less well defined than in animal cells. Microfilaments are not found in every cell type in higher plants, and in particular seem to be absent from meristematic cells. These observations suggest the caution that microfilaments may not represent part of a universally needed 'cytoskeleton', but rather are related to a specialised function, of which the cytoplasmic streaming of giant algal cells or elongated cells in higher plants are examples.

Equally, however, it is certain the present methods of electron microscopy underestimate the frequency of occurrence of microfilaments, particu-larly if they are not organised into large and easily visible bundles. The whole area of cytoplasmic organisation is one of active current research, and the status of microfilaments, at present something of a curiosity, may quickly change.

The cytoplasm

The cytoplasm may be defined as everything that is left within the plasma-membrane-enclosed space which is not part of one of the components which have been described above. The cytoplasm of meristematic cells appears to contain a random population of ribosomes, together with a number of lipid droplets within a background which may show a fine granular substructure. The impression given by the electron microscope is that there is no order at all in this arrangement. This is undoubtedly an error; the briefest glance at a living cell under the light microscope will convince a viewer that the cytoplasm is highly ordered in terms of the movements which occur within it. Quite large organelles such as plastids can be seen to move rapidly in defined directions, particularly in vacuolate cells. Even in meristematic cells, where organelle movements are less obvious, it is clear that the enormous complexity of the biochemistry of the cell could not possibly take place in a random soup of the various ions, organic molecules, proteins and nucleic acids which are found in the cytoplasm. In animal cells, fluorescent staining with antibodies to various structural proteins shows that the cells contain a 'cytoskeleton' of fibrous or filamentous elements. This idea was propounded many years ago, before the invention of the electron microscope. Evidence for such a cytoskeleton in plant cells is as yet very meagre, but this is again an area of active research where no doubt progress will be rapid. As with the nucleoplasm, it is certain that the cytoplasm must be highly organised in spatial terms in order to be able to carry out its wide range of biochemical activities.

Cell division

The meristematic cell emerges from this brief consideration as a complex unit containing many compartments, and with all the elements which go to make up differentiated cells, but in an unspecialised form. The structure of a meristematic cell and its inclusions gives no clue as to the likely fate of its progeny cells. The function of the meristematic cell is above all to divide and contribute new cells to the tissue; cells which may themselves divide again, possibly several times, but which will eventually take on a special function which will be reflected in biochemical and structural differentiation. The nature of the meristem is retained through this process, implying that the divisions within the meristem are in a sense asymmetric. One of the daughter cells remains meristematic, whilst the other, however tenuously at first, embarks upon a course of development.

It is certainly not meaningful to discuss cell division as if it were a process separate from considerations of growth and development as a whole. If cells are prevented from dividing, a certain degree of growth can ensue. For example, if seeds or embryos are irradiated with gamma rays, limited germination and growth can still follow. Indeed, there is evidence to suggest that the individual cells in such plantlets are finally larger in size than those in normal plants, as though the cessation of growth were itself under nuclear control. Nevertheless, it is quite clear that such growth as is observed in the absence of cell division is due to pre-existing ability. New cell types are not formed, and eventually the whole course of development is arrested.

Cell division serves to create new compartments within the plant body. In a sense, cell division and differentiation are antagonistic. It may be said that, in higher plants, the great diversity of cell types which is found is due to the ability of growth to be sustained within areas of meristematic activity, which in turn allow a high proportion of mature cell types to forego the need to continue to divide. As will be seen in Chapter 4, cell divisions form an integral part of many developmental pathways, and it is certainly true to say that new cell forms can usually only be generated by the stimulus to growth which cell division provides.

Mitosis and cytokinesis

Cell division comprises a cycle of different processes. The idea of a cycle is an important one, and it contrasts with the more linear series of changes which characterise progress along a developmental pathway. The visible events of cell division are mitosis and cytokinesis. Mitosis involves the condensation and ordered movement of chromosomes, and has been well studied both at the light microscope level and using the electron microscope. Cytokinesis is the partitioning of the cytoplasm which takes place by the formation of a new cell wall. In meristems, these two events are invariably linked together in an ordered time sequence; the chromosomes are partitioned and, immediately after this, the new cell wall (the cell plate) is formed between the two new daughter nuclei. The result of this orderly sequence is that every cell compartment contains a single nucleus. However, mitosis and cytokinesis are not always linked together in this way. Several mitoses can occur without the intervention of cytokinesis, in which case the result is the formation of a multinucleate cell; this is seen in the development of the free nuclear embryo in gymnosperms, and in nuclear endosperm. Similarly, a series of new cell walls may form without immediately preceding mitosis; this is seen in gymnosperm embryos when a large number of cross walls is formed following a phase of repeated nuclear division. It is also not necessary for the elaboration of a differentiated plant body that cytokinesis should occur at all. In the coenocytic algae, of which *Caulerpa* is an example which will be studied later (p. 187), the entire plant is devoid of cross walls.

Cytokinesis partitions the cytoplasm between the two daughter nuclei which are formed by the mitotic process. This partitioning is perhaps the least understood aspect of the cell division cycle. The organisation of the cytoplasm is certainly at present underestimated, and from this it follows that the apparently random nature of the distribution of its components at cytokinesis is also probably illusory. What is certain is that each new cytoplasmic compartment which is formed at cytokinesis must contain all the necessary factors and structures which are needed to sustain the nucleus. It is necessary therefore that during the mitotic cycle all of these factors and structures must themselves show a net increase, since at the beginning of a cycle they are sufficient to support one nucleus and, by its end, they must support two. Certain organelles replicate themselves by a fission process; this is true of mitochondria, and also of plastids, which will be considered in detail in Chapter 2. This replication appears to occur independently of events within the nucleus, and the partition of the products at cytokinesis also appears to be at random. Less autonomous structures such as the cytoplasmic membrane systems seem to increase throughout the mitotic cycle by a process of growth.

The partitioning of the cytoplasm by the formation of a cell wall implies a kind of fixation of any differences which exist between the two parts. The importance of this is most clearly seen in highly asymmetric divisions such as that of the fertilised egg, where the two daughter cells of the first division are obviously different in size and structure, and go on to completely different developmental fates. This is exemplified by the development of the *Fucus* egg (Chapter 7), but it is equally true of the fertilised zygote of higher plants. However, as has already been mentioned, even apparently symmetrical divisions within meristems give rise to daughter cells with different fates. One will remain in the meristematic state, and in position within the meristem, whereas the other (perhaps after further divisions) will enter a developmental pathway. Since the nuclear division process delivers an identical gene complement to each of the daughter cells, it can be inferred that such differences in cell behaviour as emerge subsequently are in one sense due to the differences in the cytoplasmic environments which arise at cytokinesis.

The cell division cycle

The process of cell division is cyclic; that is, after a certain period of time, the cell is returned to a starting point. Mitosis and cytokinesis are the visible parts of this cycle, but they are not the principal part at least in terms of a time scale. Most of the time cells are at the stage of interphase. Interphase is the stage in the cycle when the chromosomes are not condensed, and when most of the net synthesis required for the formation of new cells takes place. In keeping with the general emphasis on events in the nucleus (an emphasis which may well be considered to be misplaced), the mitotic cycle is divided into stages which are named in terms of DNA replication and chromosome condensation. Thus mitosis (M) is followed by interphase; and interphase itself consists of three stages, G_1 (a gap, during which no DNA synthesis occurs), S (the replication phase of DNA) and G_2 (a second gap). The mitotic cycle is depicted diagrammatically in Fig. 1.20.

Interphase

Interphase is divided into three stages, G_1, S and G_2. The term G_0 is sometimes used to express the state of a cell which has undergone mitosis for the last time during normal growth. It is also used with respect to cells which are in a prolonged resting state prior to

being triggered back into cycle. It is important to realise that G_1 and G_2 are not equivalent in the cycle, even though they both represent periods during which no DNA synthesis is occurring. If a cell is arrested in G_1, the next event to occur when the cycle restarts is the synthesis of DNA, whereas if the arrest occurs in G_2 then mitosis can follow without DNA synthesis. When levels of DNA (p. 34) are examined in dormant embryos it is commonly found that most of the cells are 2C, with only a few 4C. This implies that the arrest of the cycle was predominantly in G_1, before DNA synthesis had occurred. Similarly when explants of a storage tissue such as the tubers of Jerusalem artichokes are stimulated into growth, the first few divisions are synchronised due to a common point of arrest – again in G_1.

The duration of the various stages of the interphase part of the division cycle is not constant; it may vary between tissues and even between cells within a tissue. Commonly interphase accounts for about 90% of the total cycle time, with S taking up half of the interphase period. The relative importance of the G_1 and G_2 stages is variable. Near the quiescent centre in maize roots, G_1 accounts for 17% of the total division cycle, with G_2 occupying 44%. However, in the cells of the quiescent centre itself, which are characterised by a very low rate of mitotic cycling, G_1 increases in length to take up 86% of the total cycle time, whereas G_2 occupies only 10%. In the cap columnella of the same roots, G_1 is entirely absent and cells embark upon DNA synthesis immediately after mitosis is completed. On the other hand, in pollen grain mitosis in many species, G_2 is reduced to zero. The total time occupied by the complete cell division cycle is also variable. In maize roots the cells of the stelar tissue cycle in about 20–30 h, whereas cells in the quiescent centre only cycle once in about 200 h.

During the S phase, the rate of accumulation of DNA appears to be more or less constant. DNA synthesis is thought to proceed from many points simultaneously and, in some cases, a degree of organisation can be detected in this arrangement. For example, labelling experiments show that heterochromatic regions tend to replicate later in the cycle than euchromatic regions. In *Tradescantia* root tips the ends of the chromosomes are replicated last, whereas in *Crepis*, it is the centromeric regions which are the last to be replicated. It is noteworthy that when cycle times are found to vary under different conditions, the variation does not stem from changing rates of DNA synthesis, but from changes in G_1 or G_2. Even when the DNA level within a tissue is raised by polyploidy to several times the 2C level, replication times (the length of the S period) are not proportionately increased. This reflects the fact that DNA synthesis takes place from many starting points, and that the number of these starting points relates to the quantity of DNA to be replicated.

Other macromolecular components do not show the same patterns of increase as the DNA. RNA, for example, is accumulated more or less continuously, with a pause in late G_1 and a rapid burst of accumulation in G_2. RNA synthesis is at a minimum during mitosis in higher plants, and this

corresponds to the period when the nucleolus is dispersed. Protein synthesis occurs during all phases of the cycle, with peaks in G_2 and prophase in root tip cells of *Vicia faba*.

These statements relate to classes of macromolecules. When particular species of molecules are considered, a more periodic pattern often appears. tRNA, for example, seems to be specifically accumulated in the latter part of G_2. Certain enzymes show a periodic rather than continuous pattern of increase during the division cycle. Thymidine kinase and DNA polymerase, for example, show a synchronous rise in activity at the onset of the S period. The histones, which form an integral part of the structure of chromatin (p. 226) are synthesised during the S period, so that newly replicated DNA is continuously incorporated into new chromatin. It is likely that one of the reasons for multiple copies of the histone genes (p. 225) is the need for the synthesis of these proteins to keep pace with the rapid rate of replication of DNA. It has been calculated that each

Fig. 1.20. The mitotic cycle. The figure shows the named phases of the DNA synthesis cycle, and of the chromosome cycle. The distribution of microtubules is indicated diagrammatically in the outer ring.

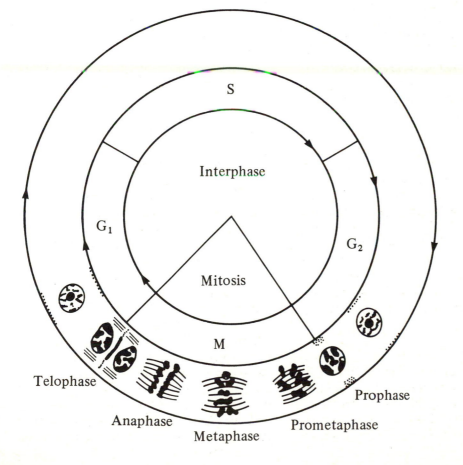

Fig. 1.21. The progress of mitosis in a cell of the endosperm of *Haemanthus katheriniae*. The pictures show the sequence of chromosome movements starting at time 0 (prophase), proceeding through prometaphase (47 min), metaphase (63 min), the start of anaphase (99 min), anaphase (109 min, 114 min, 120 min) and telophase (134 min, 143 min). Sequence of pictures courtesy of Professor Andrew S. Bajer. Bar = 10 μm.

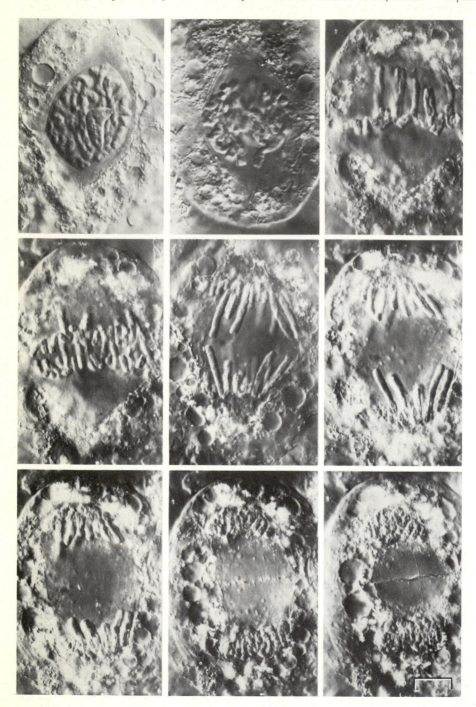

replication site in the DNA moves at a rate of about 50 nucleotides s^{-1} as DNA is synthesised.

It is likely that, as more detailed study is made of the changing patterns of enzyme activities and levels of particular proteins and other macromolecules during interphase, this period of the division cycle will emerge as being as controlled and orderly as is the period of mitosis itself. Our current appreciation of the order within mitosis stems largely from its mechanical nature and its accessibility to study with the light microscope.

The phases of mitosis

The process of nuclear division is also broken down into named phases, defined in terms of the appearance and behaviour of the chromosomes, originally described by the use of the light microscope (Fig. 1.21).

At prophase, the chromosomes condense and can be clearly seen as separate stained structures within an intact nucleus (Fig. 1.22). Prophase is also characterised by the gradual disappearance of the nucleolus or nucleoli. The timing of the disappearance of the nucleolus is slightly variable depending on tissue type – in the dividing cells of the maize

Fig. 1.22. Prophase nucleus within a cell in the meristematic region of the maize root. The chromatin is more condensed than at interphase, and the outline of the nucleolus is less regular. Bar = 5 μm.

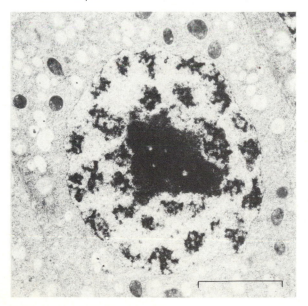

root cap, for example, nucleoli begin to disappear in late interphase. In the electron microscope the nucleolus at prophase takes on a branched appearance initially, which is often the best diagnostic feature of this phase. The end of prophase is marked by the fragmentation of the nuclear envelope and its disappearance to view in the light microscope.

The next event is the formation of the mitotic spindle and the attachment to it of the chromosomes. This is known as prometaphase and, in the light microscope, it appears as the stage at which chromosomes make apparently random movements within the spindle area. At metaphase the chromosomes pause with their points of attachment to the spindle at its equatorial plane. At anaphase the two sister chromatids separate, thus reversing physically the chemical replication of the DNA, and the chromatids move to opposite poles of the spindle. Telophase is the final mitotic stage, when the nuclear envelope reforms around each of the two masses of chromatids at the poles of the spindle, giving rise once again to two normal nuclei. The nucleolar regions are reformed from the nucleolar organisers and may quickly fuse into a single nucleolus. In most cells, the reformation of the two daughter nuclei is accompanied by cytokinesis.

This process has fascinated microscopists for a long time and in living cells it represents a remarkable spectacle; the more so in earlier times since the detailed mechanisms of the movement of the chromosomes could not be visualised. In fact advances in light microscopy allowed the discovery that the spindle is birefringent in polarised light, and that this birefringence correlates with the ability of the chromosomes to move. The more detailed description of mitosis which follows is largely based on the examination of sections of material after fixation, using the electron microscope. The interpretation of such images relies deeply on observations of living material since in mitosis, perhaps more than any other cellular process, it is the movement of the participating structures (rather than their relative positions when fixed) which leads to the greatest understanding of the events.

Prophase

Prophase was originally defined in terms of the visibility of the chromosomes. In the electron microscope it is identifiable by a marked increase in the depth of staining of the chromatin, and its condensation into a smaller number of masses, each of which corresponds to a section through a chromosome. The nucleolus in such cells often becomes highly branched at its edges, and eventually it disappears as a recognisable region within the nucleoplasm (Fig. 1.22). Whether this disappearance is real, or whether nucleolar fragments come to coat parts of chromosomes is not known for certain. It should be realised that when the nuclear envelope breaks down, the spindle area is invaded by cytoplasmic ribosomes, and these confuse the identification of the granular component of any remaining fragments of the nucleolus.

Cytoplasmic features of prophase concern the microtubules and the endoplasmic reticulum. At interphase, microtubules are arranged in so called cortical arrays along the walls of the cell (p. 23). Prophase is marked by the concentration of microtubules in the form of a discrete band which remains in the cortical cytoplasm but which can be thought of as encircling the nucleus (Fig. 1.23). The occurrence of this band of microtubules will be discussed later (Chapter 4) since it seems to have the fascinating property of predicting the position at which the cell plate will eventually fuse with the mother cell wall as cytokinesis is completed. Since the band is a very early manifestation of prophase, it has been termed the 'preprophase band' of microtubules.

The next cytoplasmic event is the formation of a clear zone around the periphery of the nucleus from which all large cytoplasmic organelles are excluded. Within this zone are found short fragments of microtubules, roughly aligned in the plane of the spindle axis. At either pole of the future spindle the zone widens and at these sites there is found to be an accumulation of endoplasmic reticulum.

The breakdown of the nuclear envelope follows this stage. This permits invasion of the nuclear space by microtubules from the spindle poles, and also by cytoplasmic ribosomes and small vesicles. The fate of the nuclear envelope is uncertain, and indeed may vary from cell to cell. In general its fragmentation

Fig. 1.23. Section through a preprophase band of microtubules in a root of the grass *Phleum pratense*. The band consists of several hundred profiles of microtubules. Bar = 0.5 μm.

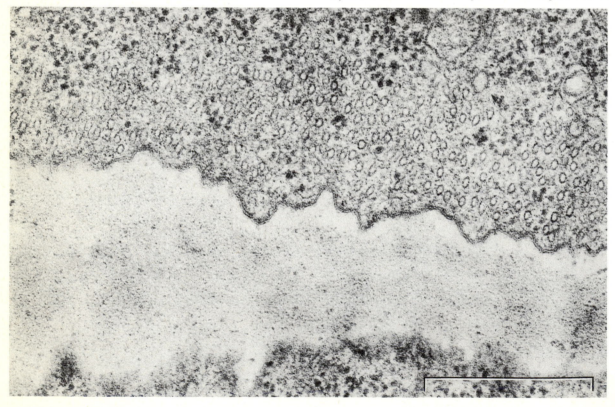

proceeds from the poles, with the equatorial regions retaining their integrity until last. There is some evidence to suggest that fragments of the membrane, together with associated nuclear pore complexes, may persist throughout mitosis. The amount of membrane surrounding the two daughter nuclei after mitosis is greater than that surrounding the original single nucleus so that, in any event, some net synthesis or incorporation of new membrane materials must occur over the mitotic stage of the cycle. Experiments with protein synthesis inhibitors suggest that this material is not synthesised *de novo* during mitosis, but rather pre-exists in the parent cell, possibly in the form of endoplasmic reticulum.

Prometaphase and metaphase

The fully formed spindle at the end of prophase consists of a system of 'continuous' fibres which run between the spindle poles, and 'kinetochore' fibres, which run from the point of attachment of the chromosome to the spindle (the kinetochore) and the spindle poles (Fig. 1.24). The birefringent 'fibres' of the light microscope correspond in the electron microscope to bundles of microtubules. The continuous fibres at least in part comprise microtubules which were moved into the spindle area from the clear zone. The kinetochore fibres consist of microtubules which have grown out from the kinetochore region of the chromosomes (Fig. 1.25). The kinetochore is a recognisable region of the chromosome, and may be regarded as a special type of microtubule organising centre (p. 23). Careful study suggests that although the continuous fibres run from pole to pole, individual microtubules within the fibres do not. 60–90% of the microtubules within the spindle occur as part of the continuous fibre system, the remainder comprising kinetochore fibres. Although these microtubules occur in bundles, and may even show occasional crosslinking, they do not form geometrically precise arrays.

The movements of the chromosomes which culminate in the pause at the metaphase plate appear to be rather random. This corresponds to a stage when the kinetochore fibres from each chromosome are interacting with the continuous fibres, resulting in apparently frantic movements of the chromosomes. Eventually, the kinetochore fibres become firmly anchored to the spindle, and the metaphase plate becomes visible due to the balancing of equal forces on either side of the kinetochore. This results from the kinetochore fibres being attached to each half spindle. The attachment can be disturbed experimentally by micromanipulation. For example, an attached chromosome can be turned round through 180°, whereupon it will reattach itself in the opposite direction. This has no consequence for the division since the sister chromatids are the same. Similarly, it is

Fig. 1.24. Microtubules in the mitotic spindle. At prophase microtubules grow from the spindle poles and the kinetochore regions of the chromosomes. At prometaphase these two systems of microtubules interact to direct the movement of the chromosomes.

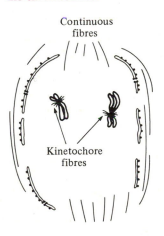

Continuous fibres

Kinetochore fibres

Prophase

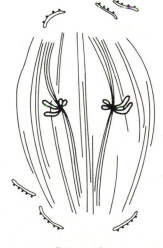

Prometaphase

possible by manipulation to force both the sets of kinetochore fibres from a single chromosome to become attached to one of the two half spindles. This will eventually result in both the sister chromatids moving into that half spindle. However, this is a highly unstable arrangement, since the two sets of kinetochore fibres usually extend in opposite directions, owing to the linear and two-sided nature of the kinetochore itself.

The metaphase plate is a still point in terms of chromosome movement. However, during pro-metaphase and metaphase, other particles within the spindle area which are not attached to the spindle are constrained to move. Thus cytoplasmic vesicles, any nucleolar fragments or fragments of chromosomes which for some reason lack kinetochores are all propelled towards the nearest spindle pole. The speed of their movement matches the normal speed of the anaphase movement of chromosomes. The arms

of the attached chromosomes are also subject to this force, so that when the metaphase plate is formed with the kinetochore regions in a single plane, the trailing arms of the chromosomes lie longitudinally in the half spindles, towards the poles. These observations are very important for the understanding of spindle movements, and are difficult to explain at present.

Anaphase

At anaphase, the poleward movement of all the chromosomes is more or less simultaneous following the separation of the sister chromatids (Fig. 1.21) Trailing arms of chromosomes can pass each other in opposite directions, showing that the movement is the result of a force applied individually to the chromosomes and not the result of a general movement of everything within a half spindle towards the nearest pole. The speed of the movement of the chromosomes is constant throughout their travel, and is the same for all the chromosomes within a given spindle. It varies between spindles in the range of 0.2–0.5 μm min^{-1}. Since large and small chromosomes move at the same rate, this must mean that the speed of their

Fig. 1.25. Section through a dividing cell in the root tip of maize. The microtubules shown are part of a kinetochore fibre. Bar = 5 μm.

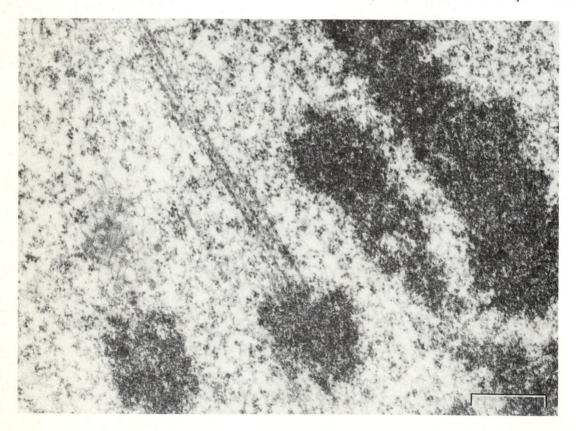

movement is not dependent upon the size of force available, but on the rate of its application. It has been calculated that the energy available from the hydrolysis of 20 ATP molecules would be sufficient to move an average chromosome from the equator of the spindle to its pole. This is a very small amount of energy indeed compared to the resources of the cell. Particulate matter other than attached chromosomes continues to move polewards at anaphase if it is ahead of the chromosomes. In the equatorial region between the separating chromosomes this movement is halted and may be reversed, leading to an accumulation of material in the central region.

Telophase

Telophase is characterised by the reappearance of the nuclear envelope around the assembled chromosomes at each spindle pole, and the gradual dispersal of the chromatin to give rise to the normal interphase appearance of the nucleoplasm (Fig. 1.26). At this time also, nucleolar organiser regions on the chromosomes begin to synthesise the products which result in the formation of a new nucleolus or nucleoli within the daughter nuclei. It may also be surmised that telophase is the time when the nuclear lamina is reformed, together with the organised attachments of the chromosomes to the nuclear envelope. Details of these processes are as yet not at all understood.

Force generation within the spindle

It is remarkable that despite intensive study, no general agreement exists on the mechanism of force generation within the mitotic spindle. That microtubules are in some way implicated in the process of chromosome movement is not disputed, since they are the only known spindle element possessing the mechanical properties which would appear to be necessary. The possibility certainly exists that there is some other as yet undetected feature of the spindle

Fig. 1.26. Telophase in a root tip cell of maize. The daughter nuclei have reformed their envelopes before completion of the new cell plate. Bar = 2 μm.

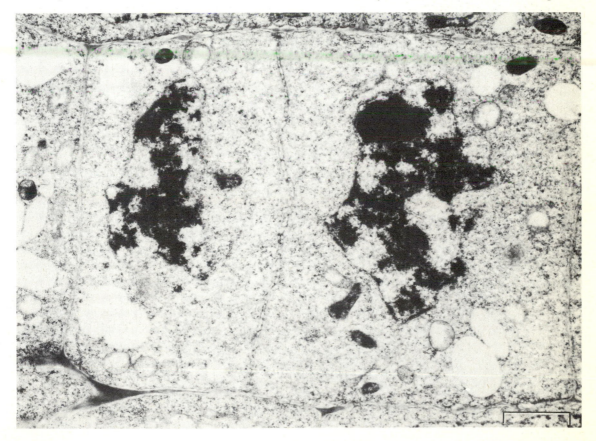

which has an important role to play. It is, for example, commonly observed that the fixatives used for electron microscopy cause a lowering of the birefringence of the spindle – in other words, they induce a degree of disorder in the structure.

There are, broadly speaking, two hypotheses to account for spindle movements. The first postulates that microtubules may move over one another by means of a series of makes and breaks in crosslinks between microtubules. This idea suffers from the weakness that crosslinked microtubules are more commonly found in situations where rigidity is required rather than mobility. The other hypothesis postulates that the microtubules of the two half spindles have different polarities, and that the force required to move the chromosomes is generated by removal of subunits from one end of the microtubules, causing them to shorten. This model relies on the kinetochore being a site of capping of the microtubules of the kinetochore fibres, which would leave the spindle-pole end of the fibres open to depolymerisation. It suffers from a general weakness in that no function is assigned to the continuous fibres. It is also true to say that any model which relies entirely on the presence and properties of microtubules in the spindle is unlikely to be convincing, if only because of the movements which occur involving particles which have no association with microtubules within the spindle area. No doubt these difficulties will be resolved in time; the mitotic spindle represents a very large structure to submit to a detailed analysis by electron microscopy, and it has only recently become possible to isolate spindles which retain some ability to function. The mitotic spindle represents the best example possible of the coordination of movements within cells.

Cytokinesis

Cytokinesis begins at telophase in the meristematic cell. The partition of the cytoplasm occurs by the formation of the cell plate between the two daughter nuclei. The growth of the cell plate is associated with a structure called the 'phragmoplast'. This consists of a region containing microtubules and small vesicles which arises first in the centre of the equatorial plane of the spindle. The cell plate forms by coalescence of the vesicles, and it grows radially outwards until it eventually fuses with the mother cell wall.

As the plate grows, the microtubules and vesicles are found increasingly only at its expanding edges, which has led to the suggestion that the microtubules are in some way directing the vesicles to their position. As with so many electron microscopic descriptions of essentially dynamic events, definitive evidence for this mechanistic explanation of the role of microtubules is lacking. It is also uncertain from which source the vesicles themselves arise. The balance of probability is that they are produced by dictyosomes. These are often, although not always, found in association with the growing edges of the cell plate (Fig. 1.27). It is also thought that at this stage elements of the endoplasmic reticulum become trapped between the fusing vesicles and mark out the positions of the future plasmodesmata which will cross the cell wall.

Cell plate formation is a process which clearly requires that membranes have specific recognition systems. The edges of the cell plate have to be able to fuse with the plasma membrane covering the mother cell wall, but they do not fuse at random with other types of membrane. The material of the cell plate is polysaccharide in nature and is recognisable as the middle lamella of the primary wall (Chapter 3). Cell plate formation is complete when the cytoplasm is partitioned. In meristematic cells, this involves only a slight extension of the phragmoplast beyond the area formerly covered by the mitotic spindle. In the very long cells of the cambium however, the phragmoplast must extend over an area many times that covered by the mitotic spindle.

There are many variations on this basic theme of mitosis and cytokinesis. In some organisms, notably lower fungi and some algae, the nuclear envelope does not break down during mitosis. The nucleolus may not disperse as has been described here, but may split into two so that each daughter nucleus receives half (*Euglena*) or it may persist and be eliminated from the spindle area altogether (*Oedogonium*). Cytokinesis may be achieved by a combination of outward growth of the cell plate and inward growth of the mother cell wall (*Spirogyra*). And as will be seen later, the geometry of the entire process may be adapted to suit particular developmental requirements (Chapter 4).

Nuclear division and ploidy

Most of the cells within the meristem of a root are diploid; that is to say, they contain two copies of each chromosome. This level of DNA is referred to as 2C. After DNA synthesis, during G_2, the cell contains a 4C level of DNA. Mitosis reduces this to 2C by the mechanical separation of the sister chromatids of each chromosome. When this series of events is linked to cytokinesis, the result is the formation of a series of cellular compartments, each with one nucleus, and each at the 2C level. As has already been mentioned, failure or delay of cytokinesis may result in the formation of cells with more than one nucleus.

However, not all the cells within a higher plant are diploid. In very many mature cells, the levels of DNA are much higher, due to cycles of replication not coupled to subsequent mitotic separation of chromatids. If one such cycle occurs, the result is a cell at 4C; after two, it will become 8C and so on. If the chromosomes retain their normal structure, this state of affairs is known as polyploidy. If however, the chromosomes exist as multistranded units, then they are said to be polytene. Polyploidy is very common in the ground tissues, the so called parenchyma, of the plant. Polyteny is less well known, and may be limited to particular cell types. Polyploidy may be induced experimentally by agents which disrupt the function of the mitotic spindle. Colchicine is the most commonly used of these; as we have seen, it acts by upsetting the equilibrium between microtubules and their subunits.

However, polyploidy is by no means an abnormal state which occurs only after experimental interference. The best explanation for the occurrence of polyploidy seems to be that it permits cell sizes to increase beyond their normal limits for the diploid state. It is as though a certain level of DNA is associated with a certain degree of growth; if the amount of DNA in a particular cell is increased, then so is the capacity of that cell for growth. In parenchyma cells, there is no particular advantage in the subdivision of compartments into small units corresponding to the diploid state; this requires the elaboration of materials associated with mitosis and

Fig. 1.27. The edge of a growing cell plate in a dividing guard mother cell of the pea (*Pisum sativum*). Microtubules are confined to the edge of the plate as it grows. Bar = 1 μm.

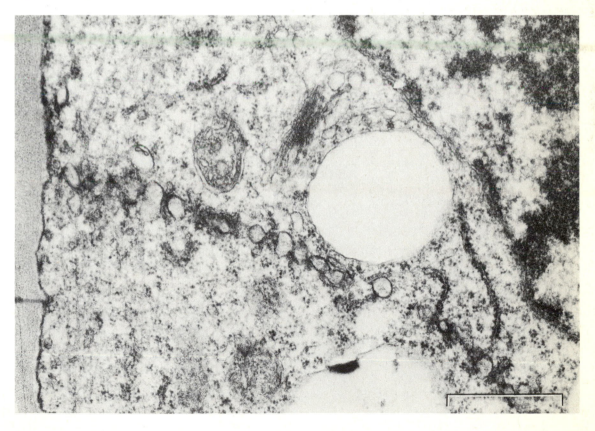

cytokinesis. Thus polyploidy can be regarded as a way of allowing tissue expansion and growth in the most efficient way possible, by minimising the occurrence of mitosis.

Increases in the levels of DNA within cells are characteristic of mature tissues. The sexual process, by contrast, involves the reduction of the amount of DNA to the haploid level (1C – half the quantity of the diploid cell). The reason for this is that fertilisation involves the fusion of two cells, the male and female gametes. This forms the zygote, which gives rise to the diploid embryo by mitotic divisions. Thus the gametes must be haploid in order that their fusion product be diploid. The process of sexual reproduction has the important advantage of allowing new combinations of genes to be formed. This occurs both at the level of

the fusion of the gametes, which may be from different individuals, and also by the process of recombination of genes which takes place during the reductive divisions which give rise to the gametes. This reductive process is known as meiosis.

Meiosis

In terms of genetic events, mitosis results in an equal partitioning of identical replicated sets of chromosomes between the two daughter nuclei. Each of these daughter nuclei is thus identical to the other genetically, and to the parent nucleus. By contrast, meiosis results in the formation of not two nuclei, but four. These arise by the division of the original diploid set of replicated chromosomes, and they are therefore all haploid. This is achieved by two nuclear cycles in succession without an intervening period of DNA synthesis. The importance of meiosis, however, lies in the events which occur during the first prophase period; events which are unique to meiosis.

The prophase nucleus contains a replicated diploid set of chromosomes; that is, two pairs each of sister chromatids which were originally derived from

Fig. 1.28. Diagram to illustrate the difference between mitotic chromosome separation and the unpairing of the meiotic bivalents. In mitosis, the kinetochores separate to give identical chromatids. In meiosis, the chiasmata separate to give paired chromatids which have undergone crossing over.

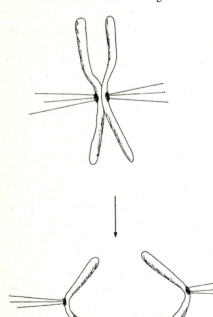

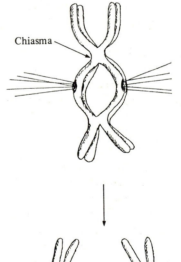

(*a*)   Mitosis   (*b*)   Meiosis

each of the parental gametes as a result of fusion during sexual reproduction. The mitotic process involves the splitting of each pair so that the daughter nuclei each receive 'half' of each chromosome. The splitting occurs at the start of anaphase, and involves the separation of the kinetochores, since this is the position at which mitotic chromosomes are joined (Fig. 1.28). The essential aspect of this mitotic arrangement is that it is cyclic; by replicating the 'half' chromosomes during interphase, the whole process can be repeated over and over again.

Meiosis is quite different. At prophase the corresponding chromosomes from each parent (called homologous chromosomes) come together to form a tight association. This is termed a bivalent. The bivalent is a lateral association of paired chromatids, one pair from each parental chromosome, and while the state of pairing lasts it is possible for genes to be exchanged at multiple sites along the length of the chromosomes. This results in the formation of new and unique combinations of genes in each chromatid. After this process is completed, the bivalents separate along most of their length, but they remain attached together at one or more points, which are known as chiasmata.

The next series of events superficially resemble mitotic metaphase, anaphase and telophase, and they result in the formation of two diploid nuclei within nuclear envelopes. However, the separation which occurs at this anaphase is not due to the splitting of the kinetochores of the paired chromosomes. It is due to the final separation of the bivalents by the parting of their point (or points) of contact at the chiasmata (Fig. 1.30). Thus each nucleus after the first meiotic division is different from its corresponding sister nucleus, since each contains paired chromatids which were originally from one parent. These paired chromatids have, of course, themselves been altered by the crossing over process which took place during prophase. This means that even if a cycle of replication were now to occur, the result would not be a return to the starting point. However, replication does not occur. The next mitotic stage is more or less conventional – anaphase separation is due to splitting at the kinetochore of each paired chromatid, and the nuclei which form at telophase from these divisions are haploid. Each of the two diploid nuclei from the first division gives rise to two haploid nuclei, and thus the

entire process of meiosis results in the formation of four nuclei from the original one. Each of these four nuclei contains a jumble of the genes present in the original nucleus, but only one such set of genes.

The stages of meiotic prophase

There are thus three major differences between meiosis and mitosis. The first is that two cycles of chromosome events follow one another without a period of DNA synthesis between. The second is that during the first anaphase, separation of the chromosomes does not involve the separation of the kinetochores of paired chromatids. The third is the crossing over process which occurs during prophase of the first division.

The first meiotic prophase involves several stages which are recognisable and named on the basis of the appearance of the chromosomes. The first appearance of the chromosomes as threads in the nucleoplasm is called leptotene. The chromosomes are at this stage attached to the nuclear envelope. The homologous chromosomes move together and pair during the stage known as zygotene. Pairing seems to proceed from the nuclear envelope inwards, and it involves the appearance of a structure called the synaptonemal complex (Fig. 1.29). The synaptonemal complex can be regarded as a sort of 'zip' which causes the homologous chromosomes to become tightly associated throughout their length. This process of association is known as synapsis.

Once synapsis is completed, the long stage of pachytene begins. This may last for days, and it is during this time that the crossing over between the homologous chromosomes takes place. The detailed mechanism of this process is mysterious, but it appears to involve the movement of large protein complexes known as recombination nodules, which are thought to mediate the exchange of genetic material between the associated DNA strands. The end of pachytene is marked by the partial separation of the bivalents, which remain attached to each other at one or more points – the chiasmata.

The stage which follows may also be very extended, and is known as diplotene. During this stage the edges of the chromosomes are often diffuse in outline, which is thought to represent the formation of open loops of DNA permitting transcription of RNA. Eventually these loops are reincorporated into the chromosomes, which undergo contraction and lose their attachment to the nuclear envelope. The point of their maximum contraction is called diakinesis. Metaphase follows, and it then leads onto the anaphase of the first division, when the bivalents finally part completely at the chiasmata and are separated into two daughter nuclei. These events are summarised in diagrammatic form in Fig. 1.30.

The significances of mitosis and meiosis

In terms of the developmental behaviour of individual plants, it might be said that meiosis is of very limited interest. The production of haploid gametes by meiosis requires a diploid cell line as its starting point; it is therefore important that the germ line, as it is known, be maintained in a diploid state. This is probably achieved within the flowering plant by a low rate of mitotic cycling at the shoot meristem (Chapter 6). However, when the significance of meio-

sis in species survival and development is considered, it is clearly very great. The ability to recombine the genes present in the diploid state means on the one hand that if beneficial mutations occur at multiple sites, there is a chance of some of the progeny receiving all the benefits. Similarly, the appearance of a lethal mutation may be less important since it can be diluted by the processes of meiosis and sexual reproduction. It is important to realise in this context that the description which has been made of meiosis is generally applicable, but its position in the life cycle of the plant depends on evolutionary position of the organism. In flowering plants, the prolonged and visible stage of the life cycle is the diploid sporophyte; the haploid gametophyte stage is short lived and much reduced in size. In lower plants the haploid gametophyte is reduced in size but nonetheless free living.

Within the life of the individual vegetative plant, the cell division cycle has several apparent purposes. The first is to permit growth by a cycle of DNA replication. In fact this may not involve mitosis and cytokinesis, but may result in the formation of a polyploid cell compartment. The second purpose is to

Fig. 1.29. The synaptonemal complex in meiotic cells in the anther of barley (*Hordeum vulgare*). Bar = 0.1 µm.

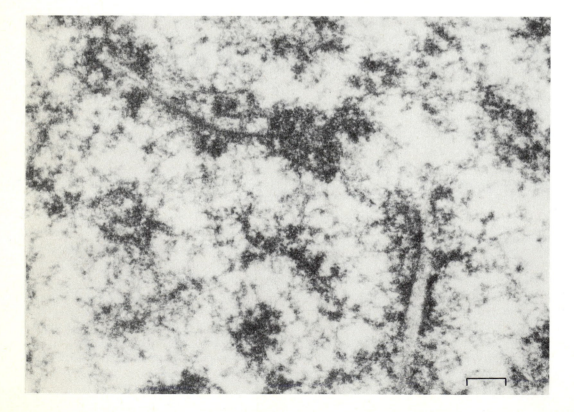

generate new cell compartments, each with an identical genetic capacity, but with a renewed capacity for growth. Most important, this renewed capacity is not necessarily uniformly expressed. Whilst the genetic material of each cell produced by mitosis is unchanged, cytokinesis allows the environment surrounding each nucleus to be different. It is this difference which gives rise to behavioural changes which we recognise as differentiation and development. The significance of cell division in the initiation of new patterns of growth and development is a theme which will recur in later chapters.

Fig. 1.30. Diagrammatic representation of the events of meiotic prophase 1. *a*, Leptotene, the chromosomes are attached to the nuclear envelope, but independent of each other; *b*, zygotene, the homologous chromosomes begin to pair by means of the formation of the synaptonemal complex; *c*, when synapsis is finished, pachytene begins (this may be extended in time); *d*, diplotene – another extended stage, with the chromosomes attached at the chiasmata; *e*, diakinesis, the point of maximum contraction of the paired chromosomes; *f*, metaphase then follows.

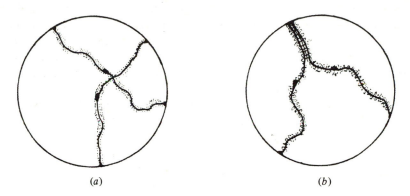

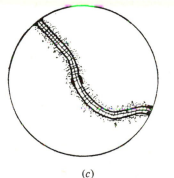

(a)

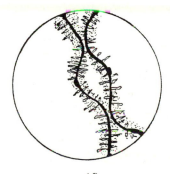

(b)

(c)

(d)

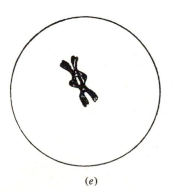

(e)

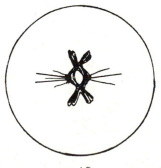

(f)

Summary

This chapter has presented an image of the idealised cell, and of its division. Cells within meristems approach this description closely, while during the course of development, a set of variations is imposed upon this basic theme. The living cell however always retains a complex series of sub-compartments, separated in space and by membranes. Within these compartments different aspects of the metabolism of the cell are contained and find their functional expression. The idealised meristematic cell contains in some form all the structures which are required for the development of a specialised role elsewhere in the plant. Variations in the organelle population, or even the type of organelles present, are common during development. On the other hand, other components of the cell, such as ribosomes and microtubules are more universally present in all living protoplasts at whatever their stage of development.

Development from this idealised state is of course not a simple matter of changing the relative importance of the components of the cell. Mesophyll tissue, for example, does not develop from the shoot meristem simply as a result of the conversion of proplastids into chloroplasts. Functional tissues do not arise in isolation. In the case of the mesophyll for example, the progeny of cells within the meristem must develop photosynthetic capacity, expressed partly in terms of the formation of chloroplasts; the cells also enlarge, become positioned within a network of air spaces, must be able to export the photosynthetic product for the benefit of the whole plant, and their environment must be controlled so as to maximise the efficiency of energy conversion. All these factors involve cooperative development within the context of neighbouring cells. Thus while the present chapter serves to emphasise the complexity of the cell environment at an individual level, this must always be regarded as only a part of the complexity of the plant itself.

The plastids

Introduction

It is convenient to regard cell development in terms of a series of changes from an original starting point, and this has been taken to be the meristematic cell. The structure and functions of any other cell in the higher plant body can notionally be traced back to this state through a series of intervening stages. The nature of the changes which accompany development is diverse, and the means of their detection equally so. At the grossest level, the cell wall represents a region which indicates the progress of differentiation along many pathways, most notably those of vascular development (Chapter 3). Within the cytoplasm, organelle populations may change in relative numbers – such variations require careful statistical analysis for their detection. Particular types of organelle may show characteristic changes within their structure. Growth, for example, in general involves the coalescence and expansion of cell vacuoles. All these changes of course imply variations in the biochemical abilities of cells as they develop. These may not be immediately detectable, but only inferred from a later and more visible effect. Modern methods of cytochemistry are beginning to approach this type of change more directly.

These processes do not occur in isolation or at random. Cell development is an ordered and, moreover, a concerted process involving all aspects of the cell's metabolism. Nonetheless, it is useful to consider one particular family of organelles in detail for the lessons which their behaviour can teach us concerning development in general. This family is the plastids: the best studied and most diverse group of organelles in the cytoplasm of higher plant cells. Plastids are of particular interest in developmental studies because the form in which they occur is closely correlated to the cell type which contains them. In meristematic cells, the representative plastid is an undifferentiated organelle called a proplastid. In the course of the development of the cell to other differentiated forms, its plastids too differentiate and develop new structures and functions. These changes are a result of the interaction between the genetic systems of the plastid and the cell, together with environmental influences. Thus the study of plastids forms a valuable introduction to the study of the development of whole cells since, in many ways, the two processes are comparable. In the present chapter we shall begin with an examination of the various forms of plastids which occur in higher plants. One of these, the chloroplast, will be considered in detail in terms of its structure and formation. Finally, in order to emphasise the flexibility of plastid behaviour in response to different genetic and environmental circumstances, we shall consider certain special aspects of plastid biology, the role of plastids in the production of variegation and the extent of their autonomy within the cell.

The structure of plastids

The plastids are a single family of organelles and, as such, have many features in common no matter what precise form they may take. Different classes of plastids are recognised on the basis of different special elements in their structure and biochemical activities. It is useful at the outset to consider plastids as though they were 'cells within cells'. This concept is a great simplification, but it is helpful initially since there are several points of superficial similarity between the plastids within a cell and the cell itself. Thus, plastids may exist as an undifferentiated type, the proplastid – corresponding to the meristematic cell. They may divide and increase their numbers by fission – comparable to cell division. They contain DNA and RNA and are capable of directing the synthesis of some of their constituent proteins. However, the concept of a cell within a cell is

just a working device; plastids are by no means autonomous or independent of the cell within which they occur, as will be seen during the course of this discussion. Plastids have no prospect of long term existence except in the environment of the cell cytoplasm.

All the plastids are surrounded by a double envelope, consisting of two membranes. The inner of these two membranes is the source of the differentiated membraneous structures which occur in the chloroplast. The central space within a plastid is called the stroma. This almost always contains ribosomes and regions in which DNA is localised, which are known as nucleoids. A variety of other inclusions may lie within the stroma, amongst which the most common are starch, lipid droplets and sometimes protein crystals. The relative development of membrane systems and stroma contents defines the different classes of plastids.

The proplastid

The simplest and least differentiated member of the plastid family is the proplastid. Characteristically proplastids are found in the zygote, the root and shoot meristems and in the reproductive tissues. Proplastids are often more or less spherical in shape, with a diameter of about 1 μm. Internal structure within the proplastid is minimal – in many sections the stroma consists merely of an amorphous mass. Usually however it contains a few scattered vesicles of indeterminate shape, together with a few lipid droplets and a small quantity of starch (Figs. 1.14, 2.1).

The cells of the shoot meristem contain in the region of 7–20 proplastids; those of the root meristem rather more, up to about 40. Proplastids are also found in the apical cells of some algae and lower plants. In general however, algae have a rather uniform population of chloroplasts, and a wide range of different plastid types does not occur on an evolutionary scale until the angiosperms are reached. It is generally accepted that, in angiosperms, proplastids are the original forerunners of all the other types of plastid in the family. Their structure suggests that they have no particular special function in a biochemical sense, but this is difficult to prove experimentally since they are not easily isolated from cells in an uncontaminated form.

The amyloplast

The next simplest plastid type is the amyloplast. This is an non-green plastid which accumulates starch. Amyloplasts are typically found in storage tissues such as cotyledons, endosperm and tubers. The familiar potato tuber is perhaps the best known example of a tissue which contains large numbers of amyloplasts. In these situations it can reasonably be assumed that the amyloplast acts as a source of energy reserve for the growth of the tissues which the storage organ supports. Amyloplasts are also found transiently in cells of the central region of the root cap. Here, they act as sensors for the direction of the gravitational field. The differentiation of the root cap will be considered in more general terms in Chapter 4, but it is interesting to note at this point that whilst the amyloplasts of the central cells act as gravity sensors or 'statoliths', as growth and division displaces the central cells to the outermost regions of the root cap, the starch within the amyloplasts is mobilised to produce an excreted polysaccharide. This illustrates a double function of the amyloplast at different stages in the life of the cell in which it appears.

Amyloplasts may be formed directly from proplastids by the deposition of starch within vesicles derived from the inner of the two plastid membranes. In other cases the starch appears to form directly within the stroma. The size of an amyloplast depends upon how much starch it has accumulated. Within the potato tuber, for example, amyloplasts with dimensions of 10–20 μm are not uncommon. The overall shape of an amyloplast depends on the number of starch grains within it, and on their disposition (Fig. 2.2). Some contain only one grain and are more or less spherical. In other cases a mass of starch grains may occur in one part of the organelle with a protuberance away from this region which contains only stromal materials. In extreme cases, it is sometimes found that the outer plastid envelope ruptures as a result of the

Fig. 2.1. A proplastid from a cell in the meristem of a maize root tip. The stroma contains starch grains, lipid droplets and a few membrane profiles. Bar = 1 μm.

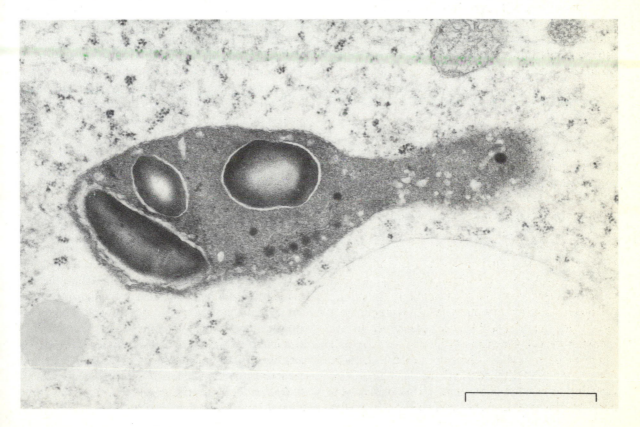

growth of the starch grains. It is, however, uncertain whether this is a real situation or simply an artefact produced by fixation methods used for electron microscopy.

During the development of amyloplasts it is sometimes observed that elements of the endoplasmic reticulum become closely associated with the outer plastid envelope. This probably reflects a role for the ER in the transport of materials into the plastid. The rate of accumulation of starch within plastids is often very high. A similar association of ER with plastids is found on a more regular basis in the green cells of some algae. It also occurs during early spring in leaves of spruce (*Picea abies*).

The accumulated starch within an amyloplast consists of two different polymers, amylose and amylopectin. Amylose is an unbranched polyglucan in which the glucose units are linked α1-4. Amylopectin has α1-6 linkage also, which results in branching and a polymer with reduced solubility. The relative amounts of these two polymers may vary according to the type of cell in which the amyloplast is found. For example, sieve tube starch tends to have a high proportion of amylopectin. On the other hand, differences within a single cell type may arise genetically;

Fig. 2.2. A group of amyloplasts in the central cells of the root cap of maize. The shape of the organelles is dictated by the presence of many starch grains in the stroma. Bar = 5 μm.

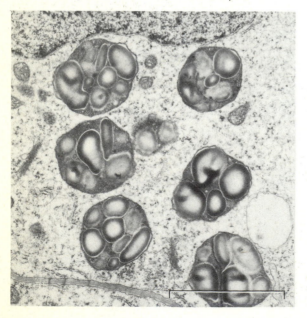

thus 'waxy' varieties of *Zea mays* (corn) almost exclusively accumulate amylopectin in the endosperm. This represents a very simple example of biochemical differentiation resulting from the local expression of genes during the course of development; this ability existing within the context of the total genetic capacity of the plant.

Other types of leucoplast

Proplastids and amyloplasts are examples of plastids which lack pigmentation. The generic name for such organelles is leucoplasts; plastids with pigments are generically known as chromoplasts. There are a few other types of leucoplast, and these can be considered together since, in a quantitative sense at least, they are of minor importance only. In developing sieve elements (p. 95), the final stages of differentiation involve the degeneration of the internal membrane systems of the plastids, the loss of ribosomes from the stroma and the accumulation of starch (predominantly amylopectin; Fig. 2.3) or protein deposits. These changes tend to be species-specific rather than linked to the course of differentiation *per se*. Indeed, it is possible to classify some plants on the basis of the types of protein crystal which are formed within their sieve element plastids. Protein crystals are also found in other situations, often in plastids which will later form chloroplasts. The transient appearance of protein deposits in this way suggests that they may represent a reserve of material held ready for later development. Protein crystals also occur in chloroplasts which are submitted to water stress, for example in mesophyll cells which have been subjected to plasmolysis.

In one situation a leucoplast provides a remarkable demonstration of the specificity of plastid differentiation linked to cell development. The plant *Cecropia* produces a highly specialised structure consisting of multicellular bodies up to 1 mm in diameter, which are provided as food for ants. The plastids within these ant-food bodies do not contain pigments, but accumulate glycogen. Glycogen is a storage polysaccharide which is normally found in liver cells in animals. The appearance of glycogen within the plastids of *Cecropia* is limited to the ant-food bodies; elsewhere the storage plastids contain starch. This clearly implies that the genes controlling glycogen production are only expressed in a single cell type in this plant.

Other plastids may accumulate essential oils, and these are sometimes called elaioplasts. Again, the occurrence of such minor groups of plastids is usually limited to particular cell types in a very striking way. These observations should alert us to the probability that plastid development is deeply linked to the development of the cell in which they occur.

Chromoplasts

The remaining types of plastid are pigment containing, and known generically as chromoplasts. Of these, by far the most widespread and important is the chloroplast, which will be considered in detail below.

Chromoplasts are the site of many of the red, yellow and orange pigments which are found in fruits

Fig. 2.3. A group of sieve element plastids within a sieve tube in a leaf of tobacco (*Nicotiana tabacum*). The contents of the plastids have more or less degenerated apart from small starch grains and a few membraneous vesicles. The double plastid envelope remains. The fibres at the outer surfaces of the plastids are P-protein. Bar = 0.5 μm.

such as tomatoes and peppers, and in roots such as carrots and sweet potatoes. This specialisation for the production of pigments is evidently a late evolutionary development; its function seems to be to attract animals. The yellow colour of many flowers resides within the plastids (blue pigments tend to be due to substances accumulated in the cell vacuole). The structure of chromoplasts is variable and depends upon the cell type in which they are found (Fig. 2.4). They have been classified according to their types of inclusion; this has some value insofar as it may reflect the way in which the chromoplasts are formed.

Globular chromoplasts are typified by the presence within the stroma of large numbers of osmiophilic droplets or 'plastoglobuli'. Plastoglobuli are a constant feature of mature green chloroplasts and, in this situation, they may represent a breakdown product of their photosynthetic membranes. However, although globular chromoplasts may arise from the breakdown of chloroplasts, they may also be formed directly from proplastids or amyloplasts without the intervention of a green stage. Floral chromoplasts of this type may arise in this direct way; in other tissues such as some leaves, globular

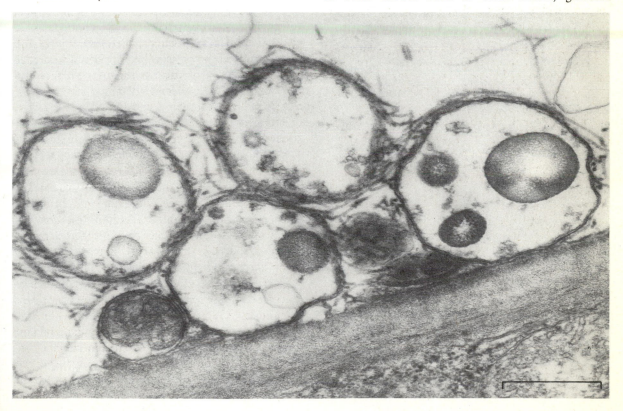

chromoplasts represent a stage in senescence.

The pigmentation of the chromoplast may also occur in the form of crystals – this is true of the chromoplasts found in carrot roots and the fruits of *Capsicum* (peppers). Again, different routes of formation are at work in these two tissues; within the root of carrot chloroplasts are not the source – the chromoplasts therefore develop more or less directly from simpler plastid types, the proplastid or the amyloplast. In the ripening fruit of the pepper, however, initially green tissue which contains chloroplasts changes to red or yellow as the constituent pigments of the plastids change their relative quantities.

Finally, chromoplasts may be classified as tubular or membraneous. Such plastids are characterised by the presence of large amounts of internal membrane material, but not in the highly organised form which is found within the chloroplast. Once again, plastids of this type may be formed from chloroplasts by a process which may be regarded as degenerative, or they may arise more directly from proplastids.

The chromoplasts as a group are highly diverse and it is difficult to draw general conclusions about their occurrence. Nonetheless, they demonstrate

Fig. 2.4. A chromoplast within a cell in the petal of a yellow *Chrysanthemum* flower. The stroma contains single thylakoid-like membranes together with spherical bodies, possibly corresponding to pigments. Bar = 1 μm.

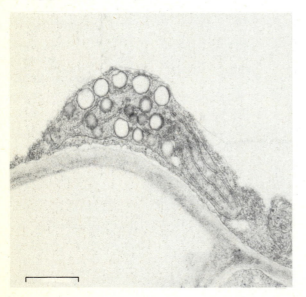

admirably several facets of plastid biology. Foremost amongst these is the plasticity of the family; its members are related to each other by a variety of different pathways, some involving processes which correlate more with senescence than with development. They also demonstrate again the way in which a cell is matched to its plastid population; a particular type of chromoplast might only occur in the petals of the flower and nowhere else throughout the entire life cycle of the plant.

The biochemistry of chromoplasts is likewise complex and highly variable. In the tomato fruit, the red pigment is predominantly lycopene, with small quantities of carotene and phytofluene. On the other hand, in the red fruits of *Capsicum* over 30 different pigment molecules have been identified, all present in relatively small amounts. Carotenoid pigments are also found in chloroplasts, but they do not show a wide range of diversity in this situation, which presumably reflects their functional importance in photosynthesis. The 'function' of pigments in chromoplasts is not directly related to the biology of the plant, but rather to the visual sense of animals.

The chloroplast

The chloroplast is the most important and widespread member of the plastid family. It is the site of the light reactions of photosynthesis and as such is the primary converter of the energy of sunlight for the whole of the plant kingdom. In a typical leaf cell there may be 20–50 chloroplasts, and it has been calculated that this represents a chloroplast concentration of about $500\,000$ mm^{-2} of leaf surface. More is known about the chloroplast than about any other type of plastid. Being so widespread it is easy to prepare pure fractions of chloroplasts and study their properties *in vitro*. Since they are green, chloroplasts can be detected visually which means that aspects of their genetics and partition between cells may be studied comparatively directly.

The typical chloroplast from a leaf of a higher plant is a lens-shaped organelle with one convex face and one face which may be convex, plane or concave (Fig. 2.5). A chloroplast is usually about 5 μm long and 2–3 μm thick. Like all plastids, it is bounded by a double envelope. Each unit membrane of the envelope is about 8–10 nm thick, and the space between them is about 10–20 nm wide. In electron micrographs this space appears to be empty. The outer membrane is very easily lost during the preparation of isolated chloroplasts; when this happens, considerable leaching of soluble proteins from the stroma occurs. The chloroplast envelope is by no means a static structure, and it is common to see infoldings of the inner membrane, in addition to those which gives rise to the formation of the photosynthetic membranes themselves. In many cases, a complex peripheral reticulum of membranes is found within the stroma beneath the inner envelope. This is thought to be concerned with the transport of materials into and out of the chloroplast.

The major protein component of the chloroplast stroma is the enzyme ribulose diphosphate carboxylase. Within the stroma are found a variety of particles, including ribosomes, starch grains and plastoglobuli (Fig. 2.6). The ribosomes of the chloroplast are smaller than those which are found in the cytoplasm – they have a diameter of 17 nm compared to that of 25 nm for cytoplasmic ribosomes. In addition, they have a lower sedimentation coefficient – 60S rather than 70–80S, and they differ from cytoplasmic ribosomes in their sensitivity to chemical inhibitors. They are usually seen as individual units, although occasional polysome-type patterns are found, suggesting that the ribosomes are active in protein synthesis. Plastoglobuli are ubiquitous in the chloroplast stroma, and usually take the form of dark-staining spherical particles with a diameter of 10–15 nm. They may be free in the stroma, or they may be found in close association with the granal membranes; it has been suggested that they represent a reserve of storage lipid. Plastoglobuli are, however, common in ageing chloroplasts; this would suggest that they arise as a result of a change in the organisation of the photosynthetic membranes.

Starch grains are a common feature of the chloroplast stroma and, as in the case of amyloplasts, the shape of the whole organelle may be distorted by the presence of several large starch grains. The starch within the chloroplast seems to lie directly within the stroma and is not limited by a membrane. The

Fig. 2.5. A chloroplast within the mesophyll of a young leaf of tobacco (*Nicotiana tabacum*). The stroma contains thylakoid membranes with grana of up to four thylakoids. A single starch grain is also visible. Bar = 1 μm.

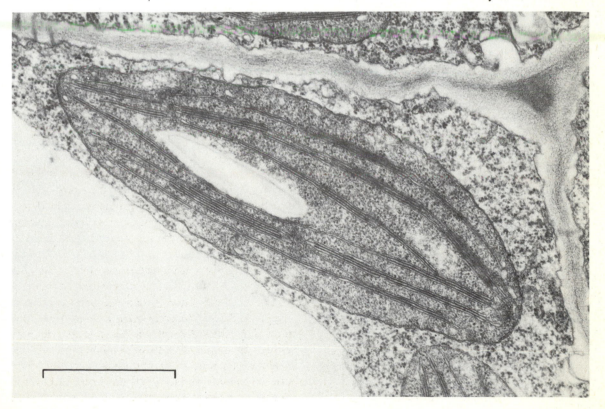

accumulation of starch may vary considerably between adjacent cells of different types within a single leaf. For example, chloroplasts within bundle sheath cells often contain more starch than the adjacent mesophyll cell chloroplasts. This may reflect the importance of the bundle sheath in the movement of photosynthetic products out of the leaf. It is also often true, particularly in grasses, that chloroplasts within the bundle sheath show less development of photosynthetic membranes than is found in adjacent mesophyll cells. This is a further example of the matching which occurs between cell type and plastid structure.

Other components of the stroma which are less frequently seen include phytoferritin and crystalline bodies composed of protein. It is also occasionally possible to see clear areas within the stroma which

contain 2.5 nm fibrils corresponding to the chloroplast DNA. Plastid DNA is discussed in more detail below (p. 62).

The lamellar system of the chloroplast

The most striking structural feature of the chloroplast is the presence within the stroma of a complex lamellar system. Using the light microscope, it is just possible to resolve individual green particles within the chloroplast; these were termed 'grana'. The reality of these green granules within chloroplasts was disputed until the advent of electron microscopy. Nowadays there is no doubt of their existence (Fig. 2.7) and of their central importance in photosynthesis.

The grana consist of stacks of membranes which are in the form of a flattened sac enclosing a space (Figs. 2.8, 2.9). These sacs are known as thylakoids, and a granal stack may consist of just 2, or up to 100, thylakoids. In sections, grana are seen to be separate within the stroma, but interconnected by paired membranes. These membranes are known as the stroma lamellae, or frets. The three dimensional

Fig. 2.6. A chloroplast within a cell of the protonema of the moss, *Physcomitrella patens*. A large quantity of starch is present in the stroma, together with thylakoid membranes and a few lipid droplets (plastoglobuli). Bar = 1 μm.

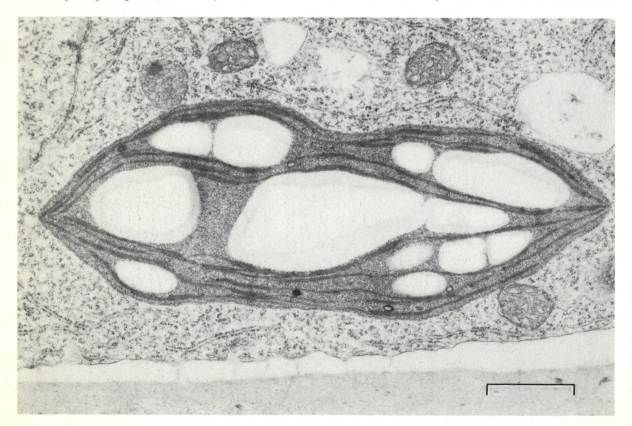

distribution of membranes is best shown by looking at shadowed preparations of fragmented chloroplasts. When this is done with chloroplasts from *Vicia faba*, it becomes clear that the grana are circular; they look rather like a pile of coins. The system of frets is also arranged as a circular array of interconnections.

The consequence of this arrangement of membranes is that the internal space enclosed within each thylakoid is in direct continuity with all the internal spaces within all the grana and frets. Considerable variations on the theme of circular grana which are aligned parallel to one another are possible. Sometimes the lengths of the thylakoids are not all the same within a given granal stack, and in other cases the branching pattern of the frets may be more complicated than a circular or spiral arrangement. However, the fundamental division of the chloroplast interior into a stroma and a continuous intrathylakoidal space, separated by a membrane, is maintained in all such arrangements.

The thickness of the membrane is about 8.5 nm, and is the same in both the frets and at the edges of the granal stacks. Thus, on morphological grounds alone, it is a reasonable assumption that the membrane is the same throughout the lamellar system. This view is reinforced by considerations of how the system develops (p. 54). However, it will be recalled that the current view of membrane structure is that components of the membrane may be mobile within its surface (p. 6), so there is no reason why different regions of an apparently uniform membrane should not be the sites of different functions. The fluidity of the system in organisational terms can be demonstrated experimentally. When spinach chloroplasts are isolated under conditions of low salt concentration, the organelles are swollen in size, and the grana disintegrate into simple sheets of membrane, but without loss of chlorophyll. When salt is added back to the medium, these sheets reorganise themselves, often into apparently quite normal grana–fret con-

Fig. 2.7. Chloroplasts within a leaf of *Coleus blumei*. The appearance of the grana depends on the plane of section. In the right-hand chloroplast, the grana are clearly circular in outline. Bar = 1 μm.

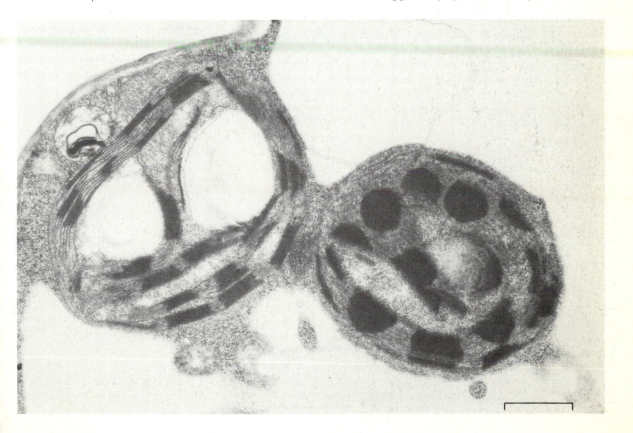

figurations. This remarkable behaviour, brought about by a simple change in the ionic conditions, serves to emphasise that the complexity of the lamellar system of the chloroplast is almost certainly derived from a single continuous membrane.

The substructure of the photosynthetic membrane

Whilst the electron microscope image of the lamellar system of the chloroplast clearly presents the

grana and frets as membraneous in character, it does not immediately demonstrate the unique character of those membranes. The membranes of the chloroplast are distinguished by the presence of pigment molecules which form an integral part of their structure. The carotenoids and the phytol chain of the chlorophyll itself are hydrocarbon entities which enter the membrane and, no doubt, substitute for other lipids which might have been present in a non-photosynthetic context.

The clearest images of the membrane substructure are obtained by the use of freeze–fracture and deep-etching techniques. These involve first the freezing of the specimen, followed either by shattering the frozen specimen with a blade (freeze–fracture)

Fig. 2.8. Diagrammatic representation of the substructure of the photosynthetic membranes. The five faces revealed by fracturing and deep etching arise from three different membrane positions.

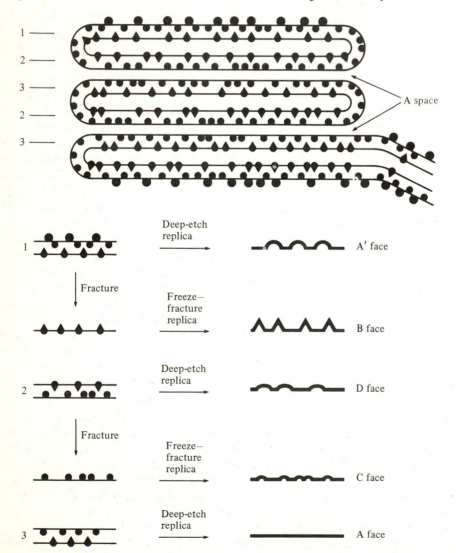

or the sublimation of water away from the surface of the specimen at low temperature and low pressure (deep etching). In both cases, the exposed surfaces are coated with a thin layer of metal, and this metal replica of the surface is the specimen which is examined in the electron microscope. It is most important to grasp a basic difference between these two techniques. When a specimen is deep etched, water is removed and the exposed surface therefore represents the 'outside' surface of any membrane in the specimen. On the other hand, when a specimen has been fractured, the exposed surfaces of membranes are 'inner' surfaces, since the fracture proceeds through the centre of the unit membrane. Thus by examining specimens prepared both by fracturing and deep etching, it is possible to analyse the distribution of particles at all four surfaces of a conventional unit membrane.

The membranes comprising the granal stack represent a complicated situation, and many terminologies have been used to describe it. The one adopted here is simply alphabetical in order to minimise confusion (Fig. 2.8). The chloroplast membrane is not symmetrical, and so freeze–fracturing reveals two different types of internal surface. The B face is that part of the membrane revealed by fracturing which has, at its back, the internal space sequestered by the thylakoid membrane (whether this is part of a disc or within a fret). The C face is also an internal surface, but is backed by the region between the discs which comprise the granal stack. The A' face is an external surface revealed by deep etching which bounds the fret channels and the outer surface of the granal stack; the corresponding outer surface within the granal stack is called the A face. Finally, the D face is the surface of the membrane which is in direct contact with the internal space enclosed by the thylakoid membrane.

The B and D faces are characterised by the presence of very large particles, 13–17 nm in diameter and 8–9 nm in height. These are present at high concentrations within the grana, but at lower concentrations in the frets. There is some evidence to suggest that within the grana these particles are ordered in semicrystalline arrays. The C face has smaller particles, 9–12 nm in diameter; these are not large enough to influence the surface morphology of the A faces, which are therefore comparatively smooth. Deep etching of the A' face shows it to be covered with particles of diameter 10–11 nm, which have the property of being removable by chelating agents.

One of the earliest attempts to correlate these structural characteristics with the presence of photosynthetic activity was based on measurements of the efficiency of light conversion into reduced carbon compounds. It was found that each molecule of carbon dioxide required at least 8 quanta of light for its reduction. When the total amount of chlorophyll present was calculated, it appeared that about 2000 chlorophyll molecules were required in the supposed photosynthetic unit (i.e. to absorb 8 quanta of light). This led to a figure of 250 chlorophyll molecules being the unit of absorption of a single quantum of light energy. When the first images corresponding to the D face were seen, it became possible to calculate approximately how many molecules of chlorophyll each large particle might contain, assuming that it was the site of the pigment. The figure reached on this assumption was 230 molecules of chlorophyll – a remarkable correspondence with the figure of 250 molecules per quantum which had been arrived at on energetic grounds. As a result of this, the large particles were given the name 'quantosomes', and each one was thought to represent a light-capturing unit. One difficulty with this idea is that the fret membranes do not have large concentrations of these particles, and yet they are known to be active in photosynthesis.

It is possible experimentally to prepare isolated fractions comprising membranes from the grana and fret membranes. When the photosynthetic capacity of these two fractions was examined, a clear result emerged at first. The frets appeared to contain only the photosystem I activity, whereas the grana contained both photosystem I and photosystem II. This suggested that the larger particles correlated with photosystem II, whereas smaller particles (present both in the grana and the frets) were the site of photosystem I. The coupling of the energy derived from light absorption to the production of ATP seems to take place at the site of the 10–11 nm particles present on the A face – that is, the face of the membrane system which is most directly in contact with the stroma. These identifications of the particles are of course only satisfactory up to a point; they fail to explain the presence of both photosystems in chloroplasts which do not form granal stacks, and they do not explain why granal stacking is necessary or at least present in higher plant chloroplasts.

The space between the discs of a granal stack is called the A space (since it is marked on either side by an A face). In fact it does not appear to be simply an empty space since, when chloroplasts are treated in such a way as to cause swelling of the grana and frets, the A space remains, as though it were the site of an adhesive between the discs. The reversible disorganisation of the grana–fret systems which is mediated by salt concentration (p. 49) suggests that the A faces of the discs may retain their identity during the disorganisation process. Possibly they represent areas on the membrane of local concentrations of charged molecules, which would be expected to have adhesive properties in the presence of solute ions. The absence of such charged molecules could then explain why certain chloroplasts do not form grana – such as the agranal chloroplasts of guard cells, or chloroplasts within algal cells. It is known that nuclear gene mutations can influence granal stacking in both higher and lower plants, and this too suggests that stacking may be mediated by the presence of particular types of molecule within the membrane.

The function of the photosynthetic membrane is thus inextricably linked to its structure. Furthermore, its chemical composition may determine the overall arrangement of the entire membrane into a system of grana and frets. The photosynthetic membrane demonstrates very clearly the importance of local variations in the composition of membranes – variations which are not apparent when conventional electron microscopy techniques are used.

Algal chloroplasts

The picture of the chloroplasts which has been drawn so far is generally true of all higher plants with certain local exceptions. For example, the limited granal development of chloroplasts within cells of the bundle sheath of grasses has already been mentioned (p. 48); guard cell chloroplasts also show a few scattered thylakoids in the stroma rather than organised grana. The variations in chloroplast structure amongst algae is even more considerable, however.

Algal chloroplasts may be lens-shaped, but they may also be star-shaped (in *Zygnema*), spiral (*Spirogyra*) or irregular in outline (*Oedogonium*). Some algal cells have only 1 chloroplast (*Chlamydomonas*), whereas others may have up to 100. All algal chloroplasts are bounded by a double membrane and all contain thylakoids and plastoglobuli. The degree of development of the thylakoid system is variable; in some cases it may be used for identification purposes. In the Rhodophyta no granal stacking occurs and the thylakoids exist as single lamellae which extend along the length of the chloroplast. In the Cryptophyta the thylakoids are found in pairs. The most common form of algal chloroplast structure has the thylakoids arranged in threes. The stacked lamellae traverse the lumen of the chloroplast without any intervening frets. In the Chlorophyta thylakoids may form dense stacks – *Nitella*, for example, has a single stack of up to 100 thylakoids within its chloroplasts. In very general terms, it can be seen that the formation of grana and frets is a later evolutionary development.

The other characteristic structure of algal chloroplasts is the pyrenoid (Fig. 2.9). In fact, not all algae possess pyrenoids, and pyrenoids are occasionally found in other plants – in, for example, *Anthoceros*, a bryophyte. In appearance, the pyrenoid is a region of densely packed fine granular material of which the major chemical constituent is protein, mostly in the form of the enzyme ribulose diphosphate carboxylase. The pyrenoid is sometimes crossed by membranes of the thylakoid system. There may be only one pyrenoid in each chloroplast, or there may be many (as in *Spirogyra*). Although, from their widespread occurrence in algae, it may be reasonably concluded that pyrenoids do have an important function, its nature is as yet uncertain. It is also not understood how higher plant chloroplasts are able to dispense with this structure altogether and yet function adequately.

An interesting feature found in many non-green algae is a marked association between the chloroplast envelope and the endoplasmic reticulum. This specialisation may be exemplified by considering the Chryosphycean alga *Orchromonas danica*. In this species the cell contains a single bilobed chloroplast and, in favourable sections, the nucleus can be seen to lie between the two lobes. The endoplasmic reticulum completely surrounds each lobe of the chloroplast, bearing ribosomes only on the surface which is distant from the chloroplast outer membrane. Direct connections can be seen between the endoplasmic reticulum and the nuclear envelope, and it has been suggested that this represents a rather direct route for the transfer of nuclear-coded proteins into the chloroplast. Why this should require a structural specialisation in only certain classes of algae is a matter for conjecture. In such cells also it is found that the photosynthetic product (corresponding to the starch within the stroma of higher plant chloroplasts) is often formed outside the chloroplast double envelope, but inside the encapsulating endoplasmic reticulum. Starch formation in other algal cells is often associated with the position of the pyrenoid (Fig. 2.9).

This diversion into alternative chloroplast structures serves to illustrate the caution which has to be used in interpreting the functions of various structural modifications. There are many factors which all chloroplasts share in common – the double outer envelope, the division of the internal space by thylakoid membranes, the presence of some machinery for protein synthesis. However, detailed understanding of the many variations which occur between different chloroplasts is certainly lacking at the present time. It must be concluded on logical grounds that particular specialisations represent only one of several possible solutions to a particular metabolic problem. The diversity of plastids within a higher plant is such that differences between them in constituent cells of a complex tissue may be as great as they are between different species at a lower point on the evolutionary scale.

The formation of chloroplasts

In the normally growing plant, chloroplasts are formed as a result of developmental changes at the shoot apex. The meristematic cells of the shoot contain proplastids which have none of the complex internal membrane system which is found within the mature chloroplast. The change from proplastid to

Fig. 2.9. A cell of the alga *Chlamydomonas asymmetrica*. The pyrenoid is visible as a dark amorphous region within the centre of the starch grain. Bar = 2 μm.

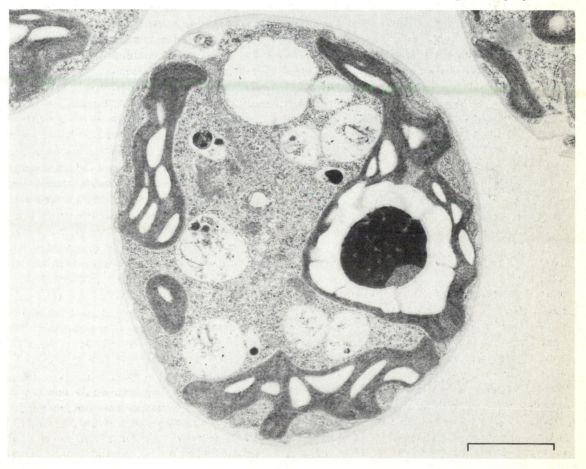

chloroplast takes place gradually as the leaf primordium is formed and expands. The change is accompanied by cell growth, and normally takes place under natural conditions of lighting. Chloroplast development along this course involves the elaboration of the inner plastid envelope to produce a series of invaginations into the stroma. These may take the form of membraneous tubules or (more probably) sheets of membrane; during the initial stages of their formation these sheets may contain numerous pores. When viewed in section this gives the impression that the stroma is crossed by strings of disconnected vesicles. At later stages of development, these pores disappear and continuous lamellar sheets give rise to the thylakoids and the grana–fret system.

It is not entirely clear how the stacked thylakoid arrangement is generated from a single membrane sheet. Two hypotheses exist to describe the process. In the first, it is envisaged that a single thylakoid-type membrane can simply fold back upon itself, giving rise to a pair of thylakoids. In the second, only one of the two leaflets of the membrane sheet is involved in the folding, and this process would give rise to a local stack of thylakoids which was inevitably connected to other stacks formed at different loci on the same membrane (Fig. 2.10). In the first model the fret connections between stacks would have to be formed by lateral extension of individual thylakoids within one

stack to make contact with an adjacent stack. Both models adequately explain the way in which an integrated internal space is formed within the granal stacks and frets. In higher plants it is evident that the internal lamellar system is generated from several initial infoldings of the inner plastid envelope; in the bryophyte *Anthoceros* it has been suggested that a single invagination of the plastid envelope gives rise to the entire internal lamellar system.

Variations on this pattern of membrane development do occur in special situations. For example, in the cereals, the meristem is at the base of the leaf and, therefore, although the plant may be growing in bright light, the meristem is in fact exposed to very dim light. In this case development from the proplastid may proceed via the initial formation of a prolamellar body, characteristic of the etioplast (see below). This type of development may also occur in the cotyledons of species which initially are underground, but which emerge into the light following germination. In seedlings of *Cucurbita*, reversible formation of a prolamellar body may occur during the night period before mature chloroplast structure has developed in the cotyledons. Finally it should be realised that in unicellular algae, chloroplasts do not develop from proplastids, but are always present in their mature form.

The etioplast

The changes which have just been described are constructed as a sequence from a series of static images of normally developing leaves growing in the light. It is, however, possible to study the formation of chloroplasts in angiosperms in a much more con-

Fig. 2.10. Possible mechanisms for the generation of stacked thylakoids. The single membrane (*a*) may either fold over itself (*b*), or one leaflet may fold over the other (*c*). A stack can be produced by a combination of these two mechanisms (*d*).

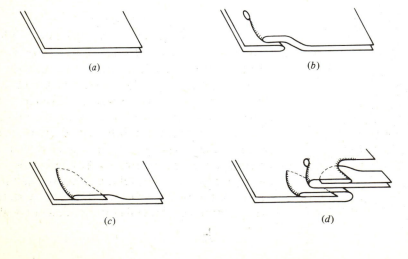

(*a*)   (*b*)

(*c*)   (*d*)

trolled manner by the use of dark-grown plants. When angiosperm seeds are germinated in the dark, they develop into seedlings which, in the great majority of cases, lack chlorophyll. Such plants are called etiolated plants. This ability to control the formation of chlorophyll in the absence of light is more or less confined to the angiosperms. Gymnosperms, mosses, ferns and most algae can produce chlorophyll in the dark. The experimental value of using etiolated plants is that when they are returned to the light, they quickly turn green and resume a normal appearance. This means that the events associated with greening can be studied in a more controlled manner than is possible with normal light-grown plants.

Etiolated plants are abnormal in many respects, but they are capable of forming leaves. When the plastids of such leaves are examined it is found that they have a characteristic structure. They have been called etioplasts in recognition of their particular identity (Fig. 2.11). An etioplast is usually ellipsoidal in shape, about 3 μm long (slightly smaller than a mature chloroplast) and bounded by a double membrane. Etioplasts may contain starch, but the structure

Fig. 2.11. An etioplast from a leaf of an etiolated maize plant. The stroma contains a small amount of starch, with a few single thylakoid membranes and a large prolamellar body. Bar = 1 μm.

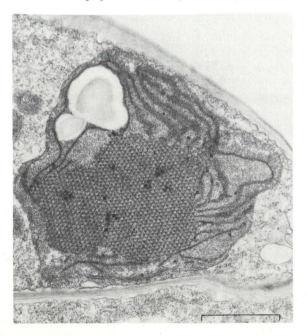

which defines the organelle is a highly ordered paracrystalline region which is called the prolamellar body. This is membraneous in character, and consists of a series of interconnected tubules on a cubic lattice. It is built up from porous invagination of the inner plastid envelope, and therefore can be regarded as having the same origin as the normal lamellar system of the mature chloroplast developing in the light. The basic building block of the prolamellar body is most commonly a tetrahedral arrangement of tubes; alternatively a six-fold symmetry may be found in the basic unit. At the edge of the prolamellar body are found a small number of flattened pieces of membrane resembling single thylakoids.

The prolamellar body represents a stage in the development of the chloroplast membrane system which is not on the normal physiological route for light-grown plants, with some exceptions (p. 54). It is the site of the protochlorophyllide, which is present in the etiolated leaf, and also of carotenoid pigments. An interesting feature of the prolamellar body is that it may take slightly different crystalline forms within etioplasts of a single cell; it is even possible that a single etioplast may contain prolamellar bodies of different structure. The form of the prolamellar body represents a very efficient way of presenting a large area of membrane surface within a small volume. The stroma penetrates between all the interconnecting tubules but is excluded absolutely from the continuous internal space which they enclose. It is probable that this arrangement is designed to allow the insertion of new membrane components from the stroma at all points of the membrane surface, rather than just at the edges (as would be the case in a less open network).

The effect of light on the etioplast

When etiolated leaves are exposed to a bright light for a few minutes, the first visible change in the etioplasts is that the highly ordered arrangement of the prolamellar body is lost. This corresponds to a rapid conversion of protochlorophyllide into chlorophyllide, and the subsequent esterification which gives rise to chlorophyll itself. These chemical changes occur rapidly – within 20 min or so of illumination in the case of leaves of *Phaseolus*. As greening continues, the prolamellar body gives rise to a series of sheets of membranes which extend out into the stroma in roughly parallel arrays. This event takes place more rapidly in bright light than it does in dim light, and it may be completed within a few minutes. Calculations show that the amount of membrane in the prolamellar body is roughly equivalent to the amount in the primary thylakoids which are formed as it disintegrates, so that this process may be regarded as one of rearrangement rather than one involving the incorporation or synthesis of new materials. The primary thylakoids formed in the prolamellar body are porous, just as are the early stages of formation of the prolamellar body itself, and the thylakoid system in light-grown plants (p. 54). These pores are relics of the lattice arrangement within the prolamellar body, and they disappear as greening continues. Once the parallel arrays of single membranes have been formed, normal chloroplast structure is achieved by the localised production of stacked thylakoids over the surface of the membrane sheets. This is thought to correspond to the second of the two models discussed for the formation of grana in normal plants (p. 54). This process takes several hours to complete and is accompanied by net chlorophyll synthesis.

The conversion from etioplast to chloroplast is to some extent reversible. Thus the breakdown of the ordered crystalline structure of the prolamellar body which accompanies a brief period of illumination is completely reversed if the plant is returned to the dark. As mentioned above, this may occur during normal growth in the cotyledons of *Cucurbita*.

Interconversion of plastids

The different types of plastids which have been described do not exist as isolated populations. The plastids of any particular cell correspond to the activity or state of differentiation of that cell. Thus meristematic cells contain proplastids, sieve elements contain sieve element plastids, and so on. Since all the cell types of a higher plant are ultimately derived from the zygote, and this cell contains proplastids, it follows that all the plastid types of the higher plant can be developed from the proplastid. The question arises as to whether the route to the formation of a particular plastid is a fixed one, and whether the process of plastid development can be reversed.

As has already been seen, some cases of plastid interconversion are relatively direct. The formation of amyloplasts within the root cap is a development which occurs very rapidly and simply by the growth of the proplastids of the originating meristem and by the

deposition of starch within the stroma. Similarly the development of chloroplasts from the proplastids of the shoot meristem is often relatively direct, although it may proceed through an etioplast stage in special circumstances. Chloroplasts are commonly regarded as an end product of plastid development; this, however, ignores the possibility of the formation of chromoplasts from chloroplasts, particularly well known in developing fruits which are initially green but which become red, yellow or orange as they ripen.

Chromoplasts are not always formed from chloroplasts, however. In the petals of *Iris germanica*, pigment is deposited within amyloplasts. The chromoplast is produced by growth following the loss of starch from the plastid stroma. In the epidermal cells of petals of *Lilium tigrinum*, the chromoplasts are produced directly from proplastids.

It is known that certain pathways of plastid differentiation are capable of reversal. Thus whilst it is common to produce chromoplasts from chloroplasts, the reverse process can take place, for example, in illuminated carrot roots. Chloroplasts can also be formed following illumination of storage tissues; the greening of potato tubers is a well known example of this, and corresponds to the conversion of amyloplasts into chloroplasts. Reversal of plastid differentiation also occurs when mature cells revert to a meristematic state. This is seen in the case of wound cambia and also in the formation of callus tissue induced by experimental treatment (Chapter 5). In such cases cycles of cell division are accompanied by the formation of proplastids from more differentiated types, such as chloroplasts or amyloplasts. Some of the pathways of plastid conversion are shown diagrammatically in Fig. 2.12.

Several points emerge from a consideration of these processes. Plastid differentiation, like cell differentiation, is a gradual process. Whilst it is convenient to talk of proplastids, amyloplasts, chloroplasts, etc, in fact there are many intermediate forms which have structural features of more than one pure plastid type. The state of the plastid population within a cell is to some extent dependent upon environmental factors. The presence or absence of light is the most obvious influence on the formation of chloroplasts in many tissues, for example. Nonetheless, it is clear that there is a marked correspondence between a cell type and its plastids. The presence of light does not inevitably cause chloroplast development – this is witnessed by the widespread occurrence of flowers which are not green; similarly, that roots do not develop chloroplasts is not simply due to the fact that they normally grow in darkness. The implication from even such general observations must be that a considerable measure of control is exerted by the cell over its plastid population. It can be inferred that the very processes of selective gene action which underlie the development of the cell itself have an important effect on the plastids which it contains.

The replication of plastids

The number of proplastids within a cell of the shoot meristem is of the order of 7–20. The number of chloroplasts within a mature leaf cell is higher, perhaps as many as 50. During leaf development there is an increase in cell numbers by division, and so it is clearly necessary that in order to increase their numbers per cell, plastids too must in some way reproduce themselves.

It is a reasonable assumption *a priori* that plastids increase in number by a replication process. The most convincing indirect evidence for this is that,

Fig. 2.12. Diagram to show some of the interconversions possible between different plastid types.

within a developing cell line, the plastid populations of increasingly mature cells are themselves at corresponding stages of their own maturation. In other words, all the plastids of a cell, no matter what its position in the plant, are normally at a similar level of differentiation. Clearly if plastids were continuously produced *de novo*, this could not be the case. A green mesophyll cell would in those circumstances show mature chloroplasts and a range of other stages corresponding to 'younger' plastids. This is not observed. Plastids contain their own DNA, which is further circumstantial evidence for an ability to replicate themselves.

The division of chloroplasts was first observed in cells of filamentous algae, and it is now possible to follow the process continuously using microcinephotography. In this way, it has been established beyond doubt that the chloroplasts of such plants as *Nitella* and *Spirogyra* replicate themselves by a fission process. In higher plants it is difficult to observe chloroplasts continuously, and so this type of direct evidence for fission is lacking. However, many electron micrographs of green leaves show chloroplasts which are dumbbell-shaped with a central constriction; this is taken to represent a stage in the division process (Fig. 2.13). The situation in angiosperms, in particular, is complicated by the fact that

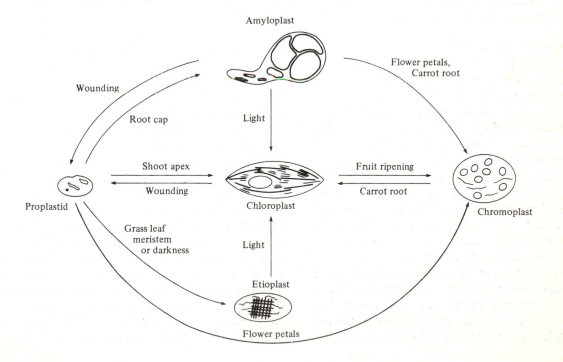

whilst mature green chloroplasts do seem to be capable of replication, most of the increase in plastid numbers which accompanies growth takes place whilst the plastids are in a partially differentiated state. In the light microscope these immature stages are difficult to identify and distinguish from other cytoplasmic particles, particularly mitochondria. Direct observation of plastid fission is therefore rendered practically impossible.

Nonetheless, the evidence from the electron microscope, even if it were the only source of information concerning plastid replication, is quite compelling. Counts of numbers of proplastids in very young developing leaves show that the replication of plastid units approximately matches the rate of cell division – in other words that the number of proplastids per cell remains more or less constant even in a dividing cell population. In the later stages of leaf development when leaf enlargement is due to cell growth rather than division, the number of plastids

per cell is seen to increase. In certain tissues it is possible to reconstruct a linear time sequence. For example, the leaves of maize develop from a basal meristem, so that if sections are made at different distances from the base of the leaf, a sample of cells of differing ages is taken. It is found that in 7 day old leaves there is a zone about 2.5 cm from the base of the leaf which shows a particularly high proportion of dumbbell-shaped plastids within the cells. If the plastids are isolated from this region they are found to be 2–3 μm long and show amoeboid movements.

More direct evidence that dumbbell-shaped plastids do indeed represent the organelle in a state of division has come from studies on isolated chloroplasts. When chloroplasts are rapidly isolated from young leaves of spinach in a medium containing inorganic salts, bovine serum albumin and an osmotic buffer, it is possible to study their replication *in vitro*. During the 100 h following isolation, each plastid divides (on average) twice. The process is signalled by the formation of a central constriction, giving rise to the dumbbell-shape which is seen in sections of fixed material. This state is fairly stable over several hours; division is then completed by a narrowing of the constriction to the point where it becomes tubular,

Fig. 2.13. A dividing chloroplast in the mesophyll of a young leaf of the pea (*Pisum sativum*). The constriction in the centre of the chloroplast is crossed by thylakoid membranes. Bar = 2 μm.

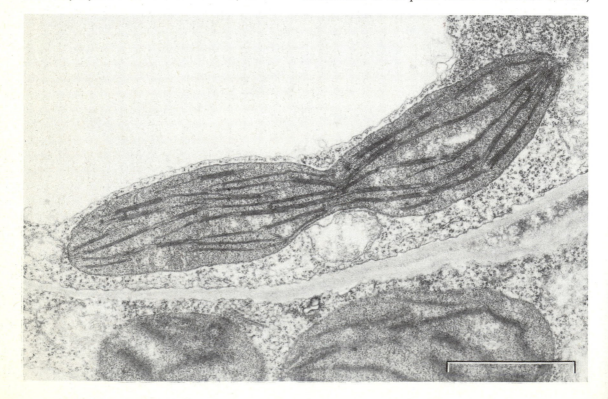

followed by separation of the two daughter plastids. After separation, the two plastids are of roughly equal size, and *in vitro*, they do not grow.

Plastid replication and cell division

Although there is no longer any doubt that plastids can divide, the detail of how this is accomplished is unclear. Particularly, it is not known how the thylakoid membranes of the chloroplast may divide. In isolated leaf discs it is found that the division of chloroplasts is light dependent, and that it is also stimulated by conditions which stimulate cell expansion. This latter observation corresponds to the pattern of behaviour of chloroplasts in intact leaves, where they replicate during cell expansion periods in order to increase their numbers per cell.

The division of the major part of the genetic material of the cell, the nuclear chromatin, is very carefully controlled by means of a complex apparatus, the mitotic spindle (p. 29). The reason for this is not hard to deduce; the mitotic spindle ensures that each daughter nucleus of a division receives an exactly equal complement of chromosomes, so that each cell formed by the division has a complete set of nuclear genes. No such precise mechanism appears to exist for the replication of plastids. This raises a number of interesting questions. Within the plastid, is there any evidence of organised or equal partition of the plastid DNA? Is there any correlation between the cell division cycle and the replication of plastids? How is the plastid population partitioned between the daughter cells at cytokinesis?

The DNA of the plastid appears within clear regions of the stroma which are known as nucleoids (Fig. 2.14). These regions contain 2.5 nm fibrils which are sensitive to DNAse treatment and become labelled by radioactive precursors of DNA. There may be many nucleoids within the stroma of a single plastid – in chloroplasts of the swiss chard it has been found that there are 18. When mature chloroplasts

Fig. 2.14. Part of a chloroplast in a leaf of *Coleus blumei*. The granal stacks consists of many thylakoid membranes. The clear area in the stroma which contains fine fibrils corresponds to a nucleoid. Bar = 0.5 μm.

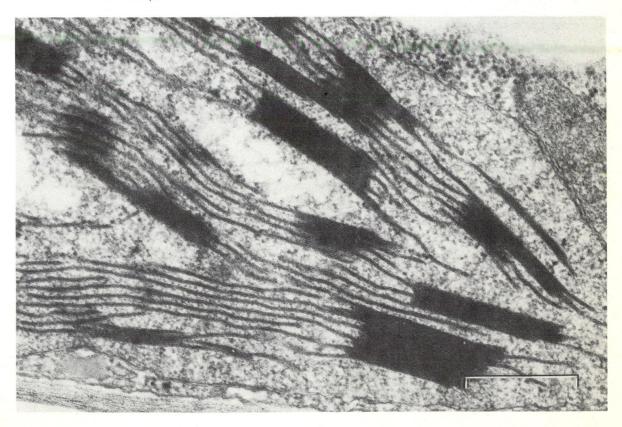

which are in process of replication are examined, it is found that there are roughly equal numbers of nucleoids in each half of the dumbbell. However, sometimes the situation is not so clear cut. It has been found for example that in swiss chard (*Beta vulgaris cicla*) proplastids only contain one nucleoid, whereas etioplasts and chloroplasts have several. This suggests that the link between the replication of the plastid DNA and the fission of the plastid itself may not be a very firm one. It is known, for example, that inhibitors of DNA synthesis do not prevent fission of plastids.

There is no visible mechanism for ensuring the equal division of nucleoids or DNA between daughter plastids. This implies that the partition may be left to chance. This is not such a risky arrangement as it might first appear however, since not only are there several nucleoids per plastid in many cases, but each nucleoid contains several copies of the plastid DNA. Therefore, the chances of a daughter plastid arising from a fission process without any DNA at all are remote. In this context it has been found that, in giant cells of the alga *Acetabularia*, the majority of the chloroplasts do not contain any detectable DNA. This may well mean that, at least in the case of mature chloroplasts, normal function of the organelle can be maintained even in the absence of nucleoids.

There is no evidence to suggest that the fission of plastids is synchronous within a single cell. This implies that the plastid fission cycle is not linked to the cell division cycle, since clearly if it were, all the plastids under the influence of a single nucleus would be expected to be at a similar stage of replication. Thus it appears that plastid DNA synthesis is controlled in a quite different manner from the replication of nuclear DNA. Where several nuclei are found within a common cytoplasm it is frequently observed that they do show a high degree of synchrony in their division cycles. This might reflect their interconnection via the endoplasmic reticulum – a pathway of communication which has no parallel when plastids are considered. It is certain, however, that some relationship does exist between nuclear cycles and plastid numbers. For example, when cells at later stages of development become polyploid (p. 35), it is found that the numbers of plastids within such cells is increased. For a single doubling of the amount of DNA, the number of plastids is increased by a factor of about 1.7. This is in contrast to the mitotic (diploid) cycling which occurs in early development, and which

is accompanied by a doubling of the plastid number for each cycle of DNA synthesis and mitosis, resulting in roughly constant numbers of plastids per cell.

Finally, there appears to be no mechanism to ensure equal partitioning of plastids at cytokinesis. The fate of an individual plastid rests rather on its position within the cell at the time of cell plate formation. It is commonly observed that, in vacuolate cells prior to division, the plastids will cluster around the nucleus. This presumably results in a more or less equal distribution of the plastids between the daughter cells of the division – however, such behaviour is not universally observed and therefore does not seem to represent a necessary part of a mechanism. The random partitioning of plastids at cytokinesis carries with it two implications, both of which are borne out by observation. The first is that if partition is random, and if the number of plastids per cell is small, then the situation should arise when one of the daughter cells receives no plastids. This has been observed in a haploid plant of *Trifolium hybridum*, where the guard cell mother cells (p. 118) had only two plastids. It was found that many of the guard cells formed by division from these mother cells had no plastids. This is good, if exceptional, evidence for random partitioning of plastids. It should be noted that in algal cells which habitually only contain one chloroplast (such as *Chlamydomonas*) each daughter cell inevitably receives one chloroplast following division. This comes about more by geometric arrangement, however, than by the existence of a specific mechanism for movement (p. 53).

The second consequence of random partitioning of plastid populations has considerable interest from an experimental and, indeed, a horticultural point of view. Suppose that a mutant plastid arises within the cytoplasm, and is able to replicate itself. Eventually by random partitioning of the mixed population of plastids so formed, a situation can arise in which one daughter cell following division has only the mutant plastids within its cytoplasm. All subsequent divisions of this cell will produce new cells which themselves are populated by mutant plastids. If the mutation is one which results in the loss or a change of pigmentation in the plastid, these progeny cells will be a different colour to the normal cells. This is an explanation for some types of variegation in plants, and is a useful tool in studying cell lineages.

These facts point to the limitations of the idea

of the plastid as a cell within a cell. Replication of plastids is a cell-like function, but it does not occur entirely independently of nuclear control. Plastids may divide at rates which are similar to cell division rates, or they may divide faster than the cell. They may even divide more slowly than the host cell, so leading to a reduction in their numbers. None of these changes occurs at random, and so many be presumed to be under nuclear control. Furthermore, although plastid fission can occur outside of the cytoplasmic environment, this process only continues for a limited time and is not accompanied by growth of the progeny.

The origin of plastids

From the above discussion, it should be clear that the presence of plastids within any plant cell is due to their presence and replication within less mature cells. Thus any particular plastid population can be traced back to its originating meristem and, from there, back to the fertilised zygote. The question then arises as to the origin of the proplastids which are present in the zygote. Since this cell is formed by the fusion of the male and female gametes, it is possible that the organelles within it might have originated from either or both parents. Alternatively, they might arise *de novo*.

The *de novo* origin of plastids has been proposed to occur during the maturation of the egg cell of the fern *Pteridium aquilinum*. In the maturing egg cells of this plant the plastids and mitochondria are seen to undergo degenerative changes. At about the same time, a series of complex evaginations of the nuclear envelope are seen to form, and these have been interpreted as newly developing precursors of pro-plastids and mitochondria. This scheme would give rise to maternal inheritance in genetic experiments (see below). However, it must be emphasised that this behaviour has only been observed in *Pteridium*; it is the only observation of its kind and has not been repeated in other ferns or higher plants despite its great theoretical importance and the fact that the original observations were made over 20 years ago. It is the opinion of most workers that either *Pteridium* represents a special and, indeed, probably a unique case, or that the interpretation of the electron microscope images was mistaken.

In the great majority of the angiosperms, the plastids of the zygote are derived solely from the mother plant via the egg cell cytoplasm. The sperm is formed from the generative cell of the pollen grain (Chapter 4) and often the plastids of the pollen cytoplasm are excluded from the generative cell during the course of the highly asymmetric division which gives rise to it. In other cases it is probable that any plastids introduced from the male parent into the zygote are eliminated by an incompatibility mechanism. The existence of this type of behaviour is an example of the matching which must occur between a cell and its plastids if the plastids are to survive.

The existence of maternal inheritance of plastids means that if a cross is made between two parents with different plastid characters, only those corresponding to the female parent will appear in the offspring. From this it follows that the offspring from the same two parent plants will differ depending on which plant was used as the male and which the female parent. Such reciprocal differences are diagnostic of a character which is inherited in a non-Mendelian way; that is, a character which is not directly controlled by nuclear genes.

In certain cases maternal inheritance of plastids is not strictly observed. The best known examples are *Pelargonium* and *Oenothera*. In the unfertilised egg of *Oenothera* there are 32 plastids. At fertilisation the sperm contributes a further 8–13 plastids to the zygote, and this enhanced plastid population may increase by fission before the zygote itself divides. The result of this is that cells which are formed by division of the zygote contain mixed populations of plastids, some from the male and some from the female parent. Since these populations are partitioned at random between daughter cells at cytokinesis, it follows that cells containing only one type of plastid might arise during the development of the plant. Once a pure line of cells has arisen in this way further divisions will not alter its purity. If the plastids of the two parents differ in their ability to produce pigments, this may result in variegation of the plant.

The presence of plastids in any plant cell is thus the result of their presence in a preceding cell, and this line of succession is unbroken through the sexual process, at least on the maternal side. The origin of plastids in an evolutionary sense is a matter for speculation. It has been suggested that the chloroplasts of present-day plants arose from sym-biotic blue-green algae. This is an attractive idea in many ways; the photosynthetic apparatus of the blue-green algae consists of thylakoids lying free in the cytoplasm. This is comparable to the situation in primitive algal chloroplasts, where thylakoids lie free in the stroma, without elaboration into grana and frets. In certain cases of endosymbiosis by blue-green algae the wall of the cynanophyte is lost, leaving it surrounded by a single or a double membrane – another striking similarity to plastid structure. However, it remains entirely possible that higher plants evolved from blue-green algae by a process entailing the encapsulation of DNA into separate compartments which in turn gave rise to the organelles which we recognise today as the nucleus, plastids and mitochondria. It is difficult to see how this matter will ever be resolved by experiment.

Plastid DNA

So far we have seen that plastids are able to replicate (and thus maintain, increase or decrease their numbers) within cells during growth and development. The early work with plastid DNA was controversial mainly because results were always subject to the doubt that the DNA found in isolated preparations of chloroplasts might represent a contaminant from the nucleus. Improvements in techniques and the demonstration of consistent differences between plastid and nuclear DNA have now removed any doubts as to the existence of plastid DNA.

The DNA of plastids is a circular molecule whose contour length depends upon the species of plant from which it is prepared. Many plastid DNAs have a contour length of about 40 μm, but the apparent uniformity which was once thought to characterise all plastid DNAs is now known to be an oversimplification. A range of lengths from about 30 to 60 μm has now been demonstrated, representing a range of molecular weights of 80–130 million. Each chloroplast contains enough DNA for several copies of the genome, perhaps up to 60; it appears very likely that all the copies of the DNA within a chloroplast are identical. Each molecule usually contains repeated regions which code for ribosomal RNA. The significance of these repeated regions is not clear, and they are not invariably found. The DNA of chloroplasts is not associated with histones and therefore does not exist in the form of chromatin as does the DNA of the nucleus. It has also been found that plastid DNA does

not contain the base 5-methylcytosine. This chemical difference between the two classes of DNA serves to emphasise the identity and reality of the plastid DNA.

The DNA of the plastid occurs in localised regions called nucleoids (p. 59). The number of nucleoids which a plastid contains depends upon the plastid's state of development. A proplastid might contain only one, whereas in a chloroplast there might be as many as 20. The number of nucleoids seems to be less than the number of copies of DNA in each plastid, which implies that each nucleoid carries more than one copy of the DNA. In *Beta vulgaris* it has been shown that the average number of copies of DNA per nucleoid is 4. The nucleoids within a plastid do not necessarily all have the same number of copies of the DNA. When the shape of the nucleoid regions is analysed it is found that they exhibit a variety of forms. In small proplastids with little internal membrane development, the nucleoid is often spherical and more or less centrally placed. In plastids with multiple nucleoids and well developed lamellar systems, the individual nucleoid is often disc-shaped. There is thus some correlation between membrane development within the plastid and the shape and number of nucleoids. In lysates of chloroplasts it is often found that the DNA is associated with membrane fragments. Taken together, these observations suggest that the DNA may be associated with membrane material within the plastid. It is possible that this association is involved in the replication of the plastid DNA.

Plastid autonomy

The presence of DNA within plastids raises the question as to the degree to which plastids are autonomous within cells. The amount of DNA in a chloroplast, allowing for known repeated sequences, would code for over 100 proteins of molecular weight 40 000. However, when the capability of plastid DNA to code for plastid materials is examined in detail, it is found that the plastid is very far from autonomous. Many of the genes which code for enzymes in the pathway of chlorophyll synthesis are located in the nucleus, as is the gene for the light-harvesting chlorophyll a/b protein. The genes for some of the ribosomal proteins of the plastid are within the nucleus, and so is the gene for the small subunit of the enzyme ribulose diphosphate carboxylase (Fraction 1 protein). The gene for the large subunit of this enzyme is within the plastid DNA. Plastid DNA is also known to code for three species of ribosomal RNA and for the transfer RNAs for amino acids. It is therefore clear that many of the major enzymes and pigments required for photosynthesis depend upon nuclear genes for their production. The same conclusion has been drawn from a study of amyloplasts in *Zea* endosperm – that successful development of the plastids depends upon the import of materials from the cytoplasm which have been synthesised under the control of genes which are not within the plastids themselves.

This means that the importance of the cytoplasmic environment is not simply that it represents an amenable medium in which the organelles can grow and divide under their own control. There is a specific dependence of the plastids upon the cell nucleus for their continued existence. This has been demonstrated most clearly in the genus *Oenothera*, which does not show maternal inheritance of plastids, but biparental inheritance. Within the subgenus *Euoenothera* there are five distinct plastid types. These were combined with six nuclear genotypes in a series of crossing experiments. Each nuclear genome was combined with each plastid type in turn, giving rise to a series of 30 different combinations of nuclear genomes and plastid types. Of these 30, only 12 gave rise to normal green plants. Other combinations produced plants which were pale green, yellow or white, and which showed other abnormalities such as poor germination and growth rates. These results emphasise the need for a degree of compatibility between the cell nucleus and its population of plastids. Conversely, they demonstrate also the fact that the genetic activity of the nucleus can influence plastid development. This in turn should alert us once again to the idea that changes in genetic activity within the nucleus which accompany development are likely to result in changes in plastids, and that this may be the underlying mechanism for the matching of the plastid type to the cell type in which it occurs.

Variegation

Variegation is a descriptive term which applies to the visual appearance of a plant, and it is as well to realise from the outset that it has many causes, some of which are not understood at all or which involve very complex interactions between the nucleus and plastids which are beyond the scope of the present discussion. However, an examination of some aspects of variegation throws considerable light on patterns of cell development and the correlations between gross morphological changes and genetic and biochemical events occurring within cells. It is these aspects which will be discussed; variegations caused by unstable nuclear or plastid genes will be, for the most part, ignored.

Physiological variegation

In the broadest terms, there are two types of patterning associated with variegation. The first of these is a figurative pattern produced by contrasting colours which is very similar from one leaf to the next on the plant. These patterns are not related to cell lines, but rather reflect specific changes in the leaf occurring due to physiological conditions which affect pigmentation. The tree *Populus lombardii*, for example, has leaves whose tip is bleached white. This is due to a physiological change in the chloroplasts of the cells at the tip of the leaf, and it does not reflect any change in gene action. There is a variety of the well known variegated plant *Coleus* which has green leaf margins and a yellow central area to each leaf (Fig. 2.15). The edge of the yellow area is ragged where it follows the line of veins into the green margin of the leaf, and this is thought to represent an effect of hormonal materials in the veins causing bleaching of the plastids. Many of the other variegations in *Coleus* are due to anthocyanin pigments and not to effects on plastids.

Fig. 2.15. A yellow/green variegated plant of *Coleus blumei*. The yellow areas (light in the picture) extend from the main veins outwards into the green.

In monocotyledons the patterns of variegation are frequently in the form of stripes, owing to the parallel nature of the venation in the leaves. In one particular yellow-striped variety of *Zea mays*, the variegation has been shown to be due to a deficiency of iron. The cells nearest to the veins contain green chloroplasts, but those in the interveinal spaces are yellow due to incomplete pigment production. This type of variegation is easily overcome by supplying the leaves directly with ferric ions.

Variegations of this type point to the fact that at a fairly gross level, nutrient status and general environmental factors do vary with the position of the cell within the tissue. Whilst variegation represents an abnormal manifestation of such variation, it nonetheless serves as visible reminder that such factors may be partially influential in cell development in general.

Cell lineage variegation

In the second major class of variegated plants, the changes in colour which are visible follow cell lineages. This means that patterns of identifiable colour change correspond to patterns of cell division within the developing leaf. This type of variegation is characterised by the fact that every leaf on a plant may show a different colour pattern (Fig. 2.16). Green parts of leaves may have small islands of white, and white areas may have small islands of green. Sometimes the variegation is only seen at certain stages of growth. In all these cases, however, the occurrence of variegation is stable and true breeding, and is therefore said to be due to stable pattern genes, to distinguish it from more complicated cases of

Fig. 2.16. A variegated tobacco plant. Each leaf is different in its pattern of pigmentation.

variegation which are due to mutable or unstable genes in the nucleus or the plastids.

Given the apparent stability of the total gene complement of nucleus and plastids in stable variegation of this type, two interesting conclusions follow. The first is that once the change which leads to say, failure of pigment production, has been initiated, it remains 'switched on' during subsequent cycles of cell division. This must be so, since colour changes follow cell lineages. The second conclusion is that, in a variegated plant which is capable of producing pure green and pure white sectors in its leaves, clearly the development of greening and normal chloroplast structure is not an inevitable consequence of the overall gene content of the nucleus and the plastids. Rather, it reflects the expression of parts of the genome. As the expression of the genome varies due to changes in developmental status of the cells, so incompatibilities between nucleus and plastids may appear, or disappear.

It should be realised that 'variegation' is a visual description of a plant's appearance, and as such expresses rather coarse control of the level of pigment production. However, there is evidence which shows that such control may be very fine indeed. For example, in a particular variegated plant of tobacco, it has been observed that leaves which appear completely white to the naked eye in fact contain green cells at the tips of the glandular hairs. This is a case of an extremely localised pigment change. It may be said to correspond to examples of normal development where the appearance of a specific cell type is matched by the appearance of specific plastid types within the cytoplasm; sieve elements and guard cells have already been mentioned in this context.

In other cases the variegation is correlated with a very much more global change in the status of the entire plant. For example, in one variegated ivy, the juvenile phase of the plant is always variegated, whilst the adult phase is always green. It is also found in other cases that whilst variegation is present in leaves, it is not found in the initial cells of growing points or the germ cells of flowering shoots. Once again, the type of interaction between nuclear and plastid genes is correlated with the developmental state of the cells in question.

Perhaps the most striking behaviour of this type is shown by the fern *Lastraea atrata*. Here the fern plant itself – the sporophyte generation – is variegated. The green tissue has a normal complement of chloroplasts and is crossed by sectors of white tissue containing small pale green plastids. It is possible to collect spores which have been formed on either white or green sectors and, when these are germinated, it is found that the gametophytes which result are always green and never variegated. These green gametophytes then produce gametes which, when fused, produced variegated sporophytes once more, and so on. A complication in this cycle is that both sporophyte and gametophyte generations are at the same level of ploidy due to abnormalities in the reproductive cycle. This means that different plastid behaviour is expressed in the two generations of the life cycle of the plant without any variation in overall gene content.

These observations stress that it is not only the total genetic makeup of an organism which determines its appearance and structure, but the way in which that makeup is expressed over a time scale. The existence of variegated plants is simply a visible expression of this effect; it need not be considered to be a unique feature of the interaction between nuclear and plastid genes. Genetic expression is thought to change with time during development, and such change has a wealth of different consequences at the structural and chemical levels. Variegation is only one small aspect of these changes.

Chimaeras and variegation

The possibility has already been mentioned several times of the existence of different populations of plastids within a single plant cell. Biparental inheritance of plastids is one cause of such an effect, and plastid mutation is another. It should be emphasised that neither of these phenomena is very common. On the one hand most angiosperms show maternal plastid inheritance and, on the other, the existence of multiple copies of the plastid DNA in each organelle means that plastid mutations are only expressed very slowly. The existence of different plastids populations can give rise to variegation if they differ in their pigmentation. One class of variegation is particularly instructive to consider from the standpoint of the organisation of cells into tissues, and this is the production of variegation by chimaeral development.

The growing point of the shoot is characterised by organisation into a series of outer layers which are termed the tunica, and a central region called the corpus. Aspects of this organisation will be discussed in more detail in Chapter 6. The basis of separation of these two cell types is that the predominant plane of division in the tunica is at right angles to the surface, whereas division within the corpus may take place in many planes. The tunica may consist of just one, or several cell layers, and these are conventionally numbered LI, LII, and so on, from the outside. LI is commonly the originating layer for the epidermis of all the leaves formed by the shoot.

It may now be considered what might happen in a plant with a mixed population of plastids. During early development, 'sorting out' can occur to give rise to pure lines of cells with only one plastid type. If a pure white cell arises in LI, subsequent development of leaves from the shoot apex will result in the formation of a white epidermis overlying green mesophyll. This alone can give rise to variegation, and is exemplified by *Pelargonium zonale*, which has green leaves with a pale margin (Fig. 2.17). The pale margin is caused by the epidermis masking the colour of the small amount of underlying green tissue at the edge of the leaf. Such a structure is known as a periclinal chimaera.

In other situations, it is possible for complete sectors of the shoot apex to become white. This may happen following plastid sorting out, or as the result of a nuclear mutation which affects all the plastids in the cell in which it occurs. Subsequent divisions of the original pure white cell will give rise to a sector in the

Fig. 2.17. Leaves of *Pelargonium zonale*, Carolyn Schmidt. The white margin of the leaf is caused by the chimeral nature of the variegation.

meristem which is genetically 'white'. In such a case, any leaves developing from this sector will show variegation or pure whiteness, depending upon the precise contribution to the mature leaf of the originating cells. Such a structure is called a sectoral chimaera. Leaves developing from other parts of the meristem will be normally green.

These simple examples serve to illustrate in a visible form the fact that complex extended tissues such as leaves originate from very localised regions of meristems comprising only a few cells. They emphasise one aspect of the organisation of meristems, which will be considered in more detail in Chapter 6.

Summary

The plastids are a particularly diverse group of organelles, which show considerable modification in their structure and activities to match the cell type in which they occur. The plastids are capable of a certain amount of autonomous action in the production of macromolecules, but they are reliant upon the import of materials from the cytoplasm to permit their continued existence. The nature of those imports is at least partially under the control of the nucleus, and the interplay between the nucleus and the plastids is the underlying cause of plastid differentiation. The plastids serve to illustrate both the idea of genetic change during development, and also the fact that cell differentiation is a matter which involves all aspects of the cell metabolism. The plastids are a valuable indicator of differentiation precisely because their structure is so variable and hence recognisable. There is no reason to doubt that the chemical diversity which this structural variation represents is just as great amongst other components of the cell which appear, from a morphological standpoint, to be more stable and therefore perhaps less interesting.

3

The cell wall and development

Introduction

In the previous chapter we saw that a cytoplasmic organelle, the plastid, responds to the changing genetic and physical environment of the cell by undergoing a series of developmental changes. In the present chapter we shall see how the cell wall also reflects the state of differentiation of the cell which it surrounds.

The cell wall is of great importance for a number of reasons, not all of which are immediately obvious. Amongst those that are obvious is its mechanical role in giving the plant the strength to stand upright and reach the light – carried to its greatest expression in the formation of forest trees. Less apparent, but equally important is the fact that the cell wall defines the interrelationships between cells. Plant cells do not migrate to their positions within the plant body; they are passively translated by the processes of growth and thickening of cell walls. Finally, the cell wall forms a continuous space within the plant which is outside the plasma membrane of each cell. In a very real sense, the cell wall is the immediate environment in which every plant cell lives. Clearly the importance of the wall in this context could be very great, since it could mediate the types of molecules found at the outer surface of the cell, as well as the physical environment there. Of course, the cell wall is in no way autonomous in this arrangement. Its chemical composition, shape and thickness are all determined by the cell which produces it. The possibility should be realised, however, that within complex tissues the behaviour of cells may be mediated by the very walls which they have themselves produced. Part of this mediation is positional, and this has been realised for a very long time; a cell's behaviour is strongly determined by its position within the plant. However, it is gradually becoming clear that the cell wall may also control important chemical parameters which affect the cell's behaviour via changes in the properties of the plasma membrane.

In this chapter we shall consider the nature of cell walls, concentrating on those aspects of most direct importance to developmental problems. Cell walls, like the cells they surround, undergo growth and differentiation. These two phases are the basis of a broad division of cell walls into two types. Primary walls are those which are still undergoing expansion. They are typically thin, plastic, highly hydrated and crossed by plasmodesmata. They are commonly regarded for convenience as being of a single uniform type, although this ignores a great diversity in chemical constituents between primary walls from different plants. Since the primary wall is growing, its major interest for development is the possible ways in which such growth is controlled in direction and extent. This is related to the formation of cellulose, the fibrillar component of plant cell walls, and a detailed consideration of cellulose forms a large part of the section on the primary wall in this chapter. Secondary walls are classified as those which have stopped expansion and are in a differentiation phase. In fact, primary and secondary characteristics can occur in the same wall simultaneously. For example, a wall may be expanding in one region and thickening in another; this occurs during xylem element differentiation (p. 101). Secondary walls are typically thickened, contain more fibrillar material (cellulose) and less water than primary walls, and often have additional chemical components not found in primary walls. Most characteristic of these is the polymer lignin, which is responsible for the comparative inertness of woody tissues to chemical or biological degradation. Secondary walls are considered in a general way in the present chapter; however more detailed descriptions of particular wall developments form part of the picture of cell specialisation which is the subject of Chapter 4. Secondary walls are characteristic of cell types in many cases. Finally in this chapter we shall consider cases of the natural breakdown of cell walls during plant growth, in order to emphasise that the wall is by no means simply an inert shell in which the plant cell lives.

Structure and composition of the primary wall

At the most general level, plant cell walls consist of a fibrillar component which may or may not be associated with a matrix material. In higher plants the wall consists of fibrils of cellulose within a matrix of other polysaccharides and proteins; in an alga such as *Chlamydomonas*, the wall consists of a complex sandwich with layers of fibrillar and crystalline components, including glycoproteins. The range of types of polymer which comprise the primary cell wall appears at first sight to be rather limited – the fibrillar component of all higher plant cell walls is cellulose, in fungi it is chitin and in some algae it is a β-1,4 xylan. The matrix materials are equally simple to describe in broad terms; in a higher plant cell wall they consist of pectins, which are water-soluble polymers, and hemicelluloses, which are insoluble in water but soluble in dilute alkalis. These classes of polysaccharides are widely distributed.

The reasons for this apparent simplicity are historical in nature; early work with plant cell walls was carried out using a sequence of fractions of the wall which was based on solubility criteria. Thus the pectins were defined as the wall component extracted by boiling water or chelating agents, whereas the cellulose was defined as being the residue left after boiling with alkalis. There were two assumptions behind this type of experimental approach to wall chemistry. The first was that the three major polysaccharide groups, pectin, hemicellulose and cellulose had some real identity, even though they were defined in terms of an arbitrary extraction procedure. The second assumption was that the extraction procedure did not alter the composition of the wall.

These assumptions are now regarded as being unsound. The most recent work with plant cell walls uses specific enzymes to degrade particular bonds within the polysaccharides comprising the matrix of the wall. Studies of this type have led to the formulation of a model for the primary wall which is characterised by a high degree of interaction between the various wall components. This interaction is thought to be important in the control of the physical properties of the wall, and hence in the control of growth. It emerges that, far from being simple mixtures of relatively homogeneous molecular types, the fractions of the wall corresponding to 'pectin' and 'hemicellulose' are complex and involve specific branching patterns and sidechains. It is also known that the primary wall contains protein.

In the present context, of growth and development, the details of the chemistry of the primary wall

Fig. 3.1. The structure of chemical cellulose. Glucose units are linked β-1,4 in long chains which are unbranched. The chains associate laterally by hydrogen bonding.

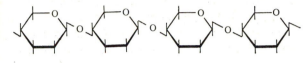

Fig. 3.2. The appearance of cellulose microfibrils when negatively stained. The microfibrils are slightly flexuous and apparently endless. These have been formed at the surface of an isolated protoplast from a leaf of White Burley tobacco (*Nicotiana tabacum*). Bar = 0.2 μm.

are to some extent peripheral to the main theme. Indeed, since chemical methods are of necessity applied to large populations of cells rather than individuals, any attempt to describe chemical changes on an individual cell basis would be misleading and premature. Nevertheless, a basic understanding of the types of polymers found within cell walls is instructive both for their possible function and role in growth, and also from the standpoint of the way in which the cell changes its biochemical activities as it differentiates. The simplicity of the wall structure which emerges from a general treatment of this type is of course illusory. The cell wall is not a passive skeleton for the plant; it is a metabolic compartment in which can occur processes of growth, development and decay.

Cellulose

Cellulose in a chemical sense is a polymer of glucose, linked β-1,4 (Fig. 3.1). Primary walls of dicotyledons typically contain 25–30% of their dry weight as cellulose. When the cellulose fraction of most plant cell walls is analysed for its component sugars, it is frequently found to contain small percent-

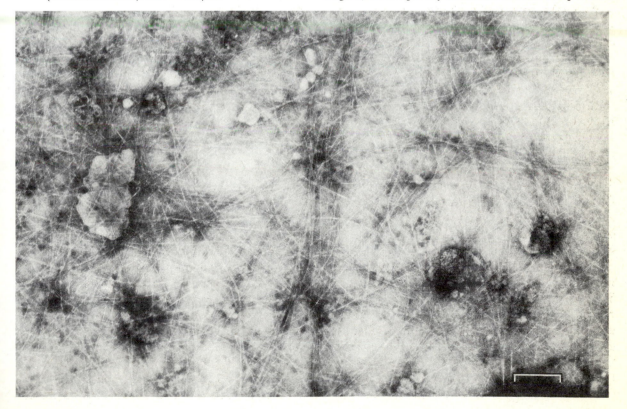

ages of monomeric units other than glucose. Xylose and mannose are the most common. X-ray data show that the cellulose of the wall is at least in part crystalline in structure, and chemical and physical measurements show that the degree of polymerisation of the glucose subunits is very high – 10 000–15 000 is the usual range. Thus individual 'molecules' of cellulose, the polyglucosan chains, may be several microns long.

In the electron microscope, cellulose appears as long fibres which are termed microfibrils (Fig. 3.2). Microfibrils are so long that it is not usually possible to identify free ends. The problem of how to reconcile this appearance of the microfibrils in the electron microscope with the facts concerning the chemistry of

cellulose has long taxed workers in this field and has yielded some fine controversy. The consensus of current opinion is that the polyglucosan chains are associated laterally and end-to-end to give rise to the microfibril. In the centre of the microfibril this association is so precise and so dehydrated that it is crystalline, and this gives rise to the X-ray behaviour of cellulose. Towards the edge of the microfibril, the association is less precise and involves the presence of water and other polysaccharides. This is the explanation for the appearance of 'contaminants' in cellulose prepared from cells walls.

The diameter of microfibrils varies both with their source and with the means used to measure it. When microfibrils from higher plants are examined by means of negative staining techniques, in which the unstained microfibril lies in a lake of electron dense stain (Fig. 3.2), the diameter of the microfibrils appears to be about 3.5 nm. On the other hand, when replica techniques are used (Fig. 3.3), this diameter appears to be about 10 nm. The cause of this discrepancy is undoubtedly the penetration of the stain in

Fig. 3.3. The appearance of cellulose microfibrils prepared as a metal replica. Their basic appearance is unchanged from that in negative staining, but the width of the individual microfibril is greater. These microfibrils are on the surface of a growing protoplast isolated from leaves of tobacco White Burley (*N. tabacum*). Bar = 2 μm.

the first technique into the outer layers of the hydrated microfibril. The best current compromise which accommodates these facts is that the cross section of the microfibril has dimensions of 4.5 nm by 8.5 nm.

To state a single set of figures in this way carries an important implication, namely that the cellulose microfibrils within the primary wall represent a uniform population. In fact, this may not be the case. If the cellulose microfibril is produced in a determinate manner by some sort of template mechanism, then it is certainly to be expected that each microfibril would be similar in dimensions to every other one. However, if the mechanism of formation of microfibrils is indeterminate, then variations could arise in the structure of microfibrils depending on the precise conditions under which they were formed. Thus the argument about the structure of the microfibril has implications for its mode of synthesis. This is a subject which will be returned to later (p. 78).

Cellulose microfibrils from cellulosic algae are quite different to those found in primary walls of higher plants. They are much larger, about 23 nm across, and they also occur in highly ordered patterns. These algae have been studied in particular with regard to the mechanism of formation and orientation of these ordered fibril arrays (p. 77).

The matrix polysaccharides

In contrast to the fibrillar content of the wall which, at least in a chemical sense, is uniform throughout higher plants and beyond, the polysaccharides comprising the matrix are found to be variable in composition. This may be a reflection on the methods of study – no matter how homogeneous a tissue may seem to be it will inevitably contain cells at different stages of development and of different ages. Since the composition of the individual wall varies with both these parameters, it is obvious that the chemical composition of a particular wall preparation will depend on the exact spectrum of cells which it contains. In addition, a wealth of different techniques has evolved for the fractionation of cell walls, and this alone can account for differences in composition of the different fractions which are isolated and analysed.

The hemicelluloses, defined as being soluble in dilute alkali, are for the most part neutral polysaccharides containing glucose, xylose and mannose. A typical primary wall of a dicotyledon contains 15–25% of its dry weight in the form of hemicellulose. The most well characterised hemicellulose is a xyloglucan. This consists of a backbone of glucose units with side branches made up from xylose units (Fig. 3.4). It is of widespread occurrence in the primary walls of angiosperms, and is known to associate tightly with the cellulose microfibril at its outer, non-crystalline surface. The hemicelluloses of gymnosperms consist of polymers with mixed chains comprised of glucose and mannose units, sometimes with galactose side chains. These polymers too are capable of forming strong associations with cellulose. This property stands out as a possible 'function' or basic characteristic of the hemicelluloses.

The pectic polysaccharides which comprise the most soluble part of the wall are a complex family of polymers. They represent up to 35% of the dry weight of the primary wall of dicotyledons. Neutral pectin components include homopolymers of arabinose and galactose, and heteropolymers built up from both

Fig. 3.4. Proposed chemical structure of the xyloglucan from sycamore callus cells. The backbone of the molecule is a chain of glucose units, with side branches consisting mostly of single xylose units. G, glucose; X, xylose; A, arabinose; F, fucose; Gal, galactose. (After P. Albersheim.)

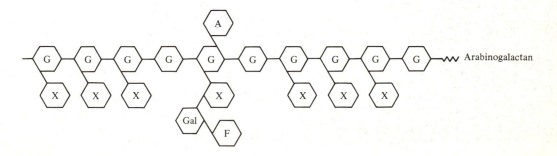

these sugars (Fig. 3.5). Acidic pectin components include polygalacturonic acid, and a mixed polymer of rhamnose and galacturonic acid which has neutral blocks of arabinan and galactan polymers attached to its backbone (Fig. 3.6).

The important property of the pectic substances, in terms of their possible function within the wall, is that they easily form viscous gels in water. This property depends upon the structure of the molecules, and the extent to which the acidic groups within them are esterified. The pectins are therefore regarded in a simple way as determining the plasticity of the cell wall. They are laid down at the very earliest stages of cell wall formation in the cell plate; the 'middle lamella' between adjacent primary walls in the tissue is a remnant of this structure, and is rich in pectin. When a tissue is treated with enzymes which dissolve pectin, it will often fall apart into its constituent cells, emphasising the role of these polymers as the cement which holds cells together.

Of course, the biological function of the pectins is no doubt a great deal more complex than simply to act as a plasticiser or as a glue in the wall. Pectins bind metal ions, notably calcium, and this can change their physical properties. Equally, pectins may change the environment for other cell wall components, especially enzymic components. Carbohydrates are very often thought of simply as structural or storage materials; however, it is clear that they dominate the extracellular environment of every protoplast within a multicellular plant, and it is likely therefore that they play an active role in many of the processes which are mediated in this compartment. The difficulty of assigning general roles to the different classes of wall components is emphasised by the fact that monocotyledons usually have only one tenth of the amount of pectin within the primary wall compared to that which is found in dicotyledons.

Protein

Besides the various polysaccharide components, the wall also contains protein. The primary walls of dicotyledons contain between 5% and 10% of protein (dry weight). The protein is distinctive in that it may contain as much as 20% of the unusual amino acid hydroxyproline. Cell wall protein also contains comparatively large amounts of the amino acids, alanine, serine and threonine, and this has led to its being compared to the structural protein of animal cells, collagen. The protein can only be extracted from the wall by methods which involve chemical degradation. This fact, together with the finding that partial hydrolysis of the protein gives rise to peptides which contain arabinose and galactose residues strongly suggests that the protein is chemically bonded to a polysaccharide constituent of the wall. The walls of monocotyledons also contain hydroxyproline-rich proteins, although the amount of hydroxyproline in these walls is only one tenth of that which is found in primary walls of dicotyledons.

In addition to the 'structural' protein, the wall also contains enzymes. The importance of turnover and modification within the space of the wall itself is only just beginning to be appreciated, and this subject will be discussed further in the context of possible mechanisms of growth of the cell wall (p. 75).

Water

Water is a universal component of the primary cell wall and its presence is necessary for a number of reasons. It is required for the operation of hydrogen bonding between wall components. It also influences the conformation of polymers and polymer aggregates – the cellulose microfibril is disorganised at its outer surface by reason of the presence of water. Finally, the

Fig. 3.5. Proposed chemical structure of the arabinogalactan from sycamore callus cells. The arabinogalactan is covalently linked to both xyloglucan and rhamnogalacturonan. Gal, galactose; A, arabinose. (After P. Albersheim.)

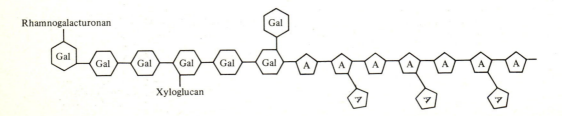

water in the wall acts as a solvent for a range of ions and small molecules which are present in the apoplasm. These ions include those which affect the plasticity of the wall itself, but other solutes include hormones and nutrients which are moving in the extracellular space. In a primary wall, 80% of the fresh weight may be water.

The chemical organisation of the primary wall – a model

The analytical classification of wall components has been made possible by a variety of extraction methods, some very drastic. Typically, in early experiments, a wall preparation obtained by solvent extraction of a tissue would be fractionated by a series of steps involving the use of chelating agents, boiling water, and hot dilute alkali or acid. Such methods not only solubilise the various components of the wall, but they also cause bond breakage within its complex tertiary structure. This means that whilst such methods may give good results in terms of the relative proportions of the various sugars within a wall, they cannot be expected to reveal much of its organisation.

More recently, detailed studies have been made of the cell walls of suspension cultures of sycamore callus tissues. In simple outline, these walls were treated with a sequence of enzymes giving rise to soluble fractions corresponding to pectin, hemicellulose, protein, and an insoluble residue. These frac-

tions were then chemically methylated, hydrolysed and reduced. This sequence of reactions leads to the elimination of any hydroxyl group which was involved in a sugar–sugar link in the original polymer. By analysing the constituent sugars and small oligosaccharides in the final mixture, it is possible to deduce the linkages which were present in the original polymeric starting material.

On the basis of such studies, a model of the primary cell wall has been proposed which has as its central feature the presence of organised linkages between the various classes of polymer. It is convenient when considering the model to begin with the cellulose microfibril. This is thought to be coated at its outer surface by the xyloglucan component of the hemicellulose fraction. This binding can be demonstrated *in vitro*; purified xyloglucan will bind strongly to cellulose powder, and the binding can be released by treatment with 8M urea, which breaks hydrogen bonds. The xyloglucan is linked covalently to a neutral 'pectin' containing galactose and arabinose. This, in turn, is linked covalently to the acidic rhamnogalacturonan of the pectin class. These linkages have been demonstrated by the isolation of fragments after enzyme treatment which contain parts of the polymers. For example, it is possible to isolate fragments of xyloglucan attached to neutral pectin, neutral pectin attached to acidic pectin, and the complete neutral pectin containing both xyloglucan and acidic pectin components. The model is represented diagrammatically in Fig. 3.7.

The model can also accommodate the presence of protein by linking it to the acidic pectin component. However, this would imply a central structural role for the protein, which is not universally accepted. The

Fig. 3.6. Proposed chemical structure of rhamnogalacturonan from sycamore callus cells. The rhamnose units cause kinks to occur in the backbone of the molecule. R, rhamnose; GAA, galacturonic acid. (After P. Albersheim.)

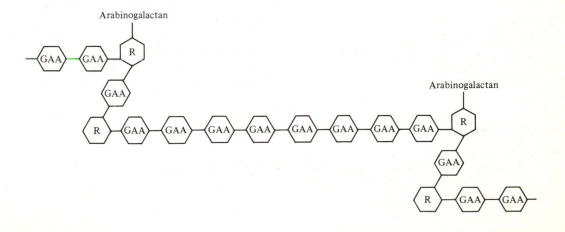

importance of the model is that all the polysaccharide components of the wall are proposed to be linked together either covalently or by hydrogen bonds. The cellulose microfibrils are considered to be completely coated with a layer of xyloglucan. It is possible that this layer is involved in the softening of the wall which accompanies growth: alternatively, growth may involve the breaking of covalent bonds elsewhere in the wall. This in turn means that it should be possible to demonstrate the presence of enzymes within the wall which would be involved in the breaking of covalent bonds. Such a demonstration has not yet been made, nor has it been possible to show differences in the linkages between the polymers isolated from walls whose growth has been greatly stimulated experimentally.

Although this model has been produced largely by the study of suspension culture cells, it does seem to have wider application to primary walls in general. Its merit is that it emphasises the likely complexity of processes involved in the loosening of the wall which

accompanies rapid growth. It also stimulates experimental investigation of these problems. However, it should be borne in mind that wide differences exist even in the gross chemical composition of walls prepared from angiosperms and gymnosperms, and monocotyledons and dicotyledons; it is, therefore, unlikely that a single model of this type will be able to account fully for the properties of all primary walls.

The physical organisation of the wall and growth

The chemical model of the wall has little to say about the way in which the polymers, and particularly the cellulose microfibrils, are organised within the wall space. This is an important problem since within the plant, the primary wall has to be able to expand in order to accommodate cell growth. This expansion does not necessarily occur equally over all the wall surface – in an elongating organ such as a root, the walls parallel to the long axis of the organ expand by a much greater amount than do cross walls. Furthermore, expansion of cell walls has to be more or less synchronised between adjacent cells. If it were not, cells would be sheared against one another, with a consequent disruption of plasmodesmata. This does not occur in the vast majority of cases, although intrusive growth of cells can occur, for example in the

Fig. 3.7. Diagrammatic representation of the interactions between the various cell wall polysaccharides. The xyloglucan is held by hydrogen bonding to the cellulose microfibril, and is linked covalently to the rhamnogalacturonan via the arabinogalactan. (After P. Albersheim.)

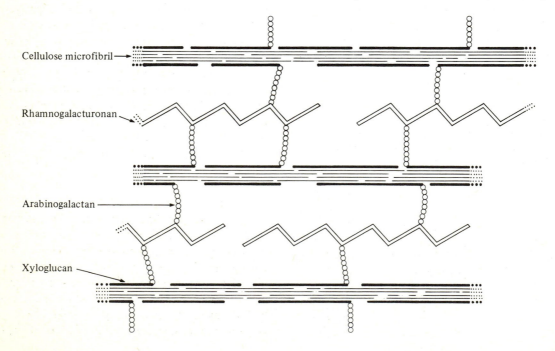

Cellulose microfibril

Rhamnogalacturonan

Arabinogalactan

Xyloglucan

development of laticifers (p. 115). Another complication concerning growth arises at points where cross walls meet expanding longitudinal walls; clearly some mechanism must exist for limiting growth at such sites, since the expansion of the longitudinal wall does not result in the thickening of the cross wall at the point of junction.

Growth of the primary wall begins as a process of addition of materials to the newly formed cell plate. Cellulose is deposited in the extracellular space and growth takes place at first both as an increase in thickness and as an increase in surface area. Subsequent to this, growth does not usually involve an increase in the thickness of the primary wall. Addition of materials to the wall does not take place equally over the entire surface, and the relative rates of expansion of the various parts of the wall determines the ultimate shape of the cell.

The first-formed part of the wall retains its identity as the middle lamella. Thus, in a general way, it might be expected that a primary wall would show a gradient of age across its structure, with the most recently formed part of the wall being next to the plasma membrane, and the 'oldest' region in the middle. This is confirmed by the observation that the majority of wall materials are incorporated at the plasma membrane surface, either because that is their site of synthesis (cellulose) or because they originate in the cytoplasm and reach the wall space in the form of membrane-bound vesicles which fuse with the plasma membrane.

The examination of the microfibril distribution within an elongating primary wall reveals an overall pattern. Near to the plasma membrane surface, the microfibrils show a preferred orientation at right angles to the direction of growth. The degree of order in the microfibrils is not high, but the pattern is recognisable. In the deeper parts of the wall, next to the middle lamella, the predominant axis of orientation of the microfibrils is along the direction of growth (Fig. 3.8). These observations have led to the so called multinet theory of growth. This states that the orientation of the microfibrils is determined by the stress laid on the wall by the process of growth itself. Thus the oldest layers of the wall, in the middle, have been subject to the greatest distortion due to growth and show a longitudinal orientation. The youngest microfibrils show a transverse orientation which is presumably due to the mechanism which controls their formation.

This simple idea elegantly describes the situation in an ideal cell which is growing in isolation or in the midst of other ideal cells without differentiation or the development of air spaces, or without even the presence of plasmodesmata. It was in fact derived in part from a study of developing cotton fibres, which are long filamentous hair cells. In three dimensional tissues, the situation is more complicated. Walls which develop against air spaces are generally thickened. In some cells, notably in collenchyma, there is marked thickening at the cell corners. There may be local variations in thickness and growth rates in the region of plasmodesmatal fields. These examples suggest that the growth of the primary wall does not occur in a passive manner once the initial synthesis of its polymeric components has taken place. It is beginning to be realised that material may be incorporated into the wall, or at least turned over, at sites far removed from the plasma membrane. This idea implies once again the presence of enzymes within the wall.

Fig. 3.8. The orientation of cellulose microfibrils in an elongating cell. In the youngest part of the wall (next to the cell interior) the orientation is predominantly transverse to the direction of growth. The process of growth itself disturbs this orientation in the older parts of the wall.

Auxin-stimulated growth and 'acid growth'

Growth can be experimentally studied in isolated segments of tissue which elongate without cell division. Such a tissue is exemplified by the oat coleoptile. Segments of coleoptiles continue to elongate after excision, and the rate at which this elongation occurs can be controlled by changes in the bathing medium surrounding the segments. In this way some of the characteristics of the growth process

can be studied, albeit in terms of the response of a large number of individual cells.

Growth is dependent upon the turgor pressure within cells; if the bathing medium contains mannitol at a concentration just insufficient to cause plasmolysis, growth ceases. Applied auxin causes an increase in growth rate after a lag period of a few minutes. This increase is known to be due to a change in the plasticity of the wall, and not, for example, to an increase in the turgor pressure within the cells. If auxin is added to the mannitol-containing medium which inhibits growth, then the change in the properties of the wall still takes place, although no growth results, since it cannot in the absence of turgor pressure.

The long-term continuation of stimulated growth requires that protein synthesis be maintained. Auxin stimulates RNA and protein synthesis in the coleoptile segments. Auxin also causes protons to be pumped outwards across the plasma membrane, so that the pH of the wall space and the medium beyond is reduced. If the pH of the bathing medium is reduced chemically, then this in itself causes an almost instantaneous increase in the elongation rate of the segment.

This is clearly a complex response. An early event in the auxin-stimulated growth is the release of non-cellulosic sugars from the wall. In oat coleoptiles this is predominantly glucose, but in pea epicotyl segments both glucose and xylose are released. This process is not inhibited by the presence of mannitol in the bathing medium; it can, however, be inhibited by calcium ions, which also have the effect of preventing the plasticity changes in the wall caused by auxin. Acidic buffers alone can cause the release of xylose and glucose from pea epicotyl segments, and it has been shown that these sugars originate from the xyloglucan which is in association with the cellulose.

These results can be interpreted as showing that turnover of polysaccharides occurs during wall softening. This is presumably enzyme mediated, and it is therefore reasonable to assume that the activity of such enzymes might be influenced by variations in pH and ion concentrations within the wall space. The acid growth effect could then be explained as a rather direct response of an enzyme (or enzymes) to a pH change. The primary auxin effect might also be pH mediated, with the longer term stimulation of growth due to increased protein synthesis.

Whilst this is a reasonably consistent picture, it should perhaps be emphasised that it is unlikely that a single explanation of this type can really account for the processes of growth in a wide variety of walls. Cell walls have different detailed compositions at different stages of growth, and certainty between the major divisions of plant types. The identification of particular enzymes responsible for the control of the physical properties of the cell wall is therefore likely to be extremely difficult. This is particularly so during growth, since it is likely that the net addition of material to the wall which growth involves will certainly mask small changes in sugar content due to polysaccharide turnover. A direct demonstration of the complexity of the situation is given by the finding that a substance called nojirimycin, which inhibits hydrolytic enzymes, is capable of preventing auxin-stimulated growth in oat coleoptiles, but has no effect on acid growth. Very much more work is required before a detailed description of the growth process in terms of biochemical activities within the wall can be given.

To place these comments in context, it should be realised that the idea of enzymes within walls is a comparatively recent one. Formerly it was believed that the major control of wall plasticity was the degree of crosslinking and gel formation in the pectin component of the wall. This too can change with pH, calcium ion concentration, and the degree of esterification of free acid groups in polygalacturonic acid. The current model also ignores the hydroxyproline-containing protein. When it was first discovered, this was given the name extensin and implicated in the control of growth. Current ideas favour its existence as a more or less independent matrix within the wall, and its function is quite uncertain. This is why it is excluded from the model of the primary cell wall (p. 73).

The formation of cellulose

The plant cell wall consists of a matrix with microfibrils embedded within it, and this is often compared to the structure of reinforced concrete. This analogy is in many ways a poor one, since the wall is a fluid and dynamic structure. Nonetheless, the change in fibril orientation which occurs during growth can be rationalised in mechanical terms; it is easier to stretch a spring (the initial transverse orientation of microfibrils) than it is to stretch a bundle of rods (the final, longitudinal orientation of micro-fibrils). Thus, in a simple mechanical way, it is to be expected that the cell would need to exert control over the way in which cellulose microfibrils are deposited within the wall. This control is more obviously seen in the secondary wall (p. 89).

The great majority of cellulose synthesis takes place at the plasma membrane surface. Fibrils are not seen in any intracellular compartment, and the surface is clearly the site of fibril synthesis in the case of isolated protoplasts (p. 79). Experiments with radio-active tracers show that incorporation of sugars into materials of high molecular weight can occur throughout the wall space – however, this is generally taken to indicate processes related to wall turnover and growth rather than massive net synthesis of fibrils.

Synthesis at the plasma membrane surface, coupled with a predominant transverse orientation in the youngest parts of the wall, suggests that the controlling mechanism for cellulose microfibril for-mation should have two characteristics. It should be present in the cortical cytoplasm next to the plasma membrane, and it should show a pattern of distribu-tion which corresponds to the transverse orientation of the youngest fibrils. In the early work with plant cells using electron microscopy, the endoplasmic reti-culum (ER) was cited as the cytoplasmic component which showed these characteristics. The association of the ER with the cell plate was regarded as particu-larly significant. Nowadays we should interpret these facts in terms of the formation and distribution of plasmodesmata rather than of the cell wall itself. It was not until 1963, using aldehyde fixation (instead of the customary potassium permangate of the time) that an entirely new set of structures was found in the root tip cells of onion which seemed to fulfil all the requirements for a role in cellulose synthesis. These structures were the microtubules (p. 21).

Microtubules and microfibril orientation

In interphase cells microtubules have a dis-tribution which is highly suggestive of a function in wall formation. They are found in the cortical cytoplasm, separated from the plasma membrane by a distance of 20 nm or less. Along the longitudinal walls of an elongating tissue, such as a root, they are found at right angles to the growth axis, mirroring the preferred orientation of microfibrils in the youngest parts of the wall. It was originally assumed that microtubules completely encircled the cell in the form of hoops, but it is now thought possible that each one is only a few microns long and extends only about half-way round the circumference of the cell. The spacing between microtubules is variable, but never less than about 35–40 nm (Fig. 1.19). This has led to the suggestion that each microtubule may be func-tionally larger than its apparent diameter of 24 nm. If this is so, then the distance of 20 nm between the microtubules and the plasma membrane could rep-resent functional 'contact'. The cortical microtubules are sometimes parallel to one another in groups, but this arrangement is not strictly defined. Over the end walls (which, in an elongating tissue, are not growing as rapidly as the longitudinal walls) microtubules are present in lower numbers, and show more random orientations.

Microtubules in such a cell therefore not only reflect the orientation of the cellulose microfibril synthesis which is taking place, but their density reflects its quantity. It is not surprising therefore that they are considered strong candidates for a control-ling function in the orientation of microfibrils. However, the situation has been complicated by observations on other cell types. In some situations growth proceeds entirely from a small region confined to the tip of the cell. This is true of pollen tubes for example, and also of root hairs. Examination of both these cell types casts a doubt on a general role for microtubules in primary cell wall growth. In root hairs microtubules are present in their cortical arrays, except at the tip of the cell, which is the site of growth. In collenchyma, which is a support tissue charac-terised by thickened primary walls, microtubules have been found to occur at all orientations with respect to the most recently formed microfibrils. On the other hand, in the thickening secondary walls of xylem elements (p. 102) there is a striking positive correla-tion between the occurrence of microtubules in the cytoplasm and the positions of the secondary wall thickenings.

The mechanism of microtubule action

If it is assumed that microtubules are indeed involved in the formation of orientated structure within the wall, it follows that this effect must be mediated by the plasma membrane. Quite how this is achieved is a matter of controversy at present. There is some evidence from the cellulosic algae that micro-fibrils might be synthesised from one end by a com-plex particle within the plasma membrane (see p. 78).

This is an attractive hypothesis, and one which it is tempting to transfer directly to the situation in higher plants. If such particles were present, then their arrangement and movement within the plane of the plasma membrane could in principle be controlled by a cytoplasmic element such as a microtubule, by means of a link through the plasma membrane. Microtubules could then act as 'rails' dictating the directions of movement of the membrane particles, and hence the orientation of the cellulose microfibril.

Much effort has been invested in attempting to demonstrate the presence of particles within the plasma membrane of higher plants which could act as the synthetic source of microfibrils. The work has been remarkably unsuccessful to date. The only pictures to emerge from many years of research have been by the use of the freeze etching technique. These have been interpreted as showing microfibrils terminating on the plasma membrane at the site of a particle, but this interpretation is not widely regarded as convincing. The problem is that the particle is only required if it is assumed that cellulose is indeed synthesised in the form of microfibrils. The evidence from higher plants at least is that this may not be the case (p. 80). If the formation of the microfibril takes place through a mechanism akin to crystallisation, as is now considered possible, then the need to demonstrate a particle engaged in the synthesis of the microfibril disappears.

This situation clearly illustrates a difficulty which is frequently encountered in the study of structural relationships. The coincidence between the occurrence and distribution of two different structural elements (in this case microfibrils and microtubules) within a fixed specimen does not necessarily mean that they are functionally related to one another. The case of cellulose synthesis is a particularly difficult one, since microfibril formation has not yet been achieved using a purified enzyme complex *in vitro*, and therefore very few facts are available concerning the details of the process. It is equally possible at the present time that some undetected influence exists which affects microfibril orientation in all situations, and microtubules in only some.

The cellulosic algae

The wall of cellulosic algae is much more obviously organised than is the case in higher plant forms. The filamentous alga *Chaetomorpha mela-* *gonium* has a wall which consists of lamellae of highly ordered microfibrils. The orientation of all the microfibrils within a given wall layer is the same, and it differs from the orientation in the adjacent layers. The innermost lamella, closest to the plasma membrane, was found to be associated with groups of granular particles. This led to the formulation of the ordered granule hypothesis of cellulose synthesis, which proposes that microfibrils are synthesised as fibrils by a multienzyme complex consisting of an array of particles in three dimensions. This complexity of the particle is required because each microfibril contains many units of 'chemical' cellulose. The microfibril is envisaged to be produced by the array in a direction which depends on the activation of particular groups of particles within it.

At about the same time as this idea was put forward, the newly invented technique of freeze etching was giving the first pictures of the inner surfaces of plasma membrane leaflets, and these were also shown to be studded with particles. Freeze etching has more recently been applied to the alga *Oocystis apiculata*, which also has ordered cellulose walls. Here a terminal complex for each microfibril was identified within the plasma membrane, this time associated with a linear array of particles which were presumed to indicate the direction of formation of the microfibril.

These observations from the cellulosic algae are of interest since they do indicate a possible way in which highly ordered arrays of microfibrils might be formed. However, it is doubtful whether the ordered granule hypothesis can be applied to higher plant cell walls. The algal microfibril is much larger than the microfibril of the higher plant cell wall, and is produced in a chemical environment which is quite different from the primary wall of higher plants. It is reasonable to conclude therefore that the mechanism of production of microfibrils might be quite different in these situations.

Primary walls and protoplasts

Apart from the production of the cell plate by the fusion of vesicles, and the formation of a wall on the shed eggs of some algae following fertilisation, cell walls do not normally arise *de novo*. Rather, they are always the result of the addition of new material to a wall which was already present. This is a constant source of difficulty in the study of cell walls; the presence of a mass of pre-existing wall means that it is difficult to detect the addition of new material, and to distinguish this from turnover associated with growth.

Although walls do not normally arise on naked membranes, it is possible experimentally to produce just such a situation. The living protoplasts can be released from a wide variety of plant tissues by the following procedure. First the tissue – for example, a piece of leaf – is plasmolysed. This protects the protoplasts and prevents their bursting once the cell walls which surround them have been removed. Next the tissue is treated with a crude 'pectinase' enzyme. This dissolves the middle lamella of the wall and causes the tissue to break down into its constituent cells. These are then treated with a crude 'cellulase' enzyme, which dissolves the rest of the cell wall and releases the free protoplasts (Fig. 3.9). Isolated protoplasts are surrounded by a plasma membrane, and electron microscopy shows that there is no wall material on this surface when they are first released. The protoplasts may be cultured in a liquid medium which contains salts and simple organic molecules including growth substances. Depending on the spe-

cies of the protoplast, a new wall begins to be formed within a few minutes or a few hours of culture. It is first seen as a few scattered microfibrils on the outer surface of the plasma membrane (Fig. 3.10). This simple observation shows two things clearly. The first is that microfibrils can be produced at the plasma membrane surface. The second is that there is no requirement for a pre-existing wall structure.

Protoplasts are spherical and those from higher plants have no intrinsic polarity (Chapter 7). The wall which they produce arises over the entire surface of the plasma membrane and consists of a random meshwork of fibrils which gradually thickens until a continuous meshwork is formed. All the various polymeric constituents of the wall are synthesised simultaneously, but only the fibrillar component is retained at the protoplast surface in the initial stages. The soluble materials, especially the pectins, are lost into the medium surrounding the protoplast, at least until a considerable amount of fibrillar deposition has taken place. This behaviour is consistent with the model of the primary wall given earlier (p. 73), since the anchorage of the non-cellulosic polymers would be expected to depend on the presence of cellulose fibrils.

Since the distribution of microfibrils on the surface of the protoplast produces no pattern, it is not

Fig. 3.9. A preparation of protoplasts isolated from a leaf of *Nicotiana plumbaginifolia*. Each protoplast is spherical and bounded only by the plasma membrane. Bar = 50 μm.

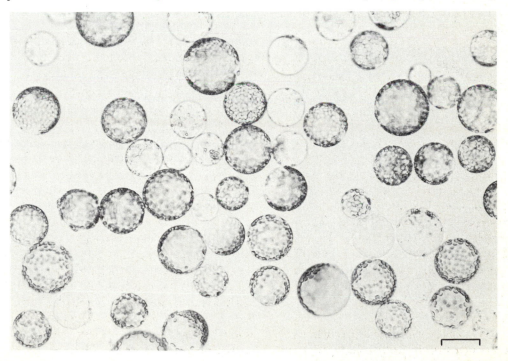

possible to draw any conclusions about the role which cytoplasmic microtubules might play in the process of fibril formation by isolated protoplasts. The microtubules are themselves randomly distributed, but even the most optimistic analyst would not assume a functional link on this basis. Protoplasts have also been examined for the presence of particles in the plasma membrane which might represent a terminating region for the production of microfibrils. Such particles have not been found.

Because the production of microfibrils by protoplasts takes place on a naked membrane, it is possible to examine early stages in the process. When this is done, it is found that microfibrils do not appear first as short lengths which progressively elongate. Rather, there is usually a lag period during which no microfibrils are observed at all, after which a few rather long microfibrils are found. This has been interpreted as evidence against the idea that microfibrils are produced from one end by a particle. It is proposed instead that during the lag period a precursor to the microfibril is synthesised and that, when it reaches a critical concentration at the surface of the plasma membrane, the microfibril is formed by a process akin to crystallisation. Such a process would not require the intervention of particles beyond the stage of synthesis of chemical cellulose or the precursor. At present this is an interesting idea which allows the absence of terminal particles within the plasma membrane of higher plants to be understood. However on its own it still fails to explain the predominant, though disorganised, arrangements of microfibrils in the primary walls of normally growing tissues.

Two other lines of evidence from work with protoplasts support the idea that the microfibril is not produced in its final form on an organising particle. Cellulose is classically defined as being insoluble in alkali. However, the very young microfibrils which appear initially on the surface of cultured protoplasts are alkali soluble, and rather easily disrupted into small fragments (Fig. 3.11). This suggests that the final stability of the microfibril is only reached after a certain period of time, which would not be expected if it were synthesised as such by a particle. The crystallisation hypothesis leads to the prediction that if something could be found which would preferentially bind to the precursor, microfibril formation would be inhibited. This behaviour is shown by the dye congo red. Protoplasts which are cultured in the presence of this substance do not produce microfibrils. However, the presence of the protoplasts causes the dye to be precipitated, and this precipitate contains glucose in a polymerised form. This is consistent with the idea that the dye preferentially binds to the microfibril precursor, and that the precursor–dye complex is insoluble.

Fig. 3.10. Scanning electron micrograph of a leaf protoplast from tobacco White Burley which has been cultured for 24 h. The surface is covered with a random meshwork of fibrous material. Bar = 5 μm.

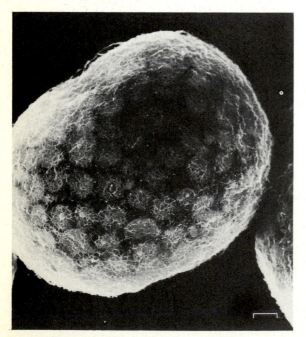

Fig. 3.11. A negatively stained preparation of microfibrils from a protoplast of tobacco White Burley, cultured for 8 h. Many of the microfibrils have fragmented into short lengths less than 1000 nm. Bar = 0.1 μm.

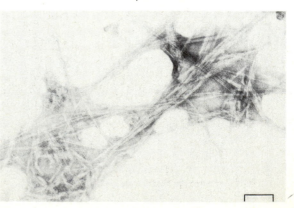

In the absence of the dye, the precursors bind to each other and the result is a microfibril; in the presence of the dye, the microfibril does not form.

Cellulose in the primary wall: a summary

This rather detailed look at cellulose synthesis is justifiable not only because of the widespread occurrence of the material and its huge biological importance, but also because the ideas in this area are in such a state of flux at present. The greatest mystery surrounds the way in which the physical unit of cellulose in the wall, the microfibril, is formed. The first idea for this is that it is synthesised as such by a multienzyme complex which appears as a particle in the electron microscope. The evidence for this is good in some algae, and it can explain how the ordered arrays of microfibrils which are characteristic of these species are organised. Alternatively, it is possible that microfibrils are formed on a more random basis by a crystallisation mechanism. The evidence for this is largely negative; that particle arrays are not encountered on a regular basis in higher plant plasma membranes, and this hypothesis does not require them.

There is no reason at present to abandon either of these ideas, within their limitations. It remains for biochemical work to establish the conditions for microfibril synthesis *in vitro*. Until this is achieved the distinguishing features of the two hypotheses cannot be independently tested, although current work with a cellulose-producing bacterium has claimed to have isolated a soluble precursor material, the basic requirement for the crystallisation hypothesis to be considered correct.

Synthesis of the matrix

The matrix of the primary wall contains polysaccharides and protein. In contrast to cellulose, which is rather uniform chemically and has a complex physical form, the polysaccharides of the matrix are chemically complex; their physical relationships within the wall are not understood at all (Fig. 3.12). The importance of the matrix materials is that they

Fig. 3.12. The primary wall at the surface of a cultured protoplast of the moss *Physcomitrella patens*. The wall clearly consists of several different materials of different staining properties and morphology, but their interrelationships are quite unclear. Bar = 0.2 μm.

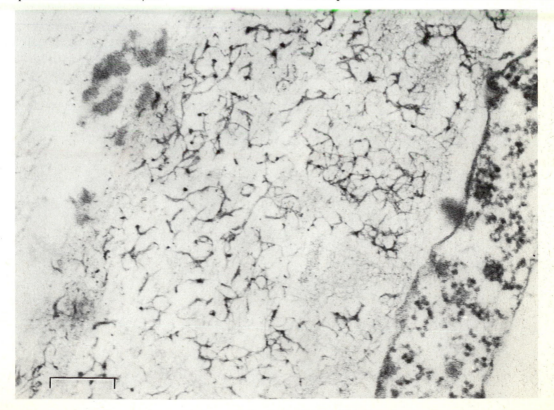

probably control the plasticity of the wall and hence its potential for growth; from a developmental point of view, their relative quantities change as the wall ages, ceases to expand and becomes thickened.

The matrix polysaccharides are derived from glucose by a complex series of reactions involving rearrangements of the sugar ring to give the other monomers – mannose, xylose, arabinose and so on. This intermediary metabolism need not concern us but the evidence suggests that changes in cell wall composition during development are mediated at the level of the polysaccharide synthetase enzymes, rather than by the availability of particular substrates.

Both the polysaccharides of the hemicellulose and pectin groups are synthesised from low molecular weight substrates which take the form of a sugar

linked to a nucleotide. For example, the polygalacturonic acid of the acidic pectin is produced by an enzyme which uses UDP-galacturonic acid as its substrate. This has been demonstrated using particulate fractions prepared from a number of different plant species. In such experiments *in vitro*, the UDP derivative is not the only possible substrate, but other nucleotide sugars are utilised less efficiently, and so it is considered that the UDP derivative is the natural substrate *in vivo*. To take an example from the hemicellulose group, it is known that xyloglucan is synthesised from UDP-xylose and UDP-glucose.

A feature of the pectins in particular is their variability in terms of fine chemical structure. Thus the carboxylic acid groups of galacturonic acid may or may not be esterified with methyl groups. This esterification takes place in synchrony with the polymerisation process, or shortly afterwards. It does not occur before polymerisation – the methylated UDP-galacturonic acid nucleotide does not act as a substrate *in vitro*. The neutral blocks of arabinans and galactans of pectin are synthesised separately and attached to the rhamnogalacturonan backbone as polymers. The

Fig. 3.13. The tip of a regenerating chloronemal filament of the moss *Physcomitrella patens*, grown from an isolated protoplast. Dictyosomes in the vicinity of the wall at this position are particularly active in the production of vesicles. Bar = 0.5 μm.

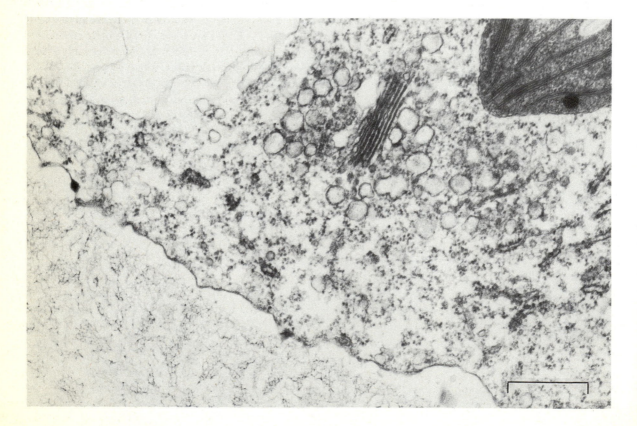

question of the variability of polysaccharide structure sets an important limitation on the biochemical study of the synthesis of matrix materials. It is clearly unsatisfactory for the biochemist to be uncertain whether a product synthesised *in vitro* has the identical structure to the natural product *in vivo*. Doubts of this type represent a major barrier to progress in the biochemistry of matrix polysaccharide synthesis.

The hydroxyproline-containing protein of the wall is synthesised from the amino acid proline, and then hydroxylated by a soluble hydroxylase enzyme. The protein is then glycosylated so that in its final form it has arabinose and galactose side chains.

Sites of synthesis of the matrix

The site of synthesis of the matrix materials of the wall is known to be on cytoplasmic membranes since, when cells are disrupted and fractionated, the synthetic enzymes are found within membrane fractions. By careful use of small amounts of aldehyde fixatives, it is possible to isolate intact dictyosomes from cells. When this is done it is found that such fractions contain polysaccharides which, after hydrolysis, yield sugars corresponding to both the pectin and hemicellulose polymers of the wall. This material can be labelled with externally applied sugars before isolation. Electron microscopy also suggests that it is the vesicles produced by dictyosomes which are involved in the delivery of matrix materials to the wall space. The best evidence for this has come from studies on the secretion of root cap slime (p. 109), which of course is not a normal wall component. Less direct evidence includes the frequent observations that cells which are growing rapidly at the tip, such as pollen tubes or the protonema of mosses, show the presence of high concentrations of dictyosomes and vesicles in this growth region (Fig. 3.13, see also Fig. 7.7). It has also been found that a fraction prepared from pea tissue which contains dictyosome membranes and vesicles demonstrates activity of enzymes capable of catalysing the production of xyloglucan from UDP-xylose and UDP-glucose.

These facts suggest strongly that the Golgi apparatus is not only the site of packaging of the polysaccharides of the matrix, but also of synthesis. Whether this is an exclusive role is less certain. It is frequently found in experiments where intact cells are fed radioactive precursors before fractionation, that other cytoplasmic membranes become labelled,

most notably the endoplasmic reticulum. However, the general conclusion stands that the matrix polysaccharides and indeed the structural protein of the wall pass through the Golgi apparatus, and are probably synthesised there. The Golgi apparatus also acts as a site of membrane transformation so that the vesicles it produces may fuse with the plasma membrane (p. 17).

The way in which the vesicles from the dictyosomes reach their destinations at the plasma membrane is not resolved. This is an important consideration since wall growth is rarely uniform over the entire surface of a cell, and so the migration of vesicles is clearly not a random process. It has been suggested that microtubules might act as direction markers for the movement of vesicles. This idea stems from the coincidence between the occurrence of microtubules and vesicles in some situations, most notably the phragmoplast (p. 34). Whether such a coincidence represents compelling evidence for a functional link between the two structures is a matter of opinion.

The secondary wall

The secondary wall is formed when growth ceases. Secondary wall development takes many forms, and certain aspects of it will not be considered in this chapter since they fit more logically into a description of developmental pathways to specialised cells. This applies to the secondary wall thickening during xylem differentiation (p. 101) and the secondary wall changes associated with sieve element development (p. 97).

Secondary wall development is not obligatory; some parenchymatous cells retain thin primary walls throughout their life. The deposition of secondary wall materials may relate rather obviously to a special function; this is true in tissues such as schlerenchyma, which is a rigid structural tissue, and also to walls with geometrically significant positions such as the epidermis. The most obvious plant product, timber, gains many of its properties because of secondary wall development, both in general and in detail. Because of its economic importance a great deal is known about the way in which secondary walls behave in relation to stress in timber.

The secondary wall is developed from the primary wall by the addition of new materials. These are not the same as were used in the growth and expansion of the primary wall. Secondary walls contain much more cellulose than do primary walls and, correspondingly, less matrix materials. However, secondary walls also contain a number of classes of chemical substance which are not commonly found in primary walls at all.

Lignin

Lignin is the most abundant plant polymer after cellulose. It consists of large amorphous molecules built up from a variety of phenyl-propane derivatives such as sinapyl, coniferyl and *p*-coumaryl alcohols. The presence of lignin in a wall confers upon it great rigidity and resistance to chemical degradation. On an evolutionary scale the presence of lignin is correlated with the development of extensive aerial forms, where mechanical strength of the tissues is necessary. In a biological sense it is the reason for the resistance to rot which is shown by many timbers and, at a cellular level, it is deposited in a localised manner in situations where the wall has to be protected from breakdown due to the presence of hydrolytic enzymes. This occurs during xylem development, for example (p. 104).

Lignin comprises 15–35% of the dry weight of supportive tissues in higher plants. Its chemistry is highly complex and the structure of its molecules does not appear to be determinate. Rather lignin consists of a polymerised network built up from its three principal starting monomers. The control of lignification within the cell appears to be at the level of the production of these monomers from the amino acid phenylalanine. The production of the polymer is not thought to proceed through a series of closely controlled enzymic steps, but via a free radical mechanism involving the enzyme peroxidase, present within the wall space. Despite this apparently uncontrolled polymerisation process, the distribution of lignin within walls is quite predictable and organised. Lignification usually begins within the depth of the wall near the middle lamella. Within the surface of the wall it may be generally deposited, or it may be localised to particular regions such as the thickenings of developing xylem elements (p. 104).

Lignin therefore clearly displays the two sides to developmental processes; its formation is controlled in time, and does not occur until primary wall growth has ceased. Its distribution is controlled in space, although the mechanism by which this is achieved is at present not understood.

Cutin and suberin

One of the important physiological problems presented to all aerial plants is the control of water loss from their surfaces. Cell walls are hydrated, and each protoplast is thereby maintained in an equable environment. The plant as whole however exposes a considerable area of surface to the drying effects of the moving atmosphere. If the outer surface of a plant were simply a uniform continuation of ordinary wall structure, the result would be a catastrophic loss of water, resulting in the death of the plant. This is prevented by the presence within outer walls of two families of polymers known as cutin and suberin. Throughout the plant kingdom, only algae and fungi lack these substances.

Cutin is not a single polymer but a mixture of polyesters of hydroxylated palmitic (C16) and oleic (C18) acids. Its exact composition is not yet known since current procedures for its analysis usually leave up to 30% of the material as an insoluble residue. In general it appears that the cutin of fast-growing plants has a higher proportion of C16 components than that of slow-growing plants. This may be reflected within the tissues of a single species; for example, in the apple, the flower has predominantly C16 subunits, whereas the fruit has both C16 and C18 subunits. Cutin is found on the outside surfaces of the aerial parts of all plants, where it forms a continuous layer. This layer is called the cuticle (Fig. 3.14). In general the cutin has an amorphous appearance in the electron microscope, although it may sometimes be lamellar. Cutin may also be found localised in regions at the base of secretory glands (p. 113).

Because of its common position at the outer surface of the plant, cutin does not turnover since it is not required to do so. However, it has recently been shown that germinating pollen of the nasturtium (*Tropaeoleum* sp.) contains a cutinase enzyme able to release monomers from cutin. This allows the growing pollen tube to penetrate the cutin layer on the stigmatic surface.

The chemical composition of suberin is also not well understood. It consists of a mixture of polymers containing long chain acids and alcohols with up to 30 carbon atoms, together with phenolic compounds. Suberin is present attached to the cell walls of periderms, including wound periderm formed by aerial parts of plants, in the endodermis and bundle sheath of grasses. These internal locations for suberin suggest a function in limiting water movement out of the vascular system. Suberised walls often have a lamellar appearance in the electron microscope which is thought to be due to admixture of waxes with the suberin. Enzymes to degrade suberin are not known to have any importance in normal growth, although it has been shown that some pathogenic fungi excrete such enzymes.

The mechanism of deposition of these insoluble and hydrophobic materials at the outer surfaces of walls is mysterious. It is possible that epidermal cell walls contain channels through which a flow of precursors could take place, with final polymerisation occurring at the surface. The enzymes responsible for some of the intermediate pathways in the formation of cutin are known to be localised in epidermal cells – a clear and not unexpected example of differentiation at the biochemical level.

Waxes and the cuticle

Waxes are found on the outer surfaces of many leaves and fruits (Fig. 3.15). They are non-polymeric, consisting of complex mixtures of paraffins, long-chain alcohols, ketones and acids. Apple fruit wax for example has been shown to contain 50 different molecular species, including all the paraffins from C15 to C33. Waxes, together with cutin, provide the water-retaining properties of the cuticle. It is also likely that the cuticle functions to protect leaf surfaces from abrasion by wind, and there is evidence that it offers frost protection. Up to 90% of the water loss from a leaf occurs through the stomata, and is therefore under the physiological control of the plant. If the wax on a leaf is removed by means of solvents, some species will exude droplets of liquid which contain sugars, oligosaccharides, amino acids and mineral salts. Thus the cuticle is important not only in the restriction of water loss, but in the retention of nutrients and small metabolites within the apoplasm. The wax layers on a leaf can be damaged by rain and this is an important consideration in agriculture, where plants may be made temporarily more susceptible to harmful effects of chemical sprays under such conditions.

Fig. 3.14. The external wall of an insectivorous gland of the Venus fly-trap (*Dionaea muscipula*). The light grey layer on the outer surface of the wall corresponds to cutin. The fibrous material beyond is the secreted material which captures and digests the plant's prey. Bar = 0.5 μm.

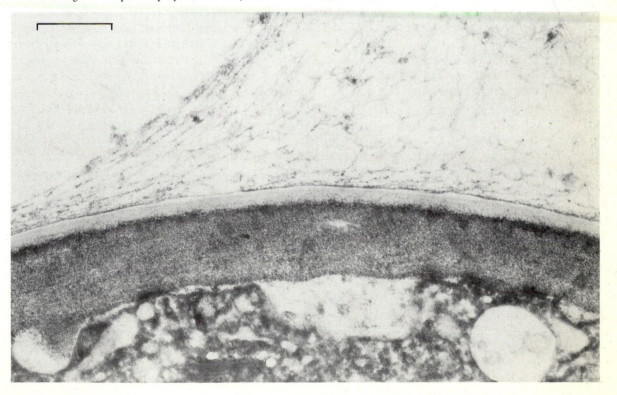

The bloom which appears on some leaves and fruits is due to the deposition of a wax layer with dimensions close to the wavelength of light. For example, the bloom on cabbage leaves is due to wax deposits in the form of rods 1–2 μm long and 0.5–1 μm in diameter. The pattern of the wax deposit stems from its chemical composition, rather than as a result of the action of a controlling template. Thus if the wax is removed from a leaf by means of a solvent, and the solvent is allowed to evaporate from a glass container, the wax layer left behind shows the same morphology as it did on the leaf surface.

The way in which wax is deposited on the outside surface of the cuticle is not known. This is no doubt partly due to the difficulty of preserving wax layers for examination in the electron microscope. Solvents used in specimen preparation remove the wax, and in addition, simple paraffins would not be stained or even stable in the electron beam. It is

known however that if the bloom is physically removed from the surface of a leaf, it can regenerate.

Finally, it should be emphasised that although there is no doubt that the cuticle does act as a resistant barrier to the loss of water and metabolites, it is by no means totally impermeable. A small water loss does occur through the cuticle, known as cuticular transpiration and, in special situations such as nectaries, secretion of aqueous solutions can apparently take place through the cuticle (p. 113).

Callose

Callose is a polysaccharide built up from glucose units linked β-1,3. It is not found in primary walls, and its appearance in secondary walls is limited to specialised situations. In sieve element development (p. 97) callose has a particular function in relation to the sieve plate, and it is also found in association with the plasmodesmata connecting the sieve element to its companion cell. Callose is remarkable for the speed at which it appears in response to wounding, especially in the phloem. This caused considerable confusion in the interpretation of

Fig. 3.15. Scanning electron micrograph of the wax on the outer surface of the leaf of the banana plant (*Musa sapientum*). Bar = 1 μm.

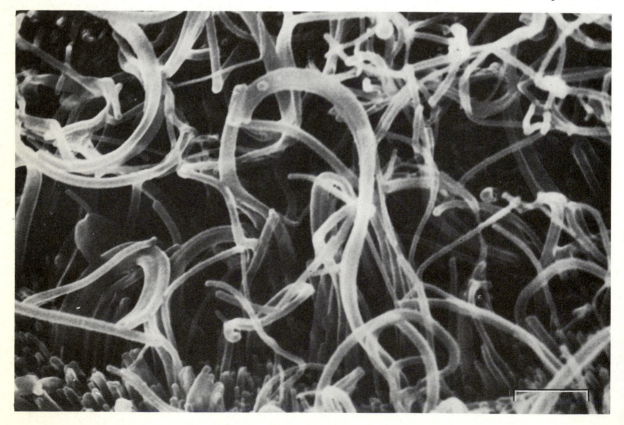

electron micrographs of developing phloem before rapid freezing techniques were introduced and confirmed callose as a normal constituent of the sieve element wall (p. 97).

Aside from its regular appearance in the phloem tissues, callose plays an important role in the development of the pollen of flowering plants. Pollen is produced by the process of reductive division called meiosis (p. 36), and arises in sets of four haploid units from each original diploid mother cell. The events of meiotic prophase ensure that each pollen grain may be genetically unique. The individual grains are initially isolated from their neighbours by a thick callose wall which is not crossed by plasmodesmata. This layer of callose remains in place until each microspore begins to develop its own sculpted wall. The callose then dissolves, releasing the pollen grains into the tapetal fluid where they complete their development. Callose, in this situation, is apparently acting as a temporary sealing material which is removed once the need for its presence passes.

The mature pollen grain commonly contains two highly differentiated nuclei. The vegetative nucleus is found within the main cytoplasmic compartment of the pollen grain, and the generative nucleus is confined to a thin-walled compartment within that cytoplasm, the generative cell (p. 120). When the pollen grain comes into contact with a suitable stigmatic surface, it germinates by producing a tube. In angiosperms at least, this tube has to penetrate the entire length of the style and carry the generative cell towards the micropyle at the base of the embryo sac. This distance may be of the order of 1 cm and, as the growth of the pollen tube proceeds from its tip, the cytoplasm is confined to the growing region by the formation of callose plugs across the diameter of the pollen tube. In this way the limited resources of the pollen tube are applied to the region of the tip where growth is occurring. Here again, callose acts as a sealing material.

Finally, callose may be produced when a pollen grain lands on an unreceptive stigma. The pollen grain wall contains proteins which act as the trigger for this incompatibility reaction. The proteins are released by contact with the stigmatic surface, and the sealing pad of callose which forms as a result of this prevents further growth of the pollen.

Callose is thus a very specialised product of differentiated cells which is found in the wall space.

Its presence implies extremely localised biochemical differentiation in the cells which produce it. Callose is also interesting from a developmental standpoint because its appearance may be transitory, or elicited by an environmental stimulus which is not necessarily a predictable part of a development programme.

Sporopollenin

Sporopollenin is another specialised cell product which, although not a general secondary wall constituent, is nonetheless an indicator of a highly differentiated state. It is found in the walls of pollen grains, where it occurs in species-specific patterns (Fig. 3.16). These patterns persist for thousands of years in pollen samples which have been preserved in peat deposits, and may be used for the identification of the types of plant present at the time of the laying down of the peat. This extreme stability of the pollen grain wall is conferred by the presence within it of sporopollenin.

The exact chemical composition of sporopollenin is not known, but it appears to be a polymer based on carotenoid derivatives. It is deposited on the outside of the pollen grains whilst they are still in the anther loculus surrounded by a layer of tapetal cells. The source of the sporopollenin is disputed; aggregates of the material are seen to line the surface of the tapetal layer at the same time as it is being deposited as an outer layer to the pollen grains. This has led to the

Fig. 3.16. Scanning electron micrograph of a pollen grain of *Bilbergia nutans*. The wall patterning of pollen grains is often species-specific. Bar = 5 μm.

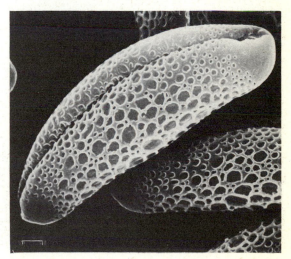

suggestion that the tapetal cells contribute to the sporopollenin coat of the pollen grains. Other workers think that the sporopollenin on the pollen grains comes exclusively from the pollen grain cytoplasm, and that the layer on the tapetal cells is formed independently as an unwettable surface to assist the release of the pollen grains when the anther finally opens. It is known that the tapetal cells supply nutrients to the pollen, in any case.

The origin of the precise patterning of the exine is also not certain. In *Ipomoea purpurea* it can be detected in negative form in the callose layer surrounding the young microspore. It has been suggested that the patterning may be controlled by the pollen grain itself in the form of localised distributions of endoplasmic reticulum and RNA within the pollen cytoplasm. It is found that if the cytoplasm of the pollen grain is fragmented experimentally, the pattern of the wall is produced correctly even over those parts of the cytoplasm which do not contain the nucleus.

Apart from the patterning of the sporopollenin, the pollen grain wall also shows differentiation in the

form of pores where the sporopollenin layer is reduced or absent (Fig. 3.17). These pores are the sites at which the emergence of the pollen tube will take place; they also contain the proteins responsible for incompatibility reactions, and for causing hay fever in humans. The positioning of the pores is not random, but relates to patterns of cell division in the developing tetrad of microspores. In *Avena sativa* (oats) for example, the single pore is produced on the face of the grain which is outermost in the developing tetrad.

The pollen grain therefore represents an extreme case of precision in the laying down of wall components, whether they be polysaccharides, proteins or hydrocarbons. At present the detailed mechanisms of this control are obscure.

Organisation of secondary walls

The primary wall is characterised by its ability to accommodate growth. The secondary wall expresses the attainment of a more stable state where specialised function is the requirement. This may be a subtle biological specialisation or it may be a mechanical one. These functions may not be mutually exclusive, and it is with the woody xylem tissue, which is the raw material of the timber industry, that very much work has been carried out on the structure and mechanical properties of secondary walls.

Fig. 3.17. Scanning electron micrographs of pollen of the black alder (*Alnus glutinosa*). The pollen grain has four pores, one of which is shown in detail. *a*, bar = 5 μm; *b*, bar = 1 μm.

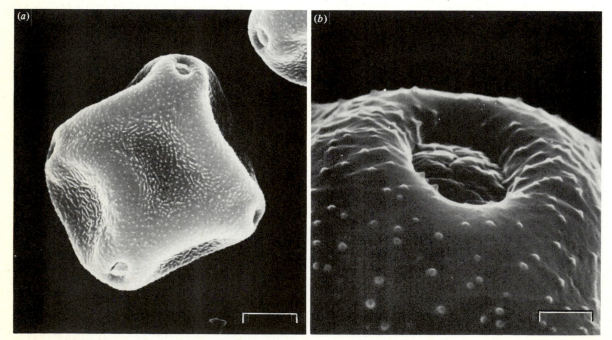

(a)   (b)

Fibrillar organisation of secondary walls

The secondary wall of the fibres and tracheids of woody tissue is prominently constructed in layers. These layers have been named S1, S2 and S3 (S for secondary). The S1 layer lies outermost with respect to the interior of the cell, in other words, next to the primary wall. The three layers can easily be seen in the light microscope and, under polarised light, they show different patterns of birefringence corresponding to differences in the predominant orientation of the microfibrils within them. The relative thickness of the three layers is variable; in cells which have thickened during summer, the S2 layer is the widest, and indeed its width may exceed that of the other two layers together. On the other hand, in early or 'spring' wood S2 may be narrower than either S1 or S3. S3 (the layer nearest the interior of the cell) is usually the narrowest layer, and may be absent in some coniferous species.

The layers differ in their microfibril orientations. The S1 layer appears to consist of perhaps two sets of crossed microfibrils, running in a slow helix around the long axis of the cell. In the S2 layer the direction of the microfibrils is predominantly at a small angle to the long axis of the cell, and therefore approaches a right angle with respect to the orientation of the S1 layer. The S3 layer, when present, is again comprised of microfibrils with a slow helical arrangement around the cell axis (Fig. 3.18). The S1

Fig. 3.18. Diagrammatic representation of the different wall layers in a thickened wall. The secondary layers are deposited on top of the primary wall after its growth has ceased.

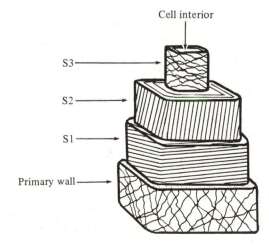

Cell interior

S3

S2

S1

Primary wall

and S2 layers, in particular, are highly laminated and this geometrical complexity no doubt serves to provide the strength and resilience which is needed in mechanical terms.

The correspondence between structure and mechanical properties is shown by a consideration of the situation which arises in cells of wood which has been formed within horizontal branches. In the vertical trunk of a tree, the normal stresses due to gravity will tend to compress all the cells at a given level to an equal degree. In a horizontal branch however, the cells in the upper part of the tissue will be in tension, whereas those in the lower part will be in compression. Tissue produced under such conditions is known as reaction wood, and shows changes in the pattern of microfibrils in the secondary wall layers.

Reaction wood is of two types. Very broadly speaking compression wood is produced on the undersides of horizontal branches of conifers, whereas tension wood is produced on the upper side of horizontal branches of angiosperms. Compression wood is characterized by a marked shortening of the individual cells, and changes in the angle of microfibrils in the S2 layer. The S3 layer is frequently absent. In tension wood there is again a shortening of cell length but, in this case, the three S layers of the wall may be replaced by a so called G (gelatinous) layer, which is unlignified. Alternatively S3 may be absent, or S2 and S3.

This situation illustrates two facts. The gross mechanical properties of extensive tissues are dependent upon the fine structure within walls of individual cells. Reaction wood is more dense and more brittle than normal wood and unsuitable for use in construction. The second point is that cells are capable of reacting to externally applied stresses by changes in their relative degree of elongation and also in the way in which they deposit their cell wall materials. Very little is known biologically of how these changes are mediated, but they do prove that development of the secondary wall is not invariant and insensitive to environmental conditions.

Chemical organisation of secondary walls

The layers of the secondary wall show chemical as well as structural differences. The walls of normal woody tissues are extensively lignified, but this does not take place evenly throughout the depth of the wall.

Lignification proceeds from the region of the primary wall inwards towards the interior of the cell, and most of the lignin occurs outside the S2 layer. Direct analysis of the different layers of the wall is not possible but, by examining walls of different ages which have thickened to varied extents, the composition of the layers can be deduced. In a study of this kind, it was found that growth of the secondary wall involved a volume increase of the wall space of more than 30 times compared to the corresponding meristematic cells (in the case of a woody stem, the cambium). Amongst the sugar monomers, it was found that arabinose was more or less confined to the middle region of the wall, corresponding to the middle lamella and primary wall. Galactose was found in the same positions but also extended into S1. As S3 was approached, the concentration of glucomannan polymers increased in the case of softwoods, with a corresponding increase in a mixed polymer of xylose and glucuronic acid in hardwoods. The cellulose concentration of the primary wall was found to be 30–40% (dry weight), increasing to 60–70% in the S1 and S2 layers. These results imply that as a cell matures, considerable changes in the metabolism of sugars and the activities of synthetic enzymes occur. In particular, the pathways for the synthesis of most of the pectic polymers are completely shut down during secondary wall formation.

Pits in secondary walls

The primary wall is crossed in many places by plasmodesmata, which are probably important channels of communication between cells (p. 4). Plasmodesmata are predominantly formed at the time of laying down of the cell plate at cytokinesis. On simple geometric grounds this would imply that their concentration along the longitudinal walls of an elongating tissue would be lower than in the cross walls (which have extended by a smaller amount). This is by and large true in primary growth, but the situation is more complicated when secondary walls are considered. Plasmodesmata are frequently found to accumulate in local regions of the primary wall, known as primary pit fields. The way in which this clustering is achieved is not known; it might result from a local inhibition of wall extension in the region of the pit field, or it is possible that plasmodesmata might somehow migrate in the wall, or perhaps be produced by a secondary mechanism at times other than at cell plate formation. However they arise, pit fields show clearly that concerted action has taken place between neighbouring cells across the walls in which they are found.

In the secondary wall the pit fields become the site of specialised areas of wall structure, and are termed pits. At the start of secondary wall deposition, thickening of the wall is restricted in the region of the pit fields. This gives rise to a channel in the secondary wall layer, at the base of which is the primary wall with

Fig. 3.19. The formation of pits. In the primary wall (*a*), plasmodesmata are restricted to localised areas, the primary pit fields. These areas are not thickened by the secondary wall (*b*). The secondary wall may form an overhang to give rise to a bordered or half-bordered pit (*c*).

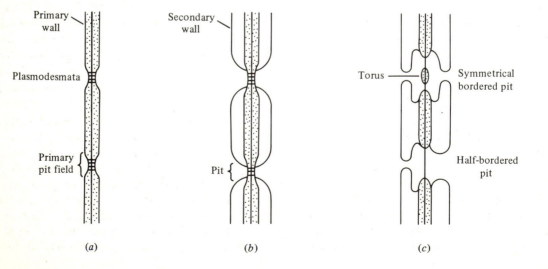

Primary wall   Secondary wall   Torus   Symmetrical bordered pit

Plasmodesmata

Primary pit field   Pit   Half-bordered pit

(*a*)   (*b*)   (*c*)

its plasmodesmata. The primary wall in this position is known as the pit membrane. Often it remains unthickened and unlignified but (in conifers) a thickened central region is formed, called a torus. This is connected to the edge of the pit cavity by radial strands in the wall. A further complication of structure may occur when the final layers of secondary wall are deposited. These may restrict the opening towards the cell interior, by forming an overhanging edge to the pit cavity. Such a structure is called a bordered pit. A simple pit is one in which the canal in the secondary wall is of uniform diameter throughout (Fig. 3.19).

Pits may or may not be symmetrical. Thus on one side of the primary wall or pit membrane, the secondary wall may develop to give a simple pit profile, whilst on the other side it may form a bordered pit. In face view it is clear that the presence of a pit seriously disturbs the predominant orientation of the microfibrils in the secondary wall layers; they curve around the outside edge of the pit. Pits are commonly arranged in files although, in vessels, their concentration in the wall may be so high that the entire microfibrillar arrangement of the wall is disturbed.

Pits represent a means of movement of materials between adjacent files of cells. The pit membrane of conifers is perforated with apertures which have been shown to allow the passage of colloidal materials from one tracheid to the next. The detailed mechanism of formation of pits is not clear; one peculiarity of the process seems to be that the plasma membrane does not follow the contour of the wall surface into the pit. Pits clearly illustrate considerable cellular control over the localised deposition of wall materials, however, and they emphasise the possibility of cooperative interaction between adjacent cells in a tissue.

The turnover and breakdown of cell walls

Development has been described as a three-stage process involving cell division, growth and differentiation. Implicit in this description appears to be the idea that development inevitably leads to increased complexity in the cell. This is in fact far from being the case. Many specialised cell types have a reduced complement of cytoplasmic organelles, and indeed, some pathways lead to the total loss of cell contents and death (p. 103). In numerical terms, this is not an uncommon event; for example, the great majority by far of cells in a woody stem of a tree are dead.

Similar considerations apply to the cell wall. Growth of the primary wall is a process which always involves an increase in the amount of material within the wall space and, likewise, thickened secondary walls are a result of the addition of newly synthesised materials. However, even primary growth, although it involves a net increase in the amount of wall material probably involves turnover of at least part of the structure (p. 76) and, in certain specialised circumstances, wholesale loss of wall materials may form a part of the development process. Such a situation has already been encountered in the transient appearance of callose around developing microspores (p. 87); similar losses of wall material occur during vascular development (p. 103), the formation of air spaces in roots of monocotyledons, and abscission (p. 127). There are two situations where wall breakdown is of great physiological and even economic importance: these are in ripening fruit and in germinating seeds.

Wall breakdown in fruit ripening

The ripening of fruits is commonly associated with a process of softening which renders them attractive to animals, whether human or otherwise. This effect no doubt assists in the dispersal of seeds and hence in the survival of the species. Fruit tissues contain a number of enzymes which could be responsible for this softening effect. Many of them are bound to the cell wall and can only be released by treating wall preparations with high concentrations of inorganic salts. The best understood breakdown mechanism concerns the pectic polysaccharides. Here, it is found that fruits contain two enzymes which together are able to solubilise the backbone of the pectin molecules. Endo-polygalacturonase attacks the α-1,4 linkage between non-esterified galacturonic acid residues in the backbone. The methyl groups which otherwise inhibit the action of this enzyme are removed by pectin methyl esterase. When the levels of these two enzymes are examined during fruit ripening, it is commonly found that the esterase activity remains more or less constant, whilst the endopolygalacturonase activity increases. In two non-ripening mutants of tomato it was found that esterase activity was present, and indeed showed an increase with the age of the fruit, but that endo-polygalacturonase activity was absent. Clearly, for the pectin to be successfully solubilised, both enzymes are necessary.

Another mechanism of softening involving the pectin has been examined in apples. Here it is found that incorporation of labelled methionine is essential if ripening is to occur. In this case, adhesion between adjacent cells is being reduced not by the hydrolysis of the pectin, but by increased methylation of the acidic groups in the galacturonic acid moieties. Interactions between adjacent polymer molecules are reduced by this change in their composition.

Turnover is also found in other polymers, although the situation is less clear here. In Conference and Bartlett pears, for example, there is considerable solubilisation of arabinose during ripening. Analysis of the products of this turnover suggests that it is not due to total hydrolysis of the neutral arabinan, but rather to cleavage within a pectin molecule which is linked to polymerised arabinose. This activity has been ascribed to endopolygalacturonase. The softening of apples is also associated with turnover in galactose. In Cox's Orange Pippin it has been shown that this is due to breakdown of neutral blocks attached to pectin at the middle lamella. The result is the loss of adhesion between cells and the appearance of a characteristic apple pectin which is deficient in neutral sugars.

All these observations strongly point to the fact that normal pectin structure is one of the determinants of cell adhesion and general tissue integrity. Maintenance of tight cell contact is obviously important in a living community of cells which are mutually dependent and which develop according to their positions relative to one another. In vegetative tissues, cell separation is usually limited to the formation of air spaces, which is most strikingly shown by mesophyll tissue. Cell separation is also a feature of the final stages of development of the root cap (p. 110).

Wall turnover during seed germination

The seed is a transportable unit containing the embryo, often with associated reserves of nutrients stored in the form of polysaccharide (e.g. starch) or protein. One of the most well studied examples of the breakdown of storage materials in seeds is that of the starch in the aleurone layer of barley seeds. The

treatment of isolated aleurone layers with gibberellic acid (p. 152) promotes the formation of a hydrolytic enzyme, amylase, which releases glucose monomers from starch. In studies on this process it was noticed that the cell walls of the aleurone layer underwent considerable breakdown following gibberellic acid treatment. This is not unreasonable, since presumably such breakdown would facilitate the diffusion of nutrients to the developing embryo.

The precise chemical cause of this wall breakdown has been difficult to determine, mainly due to the presence of large amounts of starch which contaminates any cell wall preparation from this tissue. By removing the starch enzymically it was found that the cell wall preparation which was left behind contained cellulose and a mixed polymer of arabinose and xylose based on an arabinan backbone. It was also found that the aleurone layer contained enzymes capable of breaking this mixed polymer into its constituent sugars. In a different study, staining methods were used to demonstrate the presence of callose in the cell wall, and that a β-1,3 glucanase capable of hydrolysing this was secreted from barley half seeds within 4 h of gibberellic acid treatment. Clearly this is a complex situation which is not yet fully understood, but it represents another example of biochemical specialisation fitted to a particular cell type. It is particularly interesting as an example of a biochemical activity which is elicited by the presence of a plant growth substance (see Chapter 5).

Summary

The plant cell wall presents an interesting paradox; on the one hand it is fairly regarded as the external 'skeleton' which ultimately gives the plant its form, and on the other hand it is an active participant in the life of the protoplast which it encloses. Ultimately of course many types of cell wall outlive their protoplast and remain functional, perhaps for centuries.

The development of cell walls is intimately and reciprocally related to the development of cells. Almost every example of a specialised cell has as part of its specialisation a particular wall structure, whether this be the retention of a thin primary wall in the growing regions or the parenchyma of the mesophyll, the overall development of structural strength in supportive tissues such as collenchyma, or the localised development of special structure as in sieve elements or guard cells. The position of a cell is defined by its relationship to other cells and, once division has occurred, this depends entirely on the shape and size of cell walls and their directions of growth. Consideration of this situation reveals that an extremely subtle system of control must exist on a feedback basis. An individual cell controls the deposition of its wall in a precise manner. These patterns of deposition themselves cause an alteration in the environment of the cell, and this alteration may lead to further changes within the cell itself, and so on. In addition to this, it has to be remembered that within an extended tissue in three dimensions, there is also a need for coordination of growth and expansion between cells in order to achieve the precision of form which characterises each species of plant.

This chapter has attempted to give an insight into the complexity of the cell wall and its adaptability. It should be emphasised however that most of the examples which have been considered are the result of intense study on but a few cell types. These have given rise to models of primary and secondary wall structure which should not be regarded as detailed descriptions relating to all circumstances, but as ideals from which considerable deviation will occur in any particular species or tissue.

The model of the primary cell wall may be cited specifically. Its value lies in the accuracy of its description of the chemical organisation of the wall of sycamore cells grown in suspension culture and, in fact, many of its features seem to be applicable to other wall types. However, it does not claim to show that all primary walls are the same. That this is not so is a matter of common experience and frustration in many other types of research. For example, were all primary walls the same, it would be a matter of routine to prepare protoplasts from plant tissues, since all that would be necessary would be a standard treatment with a pectinase and a cellulase enzyme. Such a regime works well in some circumstances, but not at all in others; this must imply chemical or organisational differences in primary walls. The model of the primary wall makes no attempt to explain bonding between cells. Results from the study of fruit ripening have suggested that this may be mediated by pectin polysaccharides, but once again, uniformity of behaviour cannot be assumed. This can be illustrated by reference to a technique for isolating cells from leaves. If the leaves of most plants are ground in a mortar, the result is a brei containing chloroplasts, nuclei, broken cell wall fragments and so on. On the other hand, the same procedure applied to leaves of *Zinnia* or *Asparagus* gives rise to a suspension of uncontaminated mesophyll cells, which are alive and intact. These cells will be encountered again in a subsequent chapter (p. 143), but their behaviour emphasises again that our present understanding of the properties of cell walls is not universally applicable.

Nonetheless, general conclusions can be drawn. Wall growth proceeds through two phases. Primary wall synthesis follows cell division and involves the growth of the cell. At the end of the cell extension phase, secondary wall materials are deposited on top of the primary wall. These materials differ chemically from those in the primary wall, and may be highly specialised depending upon cell type. Chemical differences imply differences in the biochemical capabilities of cells, presumably related to differential gene expression. In addition, localisation of cell wall materials implies a degree of control in spatial terms. The development of the wall, like the development of the cell itself, is not necessarily invariant, but can be changed by circumstances in the environment. These themes, of chemical change, spatial control and adaptability, will recur in later chapters.

Developmental strategies: the structure of specialised cells

Introduction

The product of the differentiation process is a cell or a system of cells which is specialised in a particular function essential to the life of the plant. Some of these cells form continuous structures which permeate the plant – the vascular system, for example. Others are relatively specific products of a sequence of developmental changes within one region – for example, the stomata on the epidermis of a leaf. Specialisation may extend in two dimensions, as in the epidermis, or in three, exemplified by the so called parenchymatous tissue of leaves, roots and stems. Some of the functions of specialised cells within a given plant are universally required by most other species – photosynthesis and the transport of soil moisture and nutrients, for example, are common activities amongst all land plants. Other specialisations may be very limited in their occurrence – the secretion of enzymes in insectivorous plants, or the secretion of latex or resin. Many specialised cells persist throughout the plant's life whilst others, such as the root cap and the floral apparatus, have a transitory appearance.

What emerges clearly from a study of this great diversity of cell types is that specialisation is the result of coordinated change; change which, for the present purposes, we have arbitrarily measured from the meristematic cell of Chapter 1. Changes during development involve the modification of semi-autonomous organelles, of which the plastids have been taken as the best studied example (Chapter 2), and also of the wall, which is in no way autonomous (Chapter 3). In the present chapter we shall consider in more detail the combinations of these changes which, together with cytoplasmic and nuclear modifications, go to define the particular class of differentiated cell. It might be envisaged that so complex is the plant cell that the number of combinations of its constituent organelles and wall structures would be almost infinite. In practice, this diversity does not occur. Only fairly limited combinations of change occur together, so that the number of recognisable cell types is remarkably small; the opinions of most plant scientists would set it below 20. The structure of any cell can in principle be rationalised in terms of its function as a fully differentiated unit, or in terms of its position along a pathway to the attainment of such a state.

The selection of cell types which is described in this chapter is based primarily on the degree of work which has been carried out in their study. Certain cell types have been popular objects for research over many years; cells within the vascular system and the root cap fall into this category. In other cases, the pathways of differentiation are less well known although the end product is clearly defined. Cells of this type are included to illustrate the principle that particular functions may be developed by a modification of some specialised cytoplasmic component – for example, the endoplasmic reticulum in secretory cells. A further important general principle is that of the importance of cell division and the control of its direction during development. This is typified by the stomatal complex in the aerial epidermis of plants. Finally, cyclical changes (typified by seasonal dormancy in the cambium) are described in order to illustrate the fact that in real plants growth and development is not necessarily continuous. Emphasis is deliberately placed on the individual cell, rather than on the functioning tissue. The behaviour of cell populations in coordinated development will be considered later (Chapters 6, 7).

The vascular system

The function of the vascular system is to supply water and nutrients to all parts of the plant. This is accomplished by two tissues. The phloem is concerned with the movement of photosynthetic pro-
ducts, principally the molecule sucrose, and the xylem with the movement of water and minerals from the soil. The terms 'phloem' and 'xylem' do not denote a single cell type, but a complex system of cells, of which some are directly concerned with the transport process, whilst others provide structural support or are involved in the loading or unloading of the transported materials. Patterns of vascular tissue development vary from tissue to tissue. In the root, development occurs in a linear sequence back from the tip. For this reason, roots are a particularly good experimental material in which to study the sequence of development. In stems with an annular meristem, the cambium, development is radial and depends on the position of the cell relative to the meristem. Internal cells give rise to xylem tissue – the 'wood' of a tree trunk, whereas cells outside of the cambium give rise to phloem. There are many variations on these two most simple patterns. The most complex patterns are found in grasses, where development may proceed in two directions simultaneously and involves the fusion of recently differentiated strands of tissue with pre-existing strands. These complications will for the moment be largely ignored, since the present concern is the modification of the individual cell rather than its part in the organisation of a tissue.

Sieve element differentiation

The characteristic cell of the phloem which is concerned directly with long-distance transport is the sieve element. This name is derived from the presence within the cell wall of regions of porous structure, the sieve areas. In lower plants and in gymnosperms the pores are rather small and uniform, whereas in angiosperms they are large and irregular. In angiosperms in particular, the end walls of the elongated sieve elements have large pores; this wall structure is known as a sieve plate. Adjacent sieve elements are connected together end to end through the sieve plates, and form a continuous tube, the sieve tube. Sieve plates may also occur on side walls connecting adjacent sieve tubes.

The mature sieve element is quite different in its internal structure from the meristematic type of cell from which it originally derived. It lacks many of the components which are typical of the meristematic cell, and it contains others which are not present at earlier stages. Mature sieve elements have proved difficult to study because of the very high osmotic pressure of their contents – as much as 30 atmospheres. This results in considerable damage to the cell contents when pressure is released, for example by cutting a piece of tissue for fixation. As a result of this, some controversy still remains concerning the way in which the mature sieve element achieves its function of the rapid transport of solutes (p. 98). However, these difficulties do not arise when differentiating sieve elements are examined, and a detailed description of this process is possible. This is best considered under three separate headings.

Cytoplasmic changes

The young developing sieve element has all the normal complement of organelles of a meristematic cell. The mitochondria appear to show no change throughout the development of the cell, and plastids remain simple in structure, although in the mature sieve element they may accumulate starch or characteristic protein deposits (p. 44). The plasma membrane becomes more densely stained at an early stage of development, and may show a marked asymmetry in this staining pattern (with the darker and thicker of the two leaflets next to the wall). The endoplasmic reticulum has a rather specific relationship to the developing sieve pores which will be considered in the next section. In addition, elements of the endoplasmic reticulum become reorganised from their usual dispersed, lamellar form into stacks of membranes without ribosomes; this complex is termed the sieve tube reticulum. These stacks may appear anywhere in the cytoplasm but may, in particular, be associated with the nuclear envelope. As differentiation proceeds, the contents of the nucleus become more disperse and eventually nothing can be seen within the nucleoplasm. The nuclear envelope then breaks down. During differentiation the vacuole membrane also breaks down, causing a dilution of the cytoplasmic ground substance. In the later stages of development, dictyosomes and ribosomes disappear from the cytoplasm.

The mature sieve element therefore contains no nucleus, ribosomes or dictyosomes; it bears very little resemblance at all to its starting condition. The remaining endoplasmic reticulum is found either as a single cisterna around the edge of the cell, or as stacks of sieve tube reticulum. The central space of the cell is occupied by the mitochondria and plastids, together

with, in angiosperms, the most characteristic biochemical product of the sieve element, P-protein.

P-protein

As differentiation proceeds, and whilst the cytoplasm still retains its synthetic ability and organisation, the sieve element in many angiosperms begins to form a characteristic cytoplasmic product which first appears as small aggregations of tubular fibrillar or granular material. This is the P-protein. P-protein appears within the cytoplasm itself – not enclosed in any membrane compartment – and for this reason its actual site of synthesis is uncertain. It is sometimes associated with the endoplasmic reticulum or dictyosomes, or even with fields of spiny vesicles; none of these coincidences represents, however, a convincing demonstration of its origin. The initially small accumulations of this material increase in size to form P-protein bodies, which are visible in the light microscope. In some genera, such as *Nicotiana* and *Coleus*, a single P-protein body is found in each differentiating cell (Fig. 4.1) and may come to equal the nucleus in size. In other genera, such as *Cucurbita*, many small peripheral P-protein bodies are formed.

The appearance of P-protein in the electron microscope is variable and this is thought to mean that the individual molecules can associate together in different ways. In soya (*Glycine max*) P-protein first appears as a granular mass, and bundles of tubules are formed within this as it grows. It has been suggested that the tubular form of P-protein is made up of two helices; when these are tightly associated, a tubular appearance results, whereas when aggregation is less tight, the appearance is that of fine fibrils. Growth of the P-protein bodies does not continue indefinitely and, in fact, ceases when the degenerative changes in the cytoplasm take place, as described in the previous section. During these changes, the P-protein disperses so that in the mature sieve element it is found throughout the cytoplasm.

The proteinaceous nature of the P-protein was initially demonstrated by staining the P-protein bodies with mercuric bromophenol blue for light microscopy. Subsequently, the filaments of the material have been chemically analysed following isolation from *Nicotiana* and *Cucurbita* species. P-proteins consist of proteins or polypeptides with molecular weights in the range 14 000–160 000. Depending on species, they may show enzyme activity, particularly acid phosphatase or nucleoside phosphatase. The chemical nature of P-protein and its ability to aggregate into tubular form has led to comparisons with the protein which comprises the microtubules. The P-protein tubules are of a similar diameter to a microtubule, but the central hole is smaller in P-protein. Furthermore, as has already been stated, the wall of the tubule of P-protein is comprised of two helices with six subunits around the tubule. This contrasts to the microtubule, which has a single helix with 13 subunits around its axis.

The sieve element wall: the sieve plate

The change in the appearance of the plasma membrane which was noted above is accompanied during early differentiation by a thickening of the wall of the sieve element. Microtubules are particularly abundant in the cortical cytoplasm of the cell during

this initial phase of thickening. The most striking changes in the structure of the wall occur in the end walls which will become the sieve plates. These arc perforated by plasmodesmata, as is normal in recently divided cells. Early on in the differentiation of the cell, cisternae of endoplasmic reticulum become closely applied to the plasma membrane around the plasmodesmata in the end walls. This is followed by the appearance of a clear region in the wall on the other side of the plasma membrane. This is a region which is, in fact, the site of deposition of the β-1,3 linked glucan callose (p. 86). The deposition of callose appears on both sides of the developing sieve plate, and eventually leads to the formation of two cone-shaped pads (Fig. 4.2). This process of deposition of callose is accompanied by local dissolving of the original wall constituents. Eventually this replacement of normal wall structure by callose leads to a continuous callose cylinder being formed through the depth of the wall. The last stage of the development of the sieve pore consists of the dissolving of this cylinder of callose. This appears to take place from the region of the middle lamella outwards, and the final result is a greatly enlarged pore in the wall which is lined with

Fig. 4.1. A P-protein body in a developing sieve element from a young leaf of *Coleus*. Bar = 1 μm.

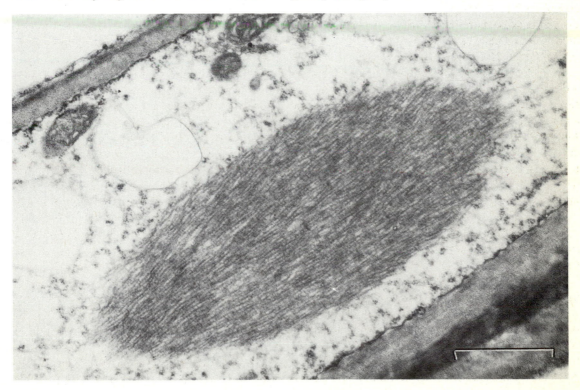

the plasma membrane (Fig. 4.3). It is usually found that a thin cylindrical layer of callose persists around the pore after its formation.

Callose is associated with sieve element walls in two other situations. Where the adjacent cell is a companion cell to the sieve element, it is commonly found that the original simple plasmodesmata which existed between the two cells take on a specialised structure at maturity. The connection on the sieve element side remains simple, but on the companion cell side it becomes branched, and associated with this branching is found a cylinder of callose within the wall (Fig. 4.4). This development is not really understood; particularly it would be interesting to know in detail how the branching of the original plasmodesmata is

achieved. Callose may be formed as a wound reaction when mature tissues are damaged. This is a very rapid response indeed, and led to considerable confusion in the early work on phloem. Nowadays it is considered that the description of a transient callose deposition at the site of the future sieve plate is a real phenomenon, and not an artefact due to wounding. Wounding also leads to changes in structure of mature sieve elements which are of considerable relevance when the possible mechanism of action of the sieve tubes in transport is examined.

The mechanism of movement of solutes

The mechanism of the movement of materials through the phloem has been a matter of controversy for a very long time. Basically two hypotheses have been advanced. The first is based on a simple physical model. If two osmometers are connected to each other by a tube, and one is filled with a solution of different concentration to the other, then mass flow of the solution takes place to equalise the concentrations. In

Fig. 4.2. A developing sieve plate in the phloem of a leaf of *Zinnia elegans*. The light-staining regions in the wall correspond to transient deposits of callose. The dark fibrillar masses in the lower cell are P-protein. Bar = 1 μm.

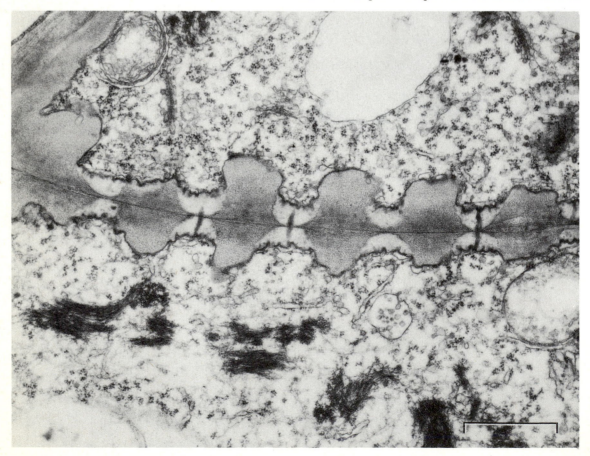

Fig. 4.3. *a*, Sequence of development in the sieve plate. The developing sieve pores are associated with endoplasmic reticulum and the deposition of callose. The mature pores are probably open. *b*, When the tissue is wounded, callose forms very rapidly to line and seal the openings in the sieve plate.

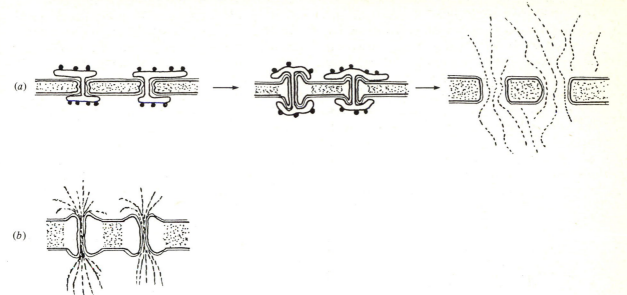

Fig. 4.4. The wall between a sieve element and its companion cell in the leaf of celery (*Apium graveolens*). The pores connecting these cells are branched on the companion cell side. Bar = 0.2 µm.

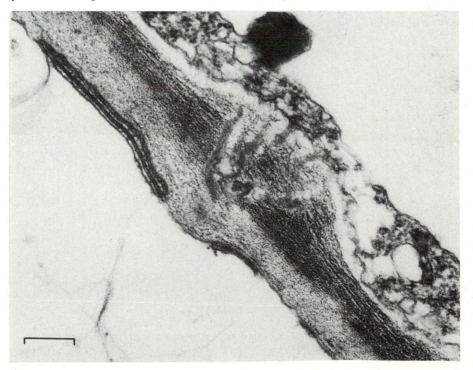

biological terms, this represents the flow of materials and water from the site of production of the solute (sucrose, in the photosynthetic tissues) to the sites of their utilisation (for example, in the root). It requires that the tube connecting the source and the sink be open, to allow for free flow of the solution.

The speed of movement of materials in the sieve tubes is very high; of the order of 50–200 cm h^{-1}. This is much higher than could be accounted for by the operation of a simple diffusion mechanism, and therefore the second hypothesis has proposed that some form of assisted diffusion acts in the phloem to achieve these transport rates. It is particularly interesting to notice that in the case of virus-infected plants, it appears that intact virus particles, which are of course very much larger than the molecules of normal phloem contents, move at the same rate as the nutrient materials.

These two hypotheses could in theory be resolved by an examination of the fine structure of the sieve tubes. In principle, if the tubes are open, then a mass flow hypothesis is not ruled out. On the other hand, if the P-protein or some other component of the remaining cytoplasm showed an organised distribution along the length of the cell, then perhaps this would indicate that an assisted transport mechanism was at work. Unfortunately, as has already been mentioned, the sieve tube is a very delicate structure. When an incision is made through one of its elements, a surge of contents takes place, resulting in the blocking of the sieve plates with a mixture of aggregated P-protein and the other cytoplasmic organelles – plastids and mitochondria. This can be regarded as a function of the P-protein; that wounding from whatever source is immediately the cause of a sealing reaction which prevents massive loss of water and

Fig. 4.5. A mature sieve plate in the phloem of a leaf of *Zinnia elegans*. The light-staining callose in this situation is formed as a reaction to wounding the sieve tube. A mass of P-protein fibrils blocks the sieve pores. Bar = 1 μm.

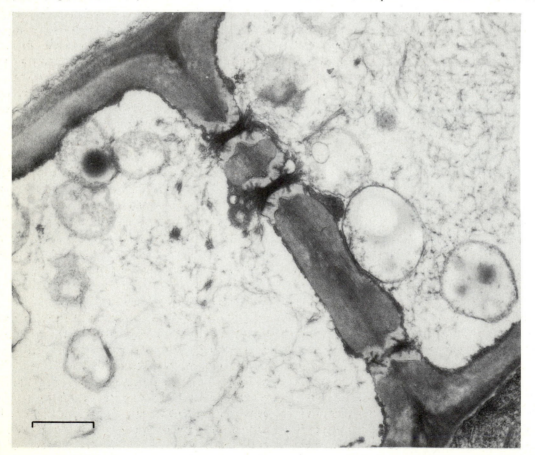

nutrients. The situation is further complicated by the rapid formation of callose at the sieve plate. In normally fixed tissue, the result is that the narrowed diameter of the sieve pore is always completely plugged with P-protein (Fig. 4.5). However, when precautions to prevent callose formation and reduce surging are taken, the picture which emerges is different. If plants are first wilted, then fixed rapidly using acrolein instead of the more usual glutaraldehyde as the fixative, a proportion of the sieve pores are found to be open. This is currently regarded as being the more likely situation in living cells. It is also thought that the P-protein may form a three dimensional network which penetrates the sieve pores and so is continuous from one element to the next.

This raises an obvious question. Does the P-protein participate actively in the transport of materials in the sieve tubes? Attractive though this idea is, the answer seems to be no. There is no evidence that the P-protein is chemically similar to any of the known active proteins, such as actin, tubulin or microfilaments. Equally, P-protein is sometimes found to occur in companion cells and even in phloem parenchyma, where rapid transport does not occur. There is also variation in the amount of P-protein in the sieve elements of different tissues; sieve elements in roots for example consistently seem to contain less P-protein than sieve elements of stems. Finally, P-protein is absent from the phloem of many monocotyledons, the gymnosperms and lower vascular plants. All these facts point to uncertainty over the role which P-protein plays in phloem transport.

Current evidence therefore suggests that movements of nutrients in the sieve tubes may indeed take place by a mass flow mechanism. The products of photosynthesis, in particular sucrose, are thought to be loaded into the sieve tube by specific carriers coupled to proton transport and utilising the energy of ATP hydrolysis. The increased osmotic pressure in the sieve tube then causes pressure-driven mass flow to occur in the direction of the 'sinks' – the sites of utilisation of the nutrients. Loading is thought to take place across the plasma membrane of the sieve element, from the extracellular space. As will be seen later (p. 105) this sometimes results in the formation of another specialised type of cell, the transfer cell.

The angiosperm sieve element provides a fine example of the processes occurring during differentiation. It is the site of a massive synthesis of a single product related to the differentiation process – P-protein. Its maturation involves the controlled loss of the genetic material of the nucleus, and also of the synthetic and secretory structures in the cytoplasm. Finally, its maturity is attained by the development of specialised wall structure; this involves spatially organised transient deposition of callose and the selective removal of normal wall constitutients. These processes clearly indicate the interplay of a changing biochemical pattern in time, and also in space.

Xylem tracheary element differentiation

Xylem tissue, like phloem, contains several different cell types. One of these, the tracheary element, has been examined in detail for its structure and development, and it corresponds to the sieve element insofar as it forms connected tubes which are the route of long-distance transport of water and minerals. The characteristic structure of the tracheary element is its thickened secondary wall. The thickenings may be annular or helical, arranged around the long axis of the cell, or they may be reticulate (Fig. 4.6). These thickenings are not required for the transport process to take place, and its seems likely that the various patterns which are formed reflect the need for mechanical strength combined with a degree of flexibility. The thickenings are lignified, and so are mechanically brittle and inextensible. It would therefore be expected that a wall with annular or spiral thickenings would be capable of a certain degree of extension without collapse, whereas if the thickenings were reticulate, the cell would be more or less inextensible. Reticulate thickenings do indeed tend to be formed towards the end of growth extension, and they are also formed in tissue culture as a response to growth-retarding treatments. The differentiation of the tracheary element can usefully be considered under two headings.

Secondary wall thickening

The development of patterned thickenings in the developing xylem element has been the object of considerable study and speculation. Before the introduction of the electron microscope, it was reported that the pattern of the thickening in the wall could be discerned within the cytoplasm. In young cells, the peripheral cytoplasm itself appeared to assume a helical configuration. This observation has not been repeated since the introduction of the electron microscope, although it perhaps should be said that this could still be due to a failure of current preservation techniques.

After the introduction of glutaraldehyde as a fixative for electron microscopy, it was quickly realised that the developing thickenings of the wall are overlaid by local concentrations of microtubules (Figs. 4.7, 4.8). The regions of the cytoplasm between the thickenings are devoid of microtubules and, instead, often contain endoplasmic reticulum profiles in close association with the plasma membrane. The dictyosomes in the developing cells are particularly active, and it appears that the vesicles produced from them are discharging their contents into the thickening regions of the wall. In glancing sections it appears that the fibrillar structure of the wall is parallel to the arrangement of the microtubules in the cytoplasm.

This situation is one of the clearest examples of a correlation between cytoplasmic microtubules and wall synthesis and fibrillar orientation; it has led to the hypothesis that microtubules are concerned with directing the orientation of wall fibrils. This hypothesis has been mentioned previously (p. 77). The differentiating xylem element has been one of the cells in which experimental tests of the idea are possible.

The drug colchicine is known to bind to microtubule subunits; this causes a shift in the monomer/polymer equilibrium, leading to loss of the microtubule structure (p. 22). When differentiating tissue is treated with colchicine, it is found that the drug can prevent the formation of thickenings if the process has not already begun. If thickenings are already present at the time of the treatment, colchicine disrupts their growth. Thickening of the wall continues, but in a far less organised manner; the bands of thickening are irregular in shape and adjacent bands may coalesce. This correlates with the disappearance of microtubules due to the drug treatment, but it is not certain in every case whether the orientation of the cellulose microfibrils within the disorganised bands is itself disrupted.

It therefore appears that whilst microtubules may in some way influence the gross organisation of the bands along the wall, they may not be directly concerned with the laying down of cellulose. The idea of their involvement in wall thickening in any case involves a slightly irrational twist of logic. During the early stages of secondary thickening it is possible for cell elongation still to be taking place; this presumably implies growth of the primary wall regions between the thickenings. Since the distribution of microtubules which supports their role in cellulose orientation is a local one, over the thickenings, it follows that some other mechanism must exist for the control of cellulose deposition in the growing primary wall parts of the same cell.

The role of the endoplasmic reticulum and the dictyosomes has also been examined since both of these components show some tendency to a distribution related to the pattern of the thickenings. It might

Fig. 4.6. Diagrammatic representation of different patterns of wall thickening found in tracheary elements. *a* annular, *b* reticulate, *c* helical, *d* pitted.

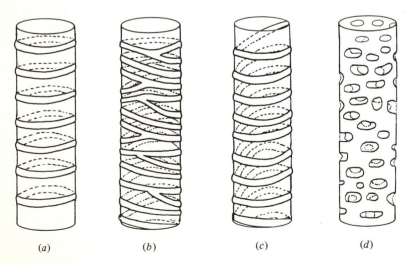

(*a*)	(*b*)	(*c*)	(*d*)

be thought that the vesicles which are seen to fuse with the plasma membrane at the site of the thickenings are contributing wall components, whereas the endoplasmic reticulum between the thickenings is inhibiting this process. In fact, when labelled glucose is fed to these cells, it is found not only in the wall of the thickenings and in the vesicles, but in the endoplasmic reticulum also.

Cytoplasmic changes and autolysis

The changes in the deposition of wall materials which have been described begin fairly early on in the differentiation process. At this time the cytoplasm does not show any change in its components, other than the distribution patterns which have been cited. As the wall reaches its final state of maturity, however, the vacuoles in the cell coalesce into a single large unit; this then ruptures, causing the degeneration of the entire cytoplasmic contents. All the membrane-bound organelles in the cell are lost, although mitochondria may persist for some time. This total breakdown of the integrity of the cell signals its death and the cessation of the development process.

The tracheary element, as its name implies, is only one part of the functioning xylem. In order for long-distance transport to be possible, the cellular tissue must be converted into a series of connected tubes. This is achieved by the hydrolysis of the end walls of each tracheary element. Clearly, such hydrolysis presents a considerable danger to the life of adjacent cells. During degeneration of the protoplast and its cell membrane, in themselves processes which are poorly understood at a biochemical level, it can be inferred that the wall of the tracheary element becomes exposed to a mixture of degradative enzymes. These are presumably acting within a non-living environment, the degenerated cell contents, and thus cannot be assumed to be under any sort of

Fig. 4.7. Diagrammatic representation of wall development during tracheary element differentiation. *a* The longitudinal walls become locally thickened, in positions which are marked by accumulations of microtubules. Active dictyosomes are a feature of these cells, and the endoplasmic reticulum may lie near to the plasma membrane between the thickenings. *b* Mature structure of the tracheary element consists of total absence of all cytoplasmic components and of the plasma membrane. Cross walls are dissolved away, and the thickenings are heavily lignified.

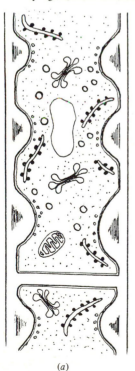

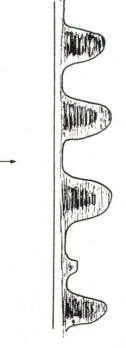

(*a*)   (*b*)

direction in the sites of their action. Why then does not the entire wall of the tracheary element dissolve?

The answer to this question is twofold. In the thickened areas, considerable lignification takes place, and this protects those regions from hydrolysis. It also appears that some component of the middle lamella is able to protect the primary wall from complete degradation. This explains how adjacent elements are connected by pits which are covered with a fibrous pit membrane (p. 90), but it leaves unresolved the problem of how – in the same cell – the end walls come to be completely removed. One suggestion which has been made is simply a mechanical one; that the comparatively large area of the end wall is so weakened, despite its protection from the inhibitory substance of the middle lamella, that it is swept away in the transpiration stream. The smaller areas of the pits between elements on their side walls would be expected to be stronger in any case, and also not to experience the degree of force felt in the end walls, since the direction of flow of the transpiration stream is preferentially along the tubes rather than between them.

Although these explanations seem to rely on a system which (to a biologist) appears to be surprisingly lacking in control, this is no doubt an erroneous impression based on a lack of understanding. The facts are that autolysis proceeds to a predictable conclusion without loss of function in either the differentiating tissue itself, or surrounding cells. This implies control – control which is exercised at the level of biochemical activities of molecules rather than in terms of structural elements which can be visualised.

Transfer cells

The vascular system does not consist solely of long-distance transport elements. Associated cells within the system have a mechanical role to play and

Fig. 4.8. An early stage in the localised thickening of a tracheary element wall in tobacco. Microtubules are localised over the thickening regions. Bar = 0.5 μm.

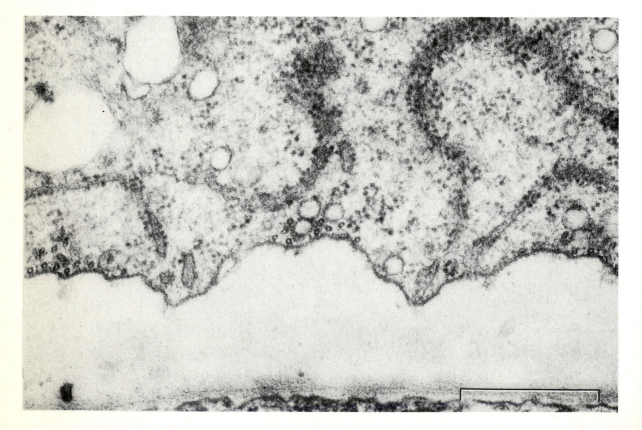

may also be concerned with the loading of the long-distance network with the products of photosynthesis, the retrieval of these products, or the retrieval of inorganic nutrients. These functions are carried out by the parenchymatous cells of the phloem and the xylem. The process of transpiration in the xylem and mass flow in the phloem may produce a problem for the plant in the following way. Water and nutrients from the roots travel in the xylem, which is effectively part of extracellular space within the plant, the apoplast. This is because the tracheary element has no remaining plasma membrane. Conversely, phloem transport is intracellular, in the symplast, since the sieve tube retains its plasma membrane even at maturity. Water and nutrients from the xylem are required inside cells of the leaf; equally, photosynthetic products of the leaf are required to be loaded into the sieve tube. In both these situations the plasma membrane of the leaf cells may be the site of very high transport rates, either of nutrients from the soil or of the products of photosynthesis. In order for such high rates of transport to be possible, it is desirable that the area of the plasma membrane be increased as much as possible. The plasma membrane of a plant cell cannot form external projections in order to increase its surface area, due to the presence of the wall, but it can form internal projections. Cells with internal projections of this type are found in many situations in plants where transport across the plasma membrane might be supposed to be unusually high. Such cells have been given the collective name of transfer cells.

Transfer cells in the vascular system

Within the vascular system, the most important type of transfer cell is associated with the xylem, and so is called the xylem transfer cell. Such cells are modified parenchyma; they are found adjacent to the conducting elements of the xylem. The individual cell often shows considerable polarisation in its wall structure. The wall which abuts the conducting element, or which is closest to it, shows extensive ingrowths (Fig. 4.9). These ingrowths, which represent secondary wall thickening, are not spatially ordered as is the case in the mechanically strengthened tracheary element. Rather they are labyrinthine and apparently random in form. They arise by local ingrowths of the wall to give first a tubular extension into the cytoplasm. This may then branch repeatedly until a network of wall ingrowths is produced which extends through the

greater part of the intracellular space. Microtubules are not implicated in the formation of this pattern of ingrowths. The result of this process is twofold. First, no part of the cytoplasm is far from the cell surface. The cytoplasm of transfer cells is often found to be particularly well endowed with mitochondria and endoplasmic reticulum, and it may be presumed that these structures function in the enhanced transfer of materials, either by supplying the energy for solute pumps across the plasma membrane, or by supplying an increased area of surface for movement within the cell or between adjacent parenchyma cells via the plasmodesmata. The second consequence of wall infolding is, of course, that the area of the plasma membrane between the conducting elements and the living tissues is greatly increased. Calculations show that the factor of this increase may be as high as 20-fold.

One problem which is consequent upon the formation of wall ingrowths is that, although the area of the plasma membrane is increased, so too is the distance between the conducting element and the plasma membrane of the absorptive cell. It appears that the wall which permeates the ingrowths is in fact unusually open in structure. Colloidal materials such as lanthanum hydroxide have been shown to be able to penetrate into the wall ingrowths, and this suggests that they effectively contain channels at least 2 nm in diameter. It is possible that such channels are lined with charged molecules which would also assist in the movement of ions along them. Thus the xylem transfer cell is a good example of a particular type of secondary wall development linked to a highly specialised function, not only at the level of the individual cell, but within a complex of cells. The apoplastic solutes gain entry to the transfer cell by means of the plasma membrane via a special wall structure; once within the symplast they may be distributed to other cells via the plasmodesmata which remain on the walls of the transfer cell in places where thickening has not taken place.

In the phloem, the situation is complicated by the occurrence of different types of transfer cell, which may be classified in terms of the position of their wall ingrowths. However, in general, they show the same overall behaviour as the xylem transfer cells. They retain a dense cytoplasm, and show a close spatial relationship to sieve elements. The commonest form of phloem transfer cell has wall ingrowths around the entire internal surface, except along the wall which abuts the sieve element. It can therefore be considered as a special type of companion cell, and its structure rationalised to the function of loading photosynthetic products into the sieve tube. It is also likely that transfer cells within the vascular system as a whole mediate a certain amount of direct cycling between the transpiration stream and the mass flow in the phloem. Phloem transfer cells only appear at the stage of growth of a young leaf when it begins to be a net exporter of nutrients, which again suggests that they have a role in the loading of the phloem. In the case of rapidly growing tissues, for example the seeds within a pea pod, the conducting systems of the xylem and phloem may be linked by transfer cells. These may function as a sort of short-

circuit mechanism, increasing the flow rate through the vascular system in a tissue which is unable to gain access to nutrient materials through the normal operations of the transpiration stream.

Adaptive formation of transfer cells

The vascular system is the most obvious example of a situation where rates of flow across the apoplast/symplast barrier might be critical in limiting the supply of nutrients for growth. Transfer cells are found in other parts of the plant where physiological conditions require their development. In general terms they are found where the flow of solvents accompanying solutes is minimal, and also where there is a large disparity in the surface-to-volume relations between the donor and recipient compartments. For example, they are found in the epidermis of aquatic plants, certain gland cells, and at the junction between the gametophyte and sporophyte in Bryophytes and Pteridophytes. They are also found in the embryo of higher plants. These all represent special circumstances in which secretion or absorption are occurring at a high rate and where solvent

Fig. 4.9. A section through a stem node of *Helianthemum*. The cells on either side of the empty xylem vessels have labyrinthine thickenings (transfer cells). Bar = 5 μm. Picture provided by Dr M. G. K. Jones. (First published in *Protoplasma*, **87**, 273–9, 1976.)

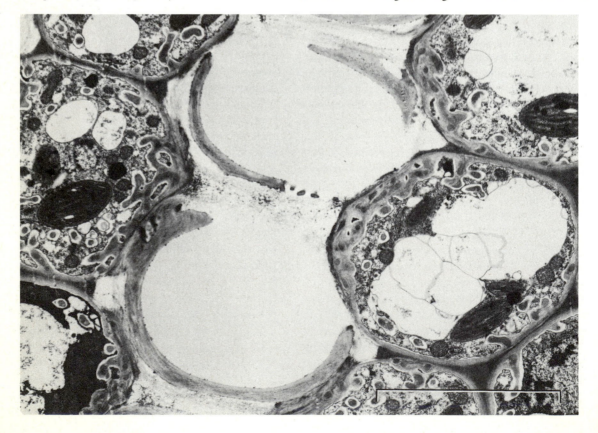

flow is either impossible or undesirable. The transfer cell in such circumstances emerges as a style of adaptive differentiation to fulfil a local physiological need. This adaptive aspect of transfer cell development is particularly well illustrated in the case of giant cells induced following invasion by the certain nematodes.

The potato cyst nematode *Heteroda rostochiensis* invades potato roots by pushing a sharp stylet through a cell wall behind the root tip or where a lateral root emerges. By cutting through cell walls with this stylet the larva migrates intracellularly until it reaches its feeding site. At this stage it is about 0.5 mm long. The first cells penetrated by the larva are cortical cells.

Fig. 4.10. A section through the edge of a giant cell induced in a root of *Helianthemum* by the root knot nematode *Meloidogyne javanica*. The wall of the giant cell develops ingrowths where it abuts the vascular system of the root. Bar = 5 μm. Picture provided by Dr M. G. K. Jones. (First published in *Protoplasma*, **87**, 273–9, 1976.)

The presence of a secretion from the stylet causes the breakdown of cell walls in a column extending away from the root tip. Expansion of this column produces a large syncytial structure containing fragments of the formerly present dividing walls. This syncytium grows in size until cells which were adjacent to xylem elements in the root are incorporated into it. At this time the cytoplasm of the cells which were adjacent to the xylem elements is filled with a variety of vesicles, small vacuoles and plastids. Wall ingrowths are then produced on the syncytium side of the common wall between the giant cell produced by the larva and the xylem elements. Wall ingrowths are also initiated when the expanding giant cell contacts a sieve tube, but such ingrowths do not expand. On walls of the giant cell which do not abut the conducting tissues, no wall growth takes place.

The multinucleate cell which is produced by the invasion of the nematode and the breakdown of the normal walls of the host has features of the typical transfer cell (Fig. 4.10). It is densely cytoplasmic with wall ingrowths adjacent to conducting elements. It is

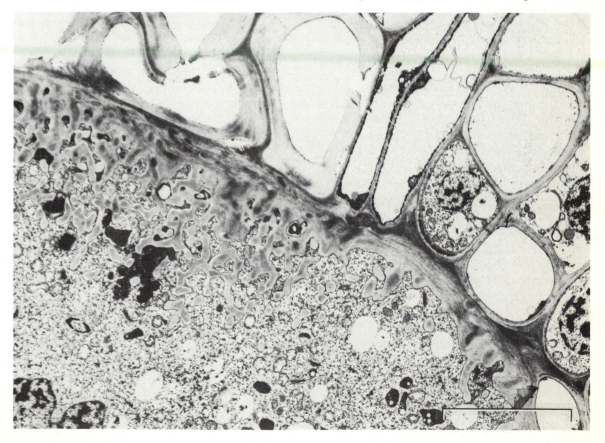

well endowed with mitochondria and endoplasmic reticulum. There is little reason to doubt that the wall ingrowths represent a specialisation of the cell which allows the larva to feed.

In such a case as this, the transfer function is elicited as a response to an external agency, and the giant cell is in no way a product of a normal developmental pathway which would be undertaken by the plant if it were uninfected. In physiological terms, the stimulus may be assumed to be the very steep gradient of nutrients between the giant cell which contains the larva and the xylem element which is the source of those nutrients. This response is specific to those walls where the gradient is steepest and most direct. The concentration gradient may therefore be in some way a cause of the differentiation of the region of the wall next to the conducting element, although how this is actually mediated is completely obscure.

This demonstrates an aspect of developmental change which has so far not been emphasised. It is tempting, when considering vascular development (or indeed any well marked differentiation pathway) to believe that a sequence of change once initiated must always proceed to its inevitable and invariant conclusion. In perhaps the majority of cases in normally growing plants, this may indeed happen. However, it is clear that cells may retain a degree of flexibility and responsiveness to their immediate environment, even after their 'differentiation' has come to an apparent end. In the example of the giant cells induced by nematode invasion, two points may be made. The cortical cells which are the primary site of invasion and which are disrupted to give rise to the syncytium are already differentiated and would probably undergo little additional change without the interference of an external agency. On the other hand, the wall ingrowths which develop do so in a manner which is not unknown in other parts of the plant. As it happens, transfer cells in the xylem are not a common feature of roots; nonetheless, their development in stem nodes and leaf traces does resemble that found in the root under these conditions.

This should perhaps alert us to the suggestion made in the introduction to this chapter that developmental changes within an apparently limitless combination of possibilities, actually occur in a comparatively small number of different ways. A further example of developmental 'switching' by external agency can be mentioned in this context. Epidermal cells of *Atriplex hastata* roots develop wall ingrowths and transfer function when the root is immersed in concentrated solutions of sodium chloride. Once again, a pattern of behaviour which is potentially present but not required during normal growth may be elicited by a suitable external stimulus.

Root cap cells

The characteristic products of the vascular system, the sieve element and the tracheary element, only achieve their final function with the irreversible loss of much or all of their biochemical machinery. This is an extreme example of development proceeding to a state from which there can be no return. The development of the root cap also represents a terminal process; this time because, having performed two quite separate types of function during its life time, the root cap cell is simply lost to the plant.

The root cap serves two purposes within the plant. It is the site of geoperception in roots. This is known from simple experiments. The root cap of grasses and cereals can be quite easily removed as a unit without damage to the rest of the root, owing to the presence of a thick wall between two meristematic regions which is called the root cap junction (p. 157). When a root without a cap is placed in a horizontal position, it fails to turn downwards. If the cap is allowed to reform, geoperception is not restored until a meristem has regenerated and a body of differentiated cells has formed as a result of the activity of this meristem. The second function of the root cap is to protect the root in its passage through the soil, and also to assist that passage. This is achieved by the secretion of a mucus from the peripheral cells of the root cap, and also by their being sloughed off as the root grows. Thus the root cap consists of three broad cell types; the meristematic cells which divide and supply new cells, the central cells which are concerned with geoperception, and the outer cells which secrete the mucus and are sloughed off. The meristem is particularly active in cell division because it has to continually replenish the cells which are being lost at the periphery.

Outer root cap cell structure

The cells of the outer root cap first drew the attention of electron microscopists because of the peculiar structure of their dictyosomes. In these cells the cisternae of the dictyosomes are swollen throughout their length, rather than compressed in the central region with peripheral vesicles at the edges. These hypertrophied cisternae eventually separate and are seen close to the plasma membrane at the outer edge of the cell. The cell wall is peculiarly thickened, also at its outer edge (Fig. 4.11). The correlation between this appearance of the dictyosomes and the secretion of a large amount of extracellular material in the form of the mucus pointed to the obvious speculation that the Golgi apparatus was functioning to produce this secretion product.

This idea was tested by the technique of autoradiography. Roots were allowed to grow in a solution containing radioactive glucose. When sections of material (fixed after 3 h continuous growth in the labelled sugar solution) were examined, radioactivity was found in the walls of the outer root cap cells, in the

dictyosomes of the same cells, and in the starch grains of the cells of the central region (see following section). When a short pulse of radioactivity was given (for 5 min, followed by a chase of up to 1 h in a solution containing unlabelled glucose) it was observed that within the outer cells label appeared first in the dictyosomes, then in the hypertrophied cisternae, and finally in the wall. Chemical analysis of the labelled material showed it to be a polysaccharide with a high proportion of galactose, galacturonic acid and arabinose – therefore, similar to pectin. This type of experiment is very good evidence indeed of a role for the dictyosomes in the secretion of a high molecular weight substance across the plasma membrane. It has subsequently often been cited as presenting evidence that the Golgi apparatus is involved directly in wall synthesis. This is not the case; the secreted product is not a normal wall product, and the outer root cap cells are not undergoing typical primary growth. Indeed, the same experiments showed that in other regions of the root tip where extension growth was occurring, label appeared in the primary wall but was not particularly associated with dictyosomes in the cytoplasm. The role of the Golgi apparatus in

Fig. 4.11. Part of an outer root cap cell in the root of maize (*Zea mays*). The dictyosomes have hypertrophied cisternae, and the contents of the cisternae are contributing to a large localised thickening of the wall space. Bar = 1 μm.

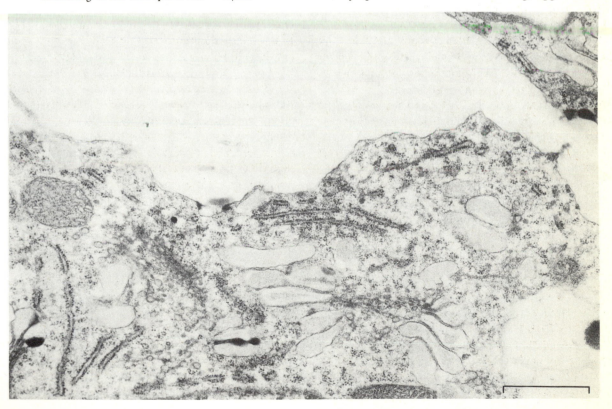

primary wall growth has been discussed previously (p. 83).

The ultimate fate of the outer root cap cell is to be sloughed off into the surrounding substrate. This results in death of the cell, although this is not an inevitable consequence. If roots are grown in culture in a medium solidified with agar for example, it is possible to observe that sloughed-off cells remain viable in this benign environment for some time, although they do not grow or divide. Cell separation is achieved by the loosening of the non-cellulosic fractions of the wall, and their partial dissolution. The control of this process is poorly understood, but its occurrence is clearly linked in some way to the arrival of the cell at its peripheral position on the outside of the root. Were this not the case, the entire structure of the root cap might be in jeopardy.

Central zone cells

The central cells of the root cap show a different structure to those of the outer regions. A principal difference lies in the plastids; in the central cells these are filled with starch and come to lie towards the bottom edge of the cell in a gravitational field. As the central cells are pushed towards the outer edges of the root cap by the continued activity of the root cap meristem, the plastids lose much of their starch and cease to be sensitive to gravitational effects. This loss of starch during maturation to the outer layers is reflected in the labelling experiments which were described in the previous section. The starch of the central cells was labelled due to its continuing synthesis at these sites; towards the outer layers of the root cap, however, the starch was not labelled, presumably because synthesis had ceased and breakdown was taking place.

Differentiation of the central zone cells does not take place in the ordered and easily sequenced manner of the vascular system. However, certain clear-cut changes in cell organelles can be detected by means of a quantitative analysis of their numbers within cells of the meristem and the central zone. During their development, the cells of the central zone undergo a 15-fold expansion in volume. This is found to be mirrored in the increase in the numbers of dictyosomes and mitochondria, suggesting that these organelles are reproduced at a rate linked to cell expansion. The area of endoplasmic reticulum also increases to match the increase in cell size. However,

it also alters its distribution within the cell. In vertically growing root tips, the endoplasmic reticulum lies parallel and close to the plasma membrane bounding all the cell surfaces. During development of the central cells, the vacuoles which were present in the meristematic region appear to disperse; in the electron microscope the impression is very strong that the cytoplasm of developing central zone cells is in fact thinner than that of the meristem (Fig. 4.12), allowing the mechanistic conjecture that this may allow more rapid sedimentation of the starch-containing plastids, or statoliths. A further point of structural interest is that because the central cells are mostly derived in columns by tranverse divisions of the meristematic cells and grow by elongation of their longitudinal walls, the numbers of plasmodesmata in the transverse walls (horizontal in a normally growing root) is far greater than the number in the longitudinal (vertical) walls. This suggested that channels of communication might exist based on the plasmodesmatal connections between the cells in vertical columns, perhaps mediated by the endoplasmic reticulum.

Geoperception

The response of the central zone cells to changes in the direction of the gravitational field has been examined by growing root tips in three different orientations – normal, horizontal and upside down. In the normal root, the amyloplasts within the central zone cells lie on top of the layered endoplasmic reticulum at the lower end of the cell. The nucleus lies above the amyloplasts, and the other cytoplasmic organelles do not appear to be distributed in a manner controlled by the direction of the gravitational field. In the upside down roots, the endoplasmic reticulum is not disturbed, but the amyloplasts lie at the end of the cell which (in a normal orientation) would be the top. This shows unequivocally that the amyloplasts of the central zone cells are sensitive to the direction of the gravitational field. In the same experiments, the starch-depleted plastids of the outer cap cells did not show this behaviour. In roots grown horizontally it was found that whilst the endoplasmic reticulum remained associated with the transverse walls (normally horizontal, now vertical, and containing plasmodesmata), it was absent between the longitudinal (now horizontal) lower wall and the amyloplasts.

This situation was found in root tips of *Zea*,

Triticum, Lens and *Vicia*. It strongly suggests that the movement of the amyloplasts is the mechanism of detection of the direction of a gravitational field, and it allows the speculation that the channel of communication between cells is the endoplasmic reticulum mediated by the plasmodesmata. The first of these two conclusions is not disputed. It has been shown, for example, that if the starch in the plastids of the central zone cells is depleted experimentally by treatment of root tips with gibberellic acid and kinetin, the geoperception is lost. However, the role of the endoplasmic reticulum and plasmodesmata in the process is uncertain – plasmolysis of root tips does not inhibit their ability to orient themselves in a gravitational field, whereas it would be expected to destroy at least a proportion of intercellular connections via plasmodesmata.

Geoperception has been studied in other cell types and the development of starch-containing plastids is not found to be a universal mechanism for detection of gravitational fields. For example, the

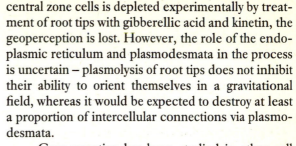

Fig. 4.12. A central zone cell from the root cap of maize. Amyloplasts lie at the bottom of the cell, and the cytoplasm is only lightly stained. Bar = 5 μm.

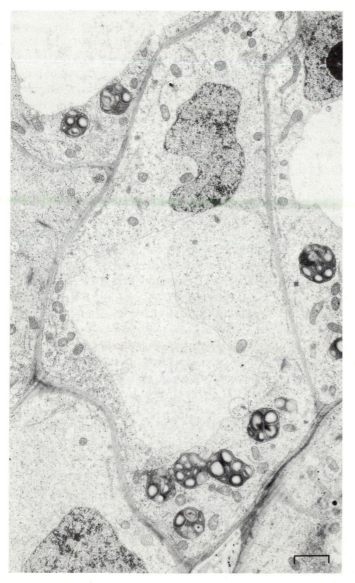

aerial roots of the orchid *Laelia anceps* do not contain starch grains. In the alga *Chara*, the rhizoid contains statoliths consisting of barium sulphate. The mechanism of geotropic movement in this cell is remarkably simple; the statoliths appear to block access to the lower part of the cell by vesicles which would otherwise contribute to its growth. The result is an increase in the relative rate of elongation of the upper part of the wall, and a downwards growth of the cell as a whole. In the roots of higher plants there is a spatial separation between the site of georeception in the central zone of the root cap and the site of reorientation of root growth on the far side of the root meristem. Thus, the stimulus has to be transmitted. This is thought to involve the movement of an inhibitory substance and will be discussed in more detail in Chapter 6 (p. 169).

The function of geoperception is a transitory one in the life of the cells of the central zone of the root cap. As the activity of the meristem pushes them towards the outer layers of the tissue, they become more vacuolate, and their carbohydrate metabolism is switched from the production of starch to the production of the mucus which the outer cap cells secrete. This illustrates the fact that specialised function within a cell can be achieved and maintained for part of its life, and is not necessarily linked to some final and irreversible state.

Gland cells

The production of a pectic mucilage by the outer root cap cells is an example of a specialist secretion. Secretion is a common activity of plant cells, since in a sense at least, the cell wall can be regarded as a secreted product. However, plants contain a wide range of specialised cells which function to secrete materials other than cell wall components. These materials vary widely in their chemical composition and include such diverse materials as the oils produced by the glands of aromatic plants such as mint, the sugar produced by nectaries, and the digestive enzymes secreted by insectivorous plants. Some secretions are of great economic importance – latex produced by rubber trees, for example, is a product of a specialised cell type. Secretory cells may be found at the surface of the plant, where they represent a further stage of development of the epidermis, involving a series of cell division stages before the actual secretory cell or cells are produced. Laticifers develop internally, either as a result of a series of nuclear divisions without cell plate formation, or by the breakdown of cell walls between adjacent existing cells.

Secretion of hydrophilic substances

In the case of hydrophilic secretions, such as mucilages and enzymes, the secretory mechanism commonly involves the passage of material through the plasma membrane in the form of vesicles which discharge their contents into the extracellular space. A particularly well studied example of this is found in the mucilage glands and digestive glands of the insectivorous plant *Drosophyllum lusitanicum*. These glands are nearly identical in their anatomy, except that the mucilage glands have a stalk which contains xylem vessels (which connect to the vascular system of the plant) whereas the digestive glands are sessile. The head of the mucilage glands consists of a triple layer of cells. The outer two layers are densely cytoplasmic, whereas the inner layer, next to the stalk, is vacuolate and has cutinised walls. The function of this cutin within the wall is to prevent general loss of water from the apoplastic space of the leaf (p. 84).

The fine structure of the cells in the outer layer is remarkable in two respects. First, the walls are irregularly thickened, in the manner of transfer cells (p. 104). Second, the cytoplasm during the secretory phase contains large numbers of active dictyosomes which produce vesicles. The role of these vesicles in the secretory process has been demonstrated by a number of simple experiments. Secretion is known to be temperature dependent; at 14°C the rate of mucilage production is one third of its value at 29°C. When the fine structure of the cells is examined at the two temperatures, it is found that the number of vesicles per unit area of cytoplasm increases by a factor of four as the temperature is raised. It was also found that the average number of cisternae in each dictyosome decreased from 4.4 to 3.1. These observations suggest strongly that the secretory product is carried in the vesicles.

If the glands on an excised leaf are subjected to a water stress, secretion stops immediately. In this situation the outer cells still contain large numbers of vesicles, suggesting that whilst production of the secretory product is not greatly inhibited by lack of water, the discharge of the vesicles across the plasma membrane is prevented. Poisoning of the glands with dilute solutions of potassium cyanide causes a decrease in the numbers of vesicles remaining in the cytoplasm, suggesting this time that fusion of vesicles with the plasma membrane is not cyanide sensitive, whereas the production of their contents is inhibited.

The product is secreted across all the cell walls of the outer layers, and discharged into the limited volume of apoplasm or extracellular space which is found between the external surface of the gland and the layer of cutinised wall material at its base. The product is forced by turgor pressure through pores in the cuticle covering the gland, and forms as a droplet on the outer surface. This pattern of Golgi-dependent secretion is common to all mucilage glands which have been studied. It is found in the Venus fly trap (*Dionaea muscipula*). Here the inner leaf surface is covered with glands which are multicellular (Fig.

4.13). The outer cells are concerned with secretion and the cytoplasm of these cells is well endowed with mitochondria and dictyosomes (Fig. 4.14). The presence of a cutinised cell wall layer at the base of glands is commonly found in aerial plants, but it is absent from submerged parts of aquatic plants. The production of wall ingrowths or transfer function is also not universal, and its development in an insectivorous plant possibly represents an adaptation based on the need for material formed by digestion of the trapped prey to be reincorporated into the leaf (Fig. 4.15).

A further example of a hydrophilic secretion is provided by nectar. This contains sugars, mainly sucrose, glucose and fructose together with small oligosaccharides and enzymes such as invertase. Not all nectar secretion requires the development of a gland; many plants simply secrete nectar from sieve elements via intercellular spaces and modified stomata. Of those that have glandular nectaries, the diversity of structure is such that a definitive mechanism for the secretion process has not emerged. Many of the secreting cells have wall ingrowths, and are

Fig. 4.13. Scanning electron micrograph of the surface of a leaf of *Dionaea muscipula* (Venus fly trap). Multicellular glands occur in small random groups. Bar = 100 μm.

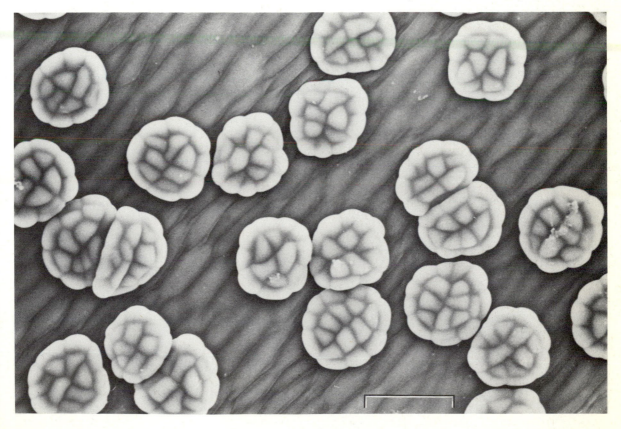

therefore classifiable as transfer cells. In these cases it seems likely that secretion may take place by direct transport of materials across the plasma membrane. In such cells, dictyosome activity is low even during active secretion phases. Nectaries commonly show marked localisation on the plant body; they are usually confined to floral parts and may function only for a few hours or days at an appropriate stage of flowering. A similar class of secretory structure is found on the leaf, where it is responsible for the process known as guttation – the formation of aqueous droplets on the margins of leaves when plants are grown in very humid environments. These are the hydathodes, and they too may make use of modified stomata as pores to control the discharge of the guttation fluid.

Secretion of hydrophobic substances

Hydrophobic secretions include a wide range of essential oils and waxes. An example of this type is given by the glandular hairs or trichomes on the leaves of *Mentha piperita*. The leaves have two types of gland,

Fig. 4.14. Section through a secretory cell in the insectivorous gland of *Dionaea muscipula*. The cytoplasm contains mitochondria with prominent cristae, together with dictyosomes. Bar =1 μm.

each consisting of a basal cell in the epidermis, a cutinised stalk cell and either one (in the hair) or eight (in the trichome) secretory cells at the top of the stalk. Activity in the glandular trichome is marked by the formation of a space between the secretory cells and the cuticle on the outside surface. During the secretion process the endoplasmic reticulum increases in extent within the secretory cells, but dictyosome activity is not seen. The secretion product accumulates in the subcuticular space and eventually may cause the degeneration of the glandular cells by breaking their contact with the stalk.

There are many variations on this theme but, in general, it appears that secretion of oils is associated with developed endoplasmic reticulum rather than with dictyosome activity (Fig. 4.16). The diversity of glandular structures is such that very few general points may be made with confidence concerning their development and function. However, they do represent an example of a type of restarted differentiation, since they are formed as a result of further development of the epidermis. This development, which may involve the production of only three or four cells, nonetheless presents in microcosm a good example of a pathway, starting with a division phase and ending with the formation of a specialised func-

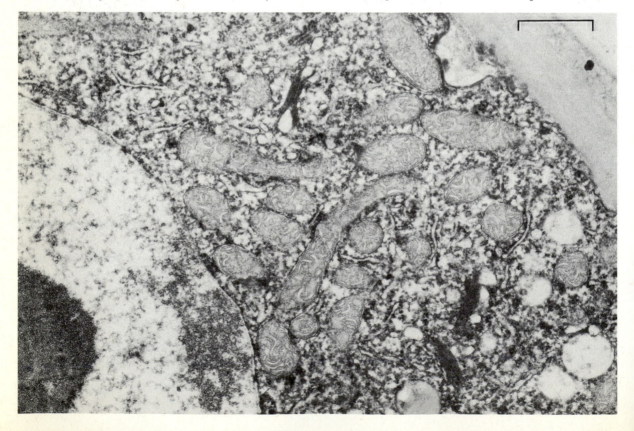

tion. In *Callitriche*, for example, the simple trichome structure first arises as an extension of one epidermal cell. This extension becomes polarised, and divides asymmetrically to give rise to the basal cell and the stalk. Further divisions of the raised cell in different planes give rise to the multicellular head of the glandular trichome (Fig. 4.17). This is but one example of the importance of planes of cell division in the determination of the shape and function of linked groups of cells. A further example which will be considered in detail below is the stomatal complex, another modification of the epidermis (p. 117).

Laticifers

To conclude this brief survey of secretory cells, mention should be made of the production of latex. Latex is the milky juice which appears as droplets on

Fig. 4.15. Section through the wall at the base of the insectivorous gland of *Dionaea muscipula*. On the gland side the wall has pronounced ingrowths, but there are no ingrowths on the leaf side. Bar = 1 μm.

the cut surfaces of many plants, the most well known perhaps being the dandelion (*Taraxacum officinale*). It may contain a variety of substances such as alkaloids, organic acids and carbohydrates as well as dispersed droplets of oils, resins and rubber. The latex of *Carica papaya* contains the enzyme papain. Latex may also contain more obvious remnants of cytoplasmic contents, such as vesicles and starch grains. These facts point to the origin of the latex as cytoplasmic (Fig. 4.18).

Latex is produced by a specialised cell type, the laticifer. Laticifers can be classified into two types depending on their pathway of development. Non-articulated laticifers arise from a single cell by a process of continual elongation and nuclear division without cell plate formation. The laticifer is thus a multinucleate cell, and the nuclei may group together, particularly in the youngest parts of the cell. Within a single cell of this type it is commonly found that the nuclei are at different stages of the division process, suggesting the movement of an unidentified mitotic stimulus along the length of the cell. Non-articulated laticifers may be simple or branched; they grow intrusively between the other cells of the plant, often

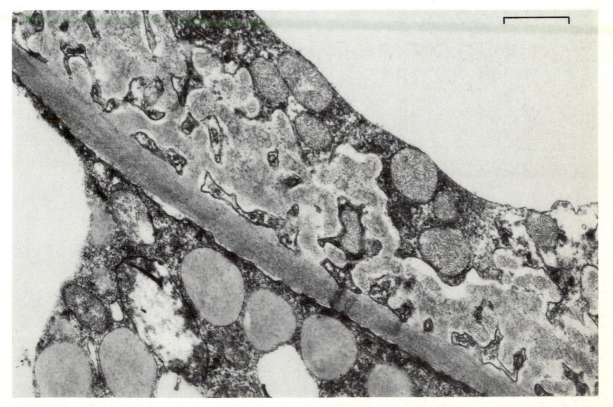

Fig. 4.16. Section through a cell in the farina gland of *Primula capitata*. This cell type secretes the white flour-like deposit ('farina') which covers the surface of the plant. The cytoplasm contains massive accumulations of smooth endoplasmic reticulum membranes. Bar = 1 μm.

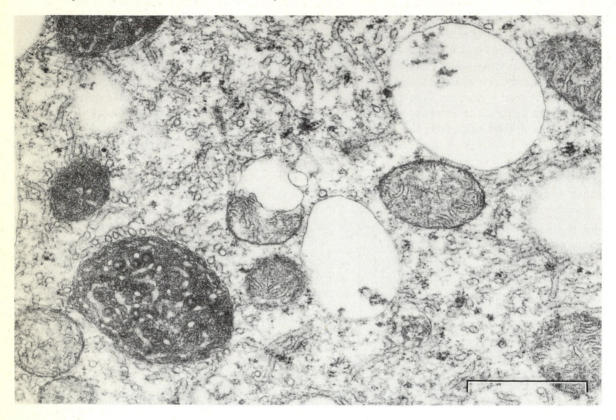

Fig. 4.17. Diagram of the sequence of development of a glandular trichomes on a leaf multicellular of *Callitriche*. A single epidermal cell first elongates at right angles to the epidermal surface, then divides asymmetrically. Further divisions of one of the daughter cells in various planes gives rise to the final structure.

very rapidly. There is some evidence that the latex of the milkweed, *Asclepsias syriacoa* contains enzymes which it is thought may be secreted to assist the passage of the laticifer through the tissues of the plant.

Articulated laticifers begin development in the form of a column of cells which become connected by the breakdown of end walls. Growth proceeds by the incorporation of other cells from the tissue into the laticifer which, once again, results in the formation of a multinucleate cell. Both types of laticifer have thickened primary walls which are not lignified. The mechanism of the production of the latex is not clear, but is thought to proceed by means of the engulfment of part or most of the cytoplasm into the vacuole. For this reason the contents of latex may be regarded both as a secretion (formation of enzymes, for example) and an excretion (the sequestering of secondary products).

Laticifers are limited to a few genera, but they provide an interesting example of an invasive style of growth, more akin to a pathogen in some of its aspects than fitting into the more usual division, elongation and specialisation pattern of development.

Stomatal development and the epidermis

Gland cells are one example of a specialisation of the epidermis. The epidermis is itself a differentiated tissue derived from its own population of initiating cells on the surface of the shoot apex and in the meristem of the root. The epidermis of roots is concerned with the absorption of soil water and nutrients. To achieve this end, the roots of many plants produce hairs which are elongated extensions of individual epidermal cells. It has been calculated that a plant of rye, *Secale cereale*, which was growing in 0.05 m^3 of soil had 14×10^9 root hairs, and a total root absorbing area of nearly 630 m^2. The epidermis which covers the aerial parts of the plant is principally concerned in the limitation of water loss. This is achieved by the production of a cuticle containing cutin, and perhaps waxes (p. 85). Epidermal cells thus

Fig. 4.18. Section through the excreted latex of *Euphorbia splendens*. Remnants of membrane-bound organelles are visible amongst the amorphous material. Bar = 1 μm.

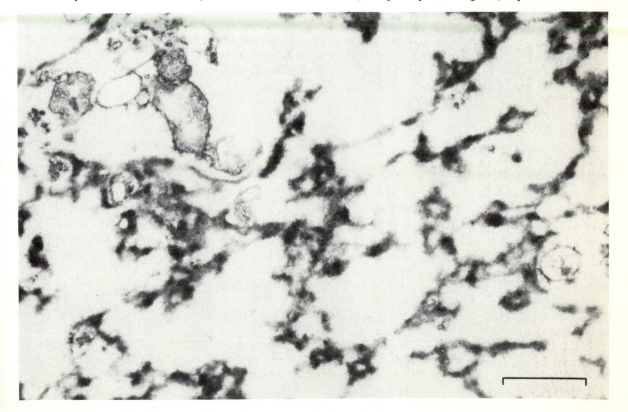

have an identifiable and specialised structure. They retain the ability to divide. This is necessary in order to allow for the growth of the plant, and also to allow reaction to wounding. Division of an epidermal cell may therefore result simply in the production of a new cell of the same type. However, as in the case of glandular hairs and trichomes, a division in the epidermis may also signal an entirely new phase of development.

A striking example of this is provided by the differentiation of the stomatal complex. This short developmental pathway is of considerable interest because it involves precise planes of cell division, and was the object of a study which first propounded a mechanism for the control of these planes. Stomata may consist of from two to eight cells. Two of the cells – the guard cells – flank the stomatal opening and control its size. Other cells may be present; these are known as subsidiary cells. The function of the stomata is to control the exchange of gas between the atmosphere and the network of intercellular spaces which is found within the photosynthetic tissues inside the leaf. This is achieved by changes in the turgor pressure inside the guard cells which result in the opening or closing of the pore which they abut. This appears to be mediated by potassium ions – loss of potassium ions from the guard cells during the dark period brings about closure of the pore.

Formation of the stomatal complex in wheat

The stomatal complex in wheat (*Triticum vulgare*) consist of four cells, two guard cells and two subsidiary cells. The epidermal cells from which the complex develops are vacuolate. Prior to the division which initiates stomatal differentiation, the nucleus of the epidermal cell migrates to one end to give a polarised structure. The cell then divides in this asymmetric configuration, to give rise to two daughter cells which are quite different in appearance and size. The future guard cell (the guard mother cell) is small and does not contain a large vacuole. The other daughter cell, which is destined to give rise to a subsidiary cell, is highly vacuolate and immediately following the division is polarised in the same sense as the original epidermal cell. As this cell progresses towards division, its nucleus migrates and comes to rest against that part of the wall which is common with the guard mother cell of the adjacent row in the epidermis (Fig. 4.19). This means that, at the start of the second division, the cell is once again very polarised, with a large vacuole and a nucleus confined to a small volume of cytoplasm at one side of the cell. The division which follows is hence highly asymmetric once more, and gives rise to a small non-

Fig. 4.19. Diagrammatic representation of the sequence of development of the stomatal complex in grasses. The pathway involves three highly predictable types of cell division.

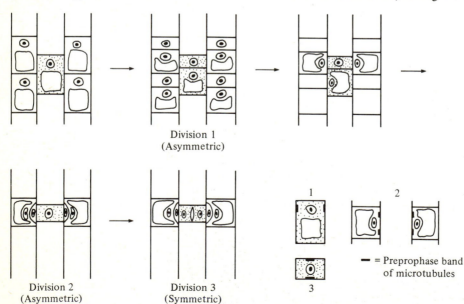

Division 1
(Asymmetric)

Division 2
(Asymmetric)

Division 3
(Symmetric)

1

2

3

— = Preprophase band
of microtubules

vacuolate cell (the subsidiary cell) and a large vacuolate cell. The cell plate which partitions the cytoplasm between these two cells at cytokinesis is curved. The final stage in the development of the complex is signalled by a symmetrical division of the guard mother cell. This gives rise to the two guard cells, which are similar in size and structure and which remain partially connected together through large spaces in the cell plate. The functioning state of the stomata is attained by thickening of this cell plate in its central region followed by splitting of the thickened wall to produce the pore (Fig. 4.20).

This sequence of asymmetric and symmetric divisions is completely predictable, and clearly indicates that cooperation is involved between adjacent cells in the epidermis. It represents a situation where the planes of cell division can be foreseen with a high degree of accuracy and, as such, was used as a model to examine possible mechanisms of control of these planes.

The control of the plane of cell division
When early prophase stages of division in the development of the stomatal complex were examined,

Fig. 4.20. The stomatal pore in *Vigna sinensis*.
Bar = 1 μm. Picture provided by Dr B. Galatis.
(First published in *Amer. J. Botany*, **67**, 1243–61.)

it was found that in each case a band of microtubules was present against the mother cell wall at the position where the future cell plate would eventually fuse with it at cytokinesis (Fig. 4.19). This band was named the 'preprophase band' of microtubules and it is now known that it begins to form during early G_2 (p. 30). Since the discovery of the band, considerable speculative and experimental effort has been devoted to the problem of its possible function.

A direct effect of the band of microtubules on the position of the cell plate is clearly unlikely, since the band disappears during the later stages of prophase. In addition, since the cell plate grows from its centre outwards, even if the band had some residual effect on the mother cell wall, this would not be a favourable way of controlling the plane of the cell plate. The band may extend along the mother cell wall for a distance which is considerably greater than the width of the cell plate; this again argues against a direct function in controlling the plate itself.

It is more likely that the function of the band is accomplished during the period of its existence prior to the breakdown of the nuclear envelope. There are two possible interpretations of the behaviour of the band. One is that it has a more or less mechanical function. In this view, the band is concerned with defining either the position or the orientation of the nucleus, or both. The best evidence for this idea

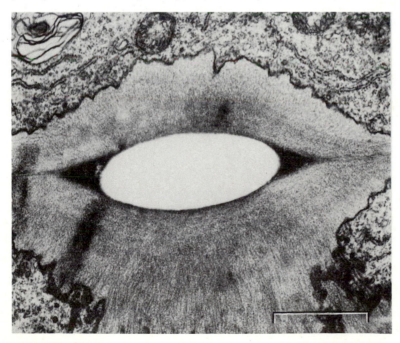

comes not from the divisions associated with stomatal development, but from asymmetric divisions in the root tip of the water fern *Azolla* (p. 160), where the appearance of the microtubule band clearly precedes migration of the nucleus to its asymmetric position.

This mechanical function has two further requirements concerning the nucleus. The first is that the very idea of orientation of a premitotic nucleus should be meaningful. Recent advances in the understanding of the complexity of the structure of the nucleus mean that this idea is not as unreasonable as was once thought. If a nucleus does indeed have skeletal elements within it which order the chromatin (p. 11) then, clearly, orientation of the entire nucleus with respect to an external axis can be envisaged. The second requirement for a mechanical function of the band to be possible is that some mechanism should exist for transferring the details of its position to the nucleus. When sections of early prophase cells are examined, it is frequently found that microtubules are present around the nucleus at the same time as the band is visible. If these two populations of microtubules were coupled together, then movement of the nucleus could be directed in this way. Once the position of the nucleus within the cell, and its orientation, are preset, then the operation of the mitotic spindle would ensure that the cell plate would be initiated in the correct plane. Of course, such a description takes no account of, for example, the curvature of the cell plate in the second asymmetric division of the stomatal complex pathway. In this case, however, it may be suggested that the position of the cell vacuole places a restraint on the final positioning of the plate.

An entirely different interpretation of these observations is possible. According to this view the band results from an accumulation of microtubular material prior to its redeployment in the mitotic spindle. When the band is present, the interphase 'cortical' microtubules are absent (p. 30). The microtubules which are seen around the nucleus during the period of the band's existence are, in this interpretation, merely being transferred from the band to the mitotic spindle, without any mechanical function.

The occurrence of the preprophase band of microtubules is by no means universal in plant cells, and this fact casts some doubt on both of the preceding suggestions as to its function. Accordingly, it has recently been suggested that the important event prior to mitosis is the localisation of microtubule organising centres (MTOCs) in an appropriate position at the cell periphery. The appearance of actual microtubules can be regarded as incidental; hence, their absence in some cells does not represent a stumbling block to this hypothesis. The importance of the clustering of MTOCs is seen to be that it would ensure their distribution between the daughter cells of the mitosis. In the young daughter cells after cytokinesis, this idea predicts that MTOCs would be concentrated in the corners where the cell plate joins the mother cell wall. Evidence for this has been presented from a study of *Azolla* root tips.

Clearly none of these ideas can be regarded as entirely satisfactory. In particular the central question is evaded: that is, of how neighbouring cells coordinate their planes of division to produce a complex of cells such as is seen in stomatal development. The preprophase band is always found in divisions with this type of requirement. Equally, it is absent in many types of filamentous tissues, such as fungal hyphae, moss protonema and algal filaments. It is also absent from dividing cells in liquid suspension cultures derived from protoplasts. The difficulty of interpreting its function no doubt stems in part from the comparative recency of its discovery. Coupled to this is the requirement for prefixation of tissues before the band can be envisaged, which means that subtle changes in its behaviour may be entirely lost to view. A further factor hampering progress in this type of study is the comparative rarity of situations in which planes of division can be confidently predicted. As will be seen later, although planes of cell division are an important factor in the development of tissues, within higher plants at least such development does not always proceed according to geometrically predictable patterns (Chapter 6).

The pollen generative cell

The stomatal complex is derived by a sequence of mitoses from the epidermis. Some of these divisions have been termed asymmetric. This is based on an obvious geometrical displacement of the mitotic apparatus within the cell, either to one end or to one side. Division from such a position produces daughter cells of widely different shape, size and developmental future. The 'symmetrical' division of the guard mother cell produces two apparently identical guard cells. Within these two types of division, symmetric and asymmetric, there is no obvious control of the division of the cytoplasmic contents, and indeed, as has been seen in the case of sorting out of plastids (p. 60), the division of cytoplasmic particles appears in many cases to be at random. This of course may simply represent a failure of observation. However, there is one example of an asymmetric division where it is clear that the two daughter cells differ not only in shape and size, but also in the nature of their organelle complement. This is the formation of the generative cell within the pollen grain. The pollen grain is the male gametophyte of a flowering plant, and contains a haploid nucleus. This nucleus divides to produce two daughter nuclei which are in quite different cellular environments and which have different fates. The vegetative cell produces the pollen tube, whilst the generative cell undergoes a further division to produce two sperm cells.

Pollen grains are independent entities, separated from their neighbours and from the parental tissue by the presence of thick and highly sculptured wall containing sporopollenin (p. 87). Prior to the division which results in generative cell formation, the cytoplasm of the pollen grain is found to contain all the normal complement of organelles associated with a meristematic cell, but with the addition of storage materials such as prominent lipid droplets, and starch in the plastids.

At prophase of the generative cell division, the nucleus migrates to one side of the pollen grain and comes to rest near to the wall. This does not appear to be accompanied by the formation of a preprophase band of microtubules although, in orchid pollen at least, a few scattered microtubules are found in the cytoplasm lying at right angles to the plane of the future spindle. The division proceeds with the formation of a highly unusual spindle structure; the pole next to the pollen grain wall (the generative pole) is flattened, and chemical studies suggest that there is a general movement of preformed protein and RNA away from this position and towards the vegetative pole. The separation of the two daughter nuclei is followed by the formation of a structure which resembles a cell plate insofar as it arises by the fusion of vesicles in association with microtubules. However, this wall does not fuse with the mother cell wall at the end of its centrifugal growth phase; rather it forms a

lining to the mother cell wall which effectively cuts off the generative nucleus entirely (Fig. 4.21). The cytoplasm enclosed by this encircling wall is limited in volume and depleted in organelle complement. It usually contains a few mitochondria, a small amount of endoplasmic reticulum, a few dictyosomes, and other miscellaneous membraneous vesicles. In very many cases it is devoid of plastids. This represents one basis for the widespread maternal inheritance of plastids (p. 61).

Continued development of the generative cell consists of its gradual separation from the mother cell wall so that, eventually, it comes to lie within the pollen grain cytoplasm. At this stage it is surrounded by a thin wall delimited on either side by a plasma membrane (Fig. 4.22). The nature of this wall is not known in detail in most species, but it is known to be polysaccharide. It may be crossed by plasmodesmata. A feature of the wall is that it has no apparent structural or mechanical rigidity, and the overall shape of the generative cell may be amoeboid. When the generative cell divides, it produces two sperm cells. In the majority of plants this happens within the growing pollen tube following its germination on the

Fig. 4.21. Diagram of an idealised pollen grain at the end of the division which gives rise to the generative cell. The generative cell is enclosed within a non-rigid wall.

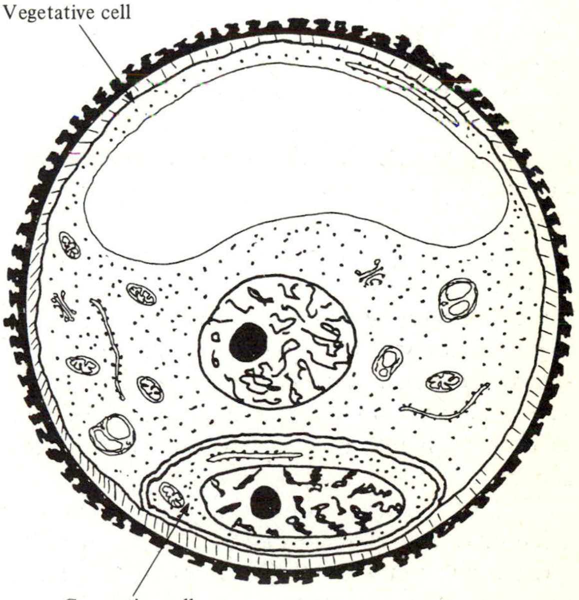

stigma. The pollen tube is very long (p. 87) and, for this reason, pictures of the sperm of flowering plants are very hard to obtain, since the small size of the sperm cells within the length of the pollen tube makes them difficult to locate. However, such evidence as there is suggests that the sperm cells contain the same types of organelles as the generative cell: a few mitochondria with very much reduced internal structure, a small amount of endoplasmic reticulum, and a small number of dictyosomes. The sperm cytoplasm contains microtubules, and the function of these is unclear. Sometimes they appear in clusters, and may be concerned with maintaining the overall shape of the cell. In other species they are not clustered, and it

has been suggested that they may function in the movement of the sperm down the pollen tube.

The pollen grain thus represents an extreme case of asymmetry in a cell division. Very rarely is the difference between daughter cells of a vegetative division so rapidly expressed as in this case. Normally, it may be inferred that even apparently symmetrical divisions are in fact asymmetric in some cryptic way, since one daughter cell may eventually differentiate into a specialised cell, whereas the other may remain as part of a meristem. In the pollen grain, one division serves to produce two completely different cells. The vegetative cell is capable of an extended period of growth by means of the pollen tube; the generative cell simply divides once, and its progeny are transported to the site at which they take part in fusion events which give rise to the zygote and the endosperm.

Dormancy and cell structure

In the examples which have so far been considered of cell specialisation and differentiation, there

Fig. 4.22. Pollen grain of *Nicotiana tabacum*, showing the large vegetative nucleus (with a nucleolus visible) and the generative nucleus within its wall lying free in the pollen cytoplasm. The difference in staining properties of the two nuclei is marked. Bar = 2 μm. (Picture by courtesy of Dr J. Dunwell.)

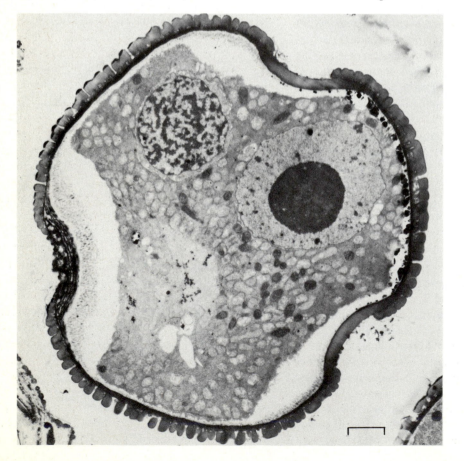

has been an implicit assumption that differentiation proceeds continuously from a meristem or undifferentiated state. For very many cells, this assumption is quite valid. In terms of the behaviour of whole plants, growth is often continuous and completed within one growing season; its termination results in the death of the complete plant body. However, there are many plants which survive for more than a single continuous growing period; they are perennial and, as well as differentiating the primary structures we have so far considered, they are able to survive periods when conditions are unfavourable to growth. The most obvious examples of this periodicity are given by trees. The suspension of active growth during cold or dry seasons involves the phenomenon of dormancy. Physiological aspects of dormancy will be considered in a later chapter (p. 215). In the current survey of cell types, dormancy is also of interest. In secondary tissues such as the stem of a tree, new cells are donated to the growing plant by an annular meristem called the cambium. This meristem shows variation in its activity and structure on a seasonal basis.

There is a second general way of surviving stress periods such as a cold season. This is by the formation of seeds or spores. In evolutionary terms, of course, it can quite validly be argued that the production of dispersive reproductive units such as seeds has survival value not only in terms of the geographical distribution of a species, but also in terms of the fact that the reproductive process itself offers the opportunity for the recombination of parental genes (p. 38). It is no doubt of considerable importance that the production of a dry, frost-resistant seed will permit the over-winter survival of a species which could not withstand the same conditions in a vegetative form. Once again, there is considerable interest both to the physiologist (p. 216) and to the student of cell structure in the production of seeds. The cells which they contain are able to survive desiccation and frost, and yet resume normal growth when favourable conditions return.

Seasonal changes in the cambium

The cambium in a woody stem is the meristem which gives rise to the vascular tissues and produces the cells which cause increase in the girth of the stem. The cambium may produce both the vertical system of the xylem and phloem, and also the rays, or horizontal system. The meristematic cells which originate these two systems are quite different in size and shape; the ray initials are more or less isodiametric, and correspond reasonably well to the ideal 'meristematic' cell which was described in Chapter 1. The cells which give rise to the vertical system are called the fusiform initials, and these are very long and narrow. In old branches of *Pinus strobus*, the fusiform initials may be as long as 4000 μm (or 4 mm)! In *Acer pseudoplatanus* (sycamore), which is the best studied species in terms of seasonal variation in structure, the fusiform initials are 320 μm long in a 2–3 year old branch, but only 2–4 μm wide and 15–20 μm across.

In the growing season, the fusiform cells contain all the usual components of a meristematic cell, although their detailed appearance is dictated by both the large volume of the cell and its extremely elongated shape. The nucleus is elongated and, in *Acer*, may reach 22 μm long, whilst in cross section it is only 2 μm by 4 μm. It is found in a more or less central position, surrounded by a thin layer of cytoplasm. The remainder of the cytoplasm is in a thin layer round the periphery of the cell, and thin strands connect these two areas of cytoplasm. The cells are very highly vacuolated, and usually contain one or two vacuoles only. When such cells are viewed under the light microscope whilst still alive, active streaming of cytoplasmic particles can be observed in the strands which cross the vacuolar space. The endoplasmic reticulum is rough, and the dictyosomes produce large, apparently empty, vesicles together with small densely stained vesicles. The plastids show very little internal development, but may have ferritin within the stroma. Starch is not present in the plastids. The walls of the active cambium are thin.

Cambial activity in temperate regions is stimulated in early spring and continues until mid or late summer, depending upon the species. The cessation of activity is accompanied by modifications in the structure of the cells. During early autumn the large vacuoles tend to break down into small units which may continue to fuse together and then break down again to give a constantly changing network within the cell. During November, many small globular vacuoles are produced; this coincides with the cessation of cytoplasmic streaming. This condition persists until about February, when the small globular vacuoles fuse together into a network, and streaming recommences. As the hydration of the cells begins in spring, this vacuolar network reforms the large vacuoles which are characteristic of the active stage.

The mitochondria of the cambial cell also show a pattern of seasonal variation in size and shape. In the active cambium they are small and more or less spherical, rarely exceeding 3 μm in diameter. As the rates of division in the cambium decrease with the progress of summer, the average size of the mitochondria increases, reaching a maximum in November, when the organelles are frequently seen to be elongated and branched. The onset of the rest period causes a reversal of this process so that, in January, the mitochondria are once again quite small (although they may retain their elongated form). Rehydration of the cells during spring, before cell division begins, causes a transient increase in the size of mitochondria; the onset of division activity restores the small spherical size to the population.

The plastids show very little seasonal variation, although in sycamore (*Acer pseudoplatanus*) they produce starch during the late summer which persists into winter. This does not seem to be a general phenomenon, however, since other species show the reverse behaviour. The dictyosomes in winter are characteristically inactive, and show no vesicle production. The endoplasmic reticulum retains its ribosomes in sycamore during the winter but, in other species, there is an increase in smooth endoplasmic reticulum during the dormant period. Winter is also characterised by the appearance of storage materials in the cytoplasm. In sycamore there is an increase in the amount of lipid found as droplets; other species may, however, accumulate crystalline protein materials within membrane-bound particles.

The cycle of cambial activity is controlled by the environment; in temperate regions it is linked to the season whereas, in the tropics, it may be linked more directly to water availability. In terms of the physiology of the cell, there is a marked increase in the osmotic pressure of the vacuole as the season proceeds; in spring it averages about 12 atmospheres in sycamore, but it may reach a level as high as 40 atmospheres during December and January if the season is a particularly cold one. In terms of the physiology of the plant, there is evidence of hormonal control of many of the processes related to resting or dormancy. This aspect will be considered later (p. 217). For the time being, the cambium serves to show that not all meristematic cells are small, densely cytoplasmic and isodiametric, and that the fine structure of cells can vary on a periodic basis related to environmental conditions, even when the function of that cell type remains the same.

The structure of storage tissues of seeds

As we have seen, cambial activity is seasonal and this correlates with a certain periodicity of changes (which are important but not particularly striking) in fine structure. The example *par excellence* of a resting period is provided by the cells of a seed where, following dispersal, viability may be retained for hundreds or even thousands of years before conditions favourable to renewed growth recur. Once again, many aspects of this behaviour have been examined on a physiological basis, and these will be considered later (p. 216). However, many seeds, particularly of cereals and leguminous plants have been the subject of structural study because of their great economic importance. The structural specialisation represented by the development of the storage tissues of seeds is of particular interest, since it results in the ability of the cells to withstand degrees of desiccation and cold which would otherwise be fatal.

The process of maturation has been followed in peas (*Pisum sativum*). Here the period from fertilisation to seed maturity takes of the order of 50 days and can be divided into four phases. The first lasts about 10 days, and consists of a cell division phase with little differentiation of the embryo into recognisable tissues. After about 10 days, the cotyledons (which are the major site of storage materials) become recognisable and expand. At about 20 days the production of storage materials begins, and growth slows down. In the final phase there is loss of fresh weight as the seed matures and dehydrates.

The phases, which can be recognised in the light microscope and by measurement of fresh weight, dry weight and protein content, are correspondingly distinctive in the electron microscope. In the cell division phase there is very little differentiation of cell types, and the cells appear to be of a uniform meristematic nature. During the second phase, vacuolation of the cytoplasm of the cotyledons takes place, first by the formation of many small vacuoles, followed by their coalescence into several large units. During this period of cell expansion, the nucleus increases in size and the nucleolus is very prominent. The endoplasmic reticulum becomes extensive and fills the cytoplasm with a network of membranes and vesicles. The end of the second phase is marked by the initial deposition of storage products; by day 20, every plastid in the cytoplasm is found to contain a starch grain. The storage phase is characterised by changes in the plastids, vacuoles and endoplasmic reticulum. Starch synthesis in the plastids proceeds at a high rate so that eventually, at the end of the phase, each plastid consists of a single large mass of starch which occupies the entire volume of the stroma – with internal membranes reduced to a peripheral layer next to the plastid envelope. Over this period, the deposition of protein begins. The precise sites of deposition of the protein are a matter for dispute, and possible variation between species. In peas (*Pisum sativum*) it appears that protein is deposited both in the vacuoles and within the lumen of the endoplasmic reticulum (Fig. 4.23). The dictyosomes seem to play no part in the storage of protein in peas.

At about the 28th day, the fourth phase of maturation beings. This is marked by a slowing down of the rate of increase in fresh weight of the seed;

eventually, after 40 days, the fresh weight of the seed actually starts to decline. In the final stages of maturation, the starch grains appear to lie within a transparent zone inside the plastid envelope, with no sign of any other internal membrane structure. The protein bodies dominate the cytoplasm, with only a few scattered vesicles and much reduced mitochondria together with a poorly stained spherical nucleus as evidence of remaining cytoplasmic organisation. In the dry seed, fat deposits tend to accumulate as droplets which line the plasma membrane on its cytoplasmic side.

These effects, which are accompanied by net loss of water, demonstrate a highly ordered series of changes which modify the organisation of all the cell organelles. They may be contrasted with the selective loss of function which occurs in the mature sieve element (p. 96) and the total loss of all structure which accompanies the final stages of differentiation of the xylem element (p. 103). The seed cotyledon is reduced in structure to large masses of dehydrated storage materials, such as lipid, protein and starch, together with minimal organisation in the form of membranes, rudimentary mitochondria, and the nucleus (Fig. 4.24). In this form, the cell is quite stable and may

Fig. 4.23. Part of the cytoplasm of a cell in a developing cotyledon of the pea (*Pisum sativum*). The cytoplasm contains many vacuoles which have a granular protein deposit within. Bar = 1 μm.

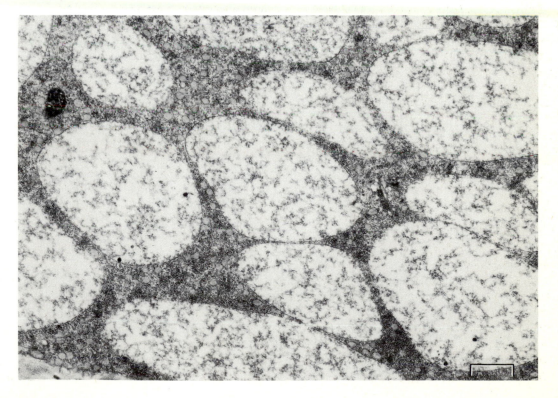

remain unchanged for long periods of time. However, within seconds of being imbibed with water, ultrastructural changes begin to be recognisable; within 1 day, the cytoplasm is the site of numerous vesicles, a few dictyosomes, and mitochondria with internal lamellae. Within 5 days the endoplasmic reticulum has reformed, mitochondria and dictyosomes have achieved their normal organisation, and the nucleus has expanded and become lobed. During normal growth of the seedling which follows this, the cotyledons fulfil their biological function of supplying nutrients to the embryo. This involves the mobilisation of storage reserves and the eventual and permanent loss of cytoplasmic organisation.

Abscission zones

Development may be continuous, or it may be periodic and, in both these situations, cell structure mirrors cell physiology and function. There is a

further type of behaviour which results in the appearance of specialised cell function: the phenomenon of abscission. Abscission is the controlled process of loss of whole parts of the plant. Leaf fall by deciduous trees is the most dramatic example of wholesale abscission; the dropping of fruits or unfertilised flowers are also abscission processes, and even 'evergreen' trees shed leaves when they reach a certain degree of maturity. The loss of these structures does not take place at random, but by the action of particular populations of cells within zones. The abscission zone can be further subdivided into the abscission layer, where cell separation takes place, and the protective layer which prevents loss of water from the plant after abscission has taken place. The abscission zone is of great physiological interest since it is recognisable within the pedicel of a flower such as that of the tomato (*Lycopersicon esculentum*) yet does not exert its final function until the fruit has ripened in the case of a successful fertilisation; on the other hand, if the flower does not set, abscission can take place without fruit development. This aspect of abscission will be considered later (p. 179).

Anatomically, the abscission zone may be recognised in many situations by the presence of a layer of cells which crosses the organ to be shed in a

Fig. 4.24. Section through a cell in the cotyledon of a dry pea seed which has been imbibed with glutaraldehyde. The cytoplasm contains many protein bodies but very little organised membrane material. Bar = 2 μm.

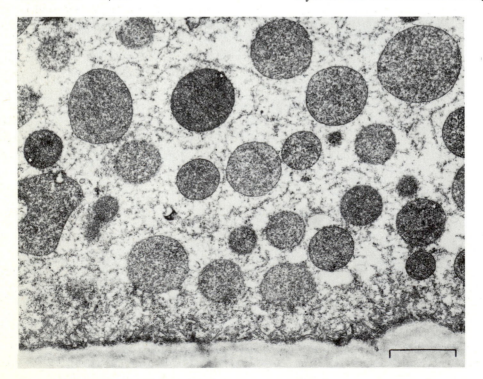

more or less flat plane. This layer may be 5–50 tiers of cells thick. Separation, when it occurs, usually involves the breakage of cell connections in a thinner layer, up to 5 cells thick at the distal end of the abscission zone (the end furthest away from the plant). The cells on the proximal side of the zone may produce specialised walls which contain cutin; these represent the protective layer. It was formerly thought that abscission resulted from the drying out of the cells in the abscission layer because of the presence of the cutinised layer, but this is now known not to be the case. The abscission zone may be recognised by eye in some situations; its position on the pedicel of a tomato flower, for example, is clearly marked by a constriction of the stalk. Ultrastructural studies of the cells of the abscission zone have shown that the cell walls are not lignified (apart from those of the vascular tissues which pass through it). Rather, many of the cells are classified as collenchyma; that is, cells in which there is a thickened wall but in which lignification has not taken place. This means that the abscission zone is not necessarily a point of mechanical weakness until induction of abscission has begun. The cells of the abscission zone are also often smaller than the surrounding parenchyma of the petiole, pedicel, etc. This implies that they have retained their capacity for division for a longer period during development. The formation of abscission zone cells has not been studied at the ultrastructural level; this reflects the general finding that there are no consistent cytoplasmic markers which define a cell in an abscission layer. In the pedicel of the tomato, it has been observed that the cells in the abscission layer contain many microbodies and also granular deposits within the chloroplasts; however, the relationship of these structures to the process of abscission has not been demonstrated. More commonly, it is found that abscission zone cells do not have specialised cytoplasmic inclusions.

Abscission itself is an active process, not a degenerative one. It proceeds by the separation of cell wall at the middle lamella, often (though not always) from the positions of branched plasmodesmata. The process is not highly organised in space, since complete dissolution of the middle lamella can occur in one region of the zone before it begins in another. The biochemistry of the separation process is certainly complex, and appears to involve at least two types of enzyme. Activities of a cellulase-type enzyme and a polygalacturonase have been shown to increase during abscission. The fact of abscission as an active process is emphasised by the finding that protein synthesis is necessary for its completion. It is also known that ethylene, a potent stimulant to abscission, increases the secretion of these enzymes into the wall space.

Final separation of the middle lamellae and partial dissolution of unlignified walls are not sufficient to guarantee physical separation of the organ which is to be shed from its parent plant. The remaining vascular tissue is lignified and hence resistant to degradation by enzymes. It appears that these cells are actively broken by forces generated within the abscission cells themselves. In very simple terms this may be described as involving the expansion of the cells of the abscission zone on its proximal side (remaining in contact with the plant), whereas those on the distal (shed) side tend to contract due to water loss. This differential swelling and contraction generates sufficient force to rupture the tissues of the vascular system. In the case of the horse chestnut (*Aesculus hippocastanum*), the severed vascular strands are clearly visible and form the 'nails' on the horseshoe shape (Fig. 4.25).

Abscission appears to be under the control of plant hormones, and this well studied aspect of it will be examined later (p. 179). In our present context it demonstrates that populations of cells may be left after the differentiation process in a physiological state which is different from that of their neighbours, even though marked cytoplasmic modification has not taken place. Visible markers of the zone are certainly present in the form of the small size of the cells and their unlignified walls. It is highly likely, however, that these modifications alone do not explain the process of abscission; abscission zone cells are probably specific targets for the hormonal messages which elicit their particular function.

Summary

This survey of differentiated cell types does not pretend to be comprehensive. It has been selective in a manner which is intended to point out certain general principles underlying the differentiation process and the origin of specialised function. Not all of these principles are demonstrated in every case and, of course, since the discussion has been limited to considering events at the level of the cell, many questions remain as yet unasked.

The simplest example of a developmental pathway is afforded by the vascular system, and this remains the most well documented area in plant cell development. The three phases of development – division, elongation and differentiation – are clearly separated in a linear sequence in this situation. Furthermore, in the case of the formation of the transporting cells, the sieve tube and xylem vessels, there is no possibility of complication due to diversion of an end product of one pathway into another. These cells are effectively dead and will not undergo further change.

This simple linear development is not always found. In the root cap, a period of division is followed by expansion and differentiation but, as we have seen, the first stage of development to a significant specialised function is not terminal; the geoperceptive cells of the root cap are inexorably moved outwards to the surface where they will assume a different role. The central zone cells are therefore transient. A different style of transience is also indicated by the differentiation of the stomatal complex from the epidermis, and also by the adaptive formation of transfer function in nematode-induced transfer cells. These phenomena should alert us to the general fact that, provided a cell retains its genetic and synthetic machinery in the form of the nucleus and cytoplasmic organelles, its differentiated state need not necessarily be regarded as fixed, even though during normal growth this may usually be the case.

It has also been seen that specialisation may involve modifications to the cell wall, or to one cytoplasmic component in particular, or to the entire contents of the cytoplasm. Often these changes occur in a highly organised manner: the formation of transfer cell ingrowths may be limited to one particular wall only within a cell, the secretion of an extracellular product such as root cap slime may be polarised within the cell, or selective breakdown of a cell wall may take place in only one plane. These observations serve to emphasise strongly the value of structural studies in stressing the importance of spatial, as well as temporal, considerations in development.

Finally, in this chapter, it has been possible to begin to glimpse a further level of complexity in cell development. Although the process can be simplified to descriptions of events within individual cells, it should be becoming clear that within the growing plant, differentiation never occurs in isolation. Cooperative involvement of adjacent cells is implied in the formation of continuous connected structures such as the sieve tubes, in the development of a two dimensional complex such as the stomata, and in the formation of limited three-dimensional zones of specialised cells such as the abscission zone. The remainder of this volume seeks to examine some of the problems of how this organisation of individual cells is achieved and regulated.

Fig. 4.25. A dormant branch of the horse chestnut (*Aesculus hippocastanum*). The 'nails' in the horseshoe below the dormant buds represent the severed vascular strands from the previously attached leaf.

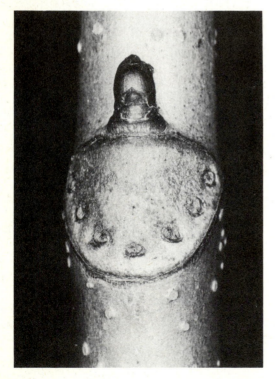

Hormones and cell differentiation

Introduction

The development of specialised functions within the cells of intact plants proceeds by a series of structural and metabolic changes from a small number of starting points, the meristematic cells. The description of these processes necessarily involves the idea of time, but it does not appear in principle to need the interference of ideas concerning cellular interaction or position, unless a complex of cells or a functioning tissue is being considered. Of course, within the intact plant, the state of differentiation of a given cell is very much dependent upon its position relative to other cells, but this would seem to be as likely to represent a particular nutritional or environmental locus as it would be to an inviolate consequence of the nature of the neighbouring cells in a tissue. In other words, it seems at least reasonable to assume that if cells could be isolated, they could be made to follow developmental pathways *in vitro* by suitable manipulations of their environment. The nature of such manipulations could then be assumed to be a reflection of the sorts of stimuli which lead to differentiation in the intact plant and, thus, some considerable insight would be gained into the way in which the plant body organises itself.

This idea occurred first to Gottlieb Haberlandt, who, at the turn of the century, was the first person to attempt the culture of isolated plant cells. His avowed aims in these experiments were to gain insight into the properties of plant cells as 'elementary organisms', and also to see what could be deduced about the interrelationships of cells in intact organisms. His chosen starting materials were always what we would now regard as differentiated cells – in particular, he used the mesophyll and palisade cells from leaves of *Lamium*. His reason for this was the sound one that photosynthetic tissues would probably survive better *in vitro*. The media which he used for the culture of these cells were simple salt solutions, or tap water, sometimes supplemented with sucrose. He did not apply rigorous aseptic techniques to his experiments and, in fact, noted in passing that the presence of bacteria or fungi in his culture did not affect the 'progress' of his cells.

In a crucial sense, his experiments were a failure. He never observed cell division and, although the cells survived for a few weeks (during which time he was able to demonstrate some growth and the deposition of starch), they eventually went into a decline followed by death. With the benefit of 80 years of hindsight, it is now possible to say that this failure was entirely predictable. Using these cells and media, no other result could have been achieved. Nevertheless, these experiments represent the starting point of a new approach to studying the problems of cell development. Haberlandt himself conjectured that 'growth enzymes' would probably be required to stimulate cell division, and he concluded his paper with the entirely speculative prediction that one might be able to cultivate embryos from vegetative cells. Forty years later, this prediction was fulfilled.

Plant growth substances

Following Haberlandt's pioneering lead, progress in the field of cell and tissue culture was slow. By 1922 it was possible to show that isolated roots could be maintained in culture, but the growth of an 'undifferentiated' callus tissue was not achieved until the newly isolated growth substance auxin was used in media during the 1930s. Tissue culture techniques led to the discovery of a second class of growth substances, the cytokinins, in 1955. At the present time it is possible to recognise several groups of substances which may be regarded as plant growth regulators or hormones.

The term plant hormone has a sense different from the word hormone in animal cell biology. In zoology, a hormone is the product of one sort of cell which is transported through the body and has its action at a site remote from its point of synthesis. Plant hormones are not like this; they may act on the cells in which they are produced, and they may show very different actions on different types of cells, and at different concentrations. In intact plants it is likely that growth regulation is the result of the combined action of growth substances, either simultaneously or in sequence. For these reasons it is perhaps preferable to refer to the plant hormones as plant growth regulators or plant growth substances. One early concept which reflects this difference between a hormone and a growth regulator is the idea of the existence of specific inducers of organ formation. It was thought possible that substances might be identified which would, for example, stimulate the formation of shoots or of flowers. The plant growth regulators which have so far been discovered do not function in this way.

In the present chapter we shall consider the nature of the known plant growth regulators and the way in which they may be used as external stimuli to influence the growth of isolated plant tissues, cells or callus tissues. The study of cultured cells has been an integral part of the growth of our understanding of plant growth substances and allows their action, at least in principle, to be examined in a more controlled way than is possible with intact plants. The problem of the role of growth regulators in whole plants will be returned to in Chapter 6.

Auxin

Auxin was discovered as a result of experiments first initiated by Charles Darwin on the growth curvature of grass coleoptiles. Unilateral lighting causes the coleoptile to bend towards the light, and Darwin showed that, for this effect to take place, an intact tip was necessary. Removal of the tip or shading it from the light abolished the curvature. It was later found that if the tip were removed, then replaced after interposing a strip of gelatin between the tip and the rest of the coleoptile, curvature was restored. It was also found that if the tip were replaced to one side of the coleoptile, curvature resulted even in the absence of a unilateral light stimulus. These experiments strongly suggested that the curvature was mediated by the diffusion of a chemical substance. In 1926 Frits Went carried out a simple experiment to prove this point. Agar blocks were placed on the cut ends of coleoptile tips. After a period of time, if the agar blocks were then used as artificial tips and placed asymmetrically on decapitated coleoptiles, curvature resulted (Fig. 5.1). Furthermore, the angle of the curvature was shown to be proportional to the amount of the diffusable substance present; it therefore became possible to develop the coleoptile curvature test as a quantitative assay of auxin activity. The amount of auxin in tissue is extremely small (of the order of tens of micrograms per kilogram of fresh weight) and, as is now known, rather easily oxidised. The chemical identification of auxin was not made with material from plants at all, but from its presence in human urine. As an example of the difficulty of any identification, it may be cited that the definitive determination of the nature of the natural auxin in coleoptiles of *Zea mays* (maize) was not made until 1972, and involved using 15 000 excised tips to obtain the material which was then examined by mass spectroscopy.

The auxin from human urine, maize coleoptiles and several other plants has been shown to be indole-3-acetic or IAA (Fig. 5.2). It should be made quite clear however that the term 'auxin' is a generic one,

not a chemical name. A substance can behave as an auxin in a variety of tests, of which the coleoptile curvature test is but one example. Auxin-like behaviour, however, does not identify that substance as IAA. A wide variety of auxins is known today; some are simple derivatives of IAA (such as indole-3-acetonitrile), whereas others are completely synthetic (such as 2,4-dichlorophenoxyacetic acid – 2,4D). Synthetic auxins of this type are frequently more stable than naturally occurring auxins, and so are often used in culture media.

The biochemistry of IAA

The biosynthesis of IAA proceeds from the amino acid tryptophan. This has been elegantly demonstrated by using the amino acid in a form double labelled with radioactive carbon and tritium atoms as a substrate for sterile pea (*Pisum sativum*) plants. The extracted IAA was also double labelled, and the ratio of the two labels in the IAA was the same as in the supplied tryptophan. The exact pathway of the synthesis may not be the same in all plants. Some experiments suggest that indole pyruvic acid is the intermediate, whilst others show that synthesis proceeds via tryptamine. Of some interest is the finding that epiphytic bacteria found in association with plants can also synthesise IAA.

The level of IAA within a tissue depends on a number of factors, of which the balance between synthesis and breakdown is the most obvious. IAA is known to be degraded by oxidative enzymes. Some of these enzymes are peroxidases, but specific IAA oxidases are also known to occur. The activities of these enzymes can be influenced by plant growth regulators themselves, so that the possibility exists for very complex control mechanisms, by which the presence or absence of other growth substances may accelerate or prevent the breakdown of IAA. The cellular sites of the enzymes which degrade IAA are not known; some occur in the wall and others are attached to membranes. Various phenolic compounds can inhibit IAA oxidases and so act as auxin protectors.

Fig. 5.1. Diagrammatic representation of the role of auxin in the light-induced curvature of the oat coleoptile. *a*, Unilateral lighting causes the intact tissue to bend towards the light; *b*, removal of the coleoptile tip prevents the curvature; *c*, replacement of the tip by means of a gelatin bridge restores the ability to curve; *d*, if the tip is placed in contact with an agar block, auxin diffuses into the agar. Placing the loaded block asymmetrically on the cut end of the coleoptile causes curvature in the absence of unilateral lighting.

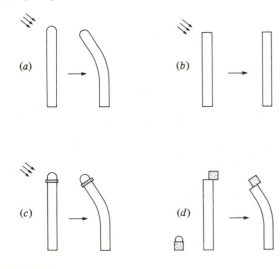

Fig. 5.2. Indole-3-acetic acid (IAA).

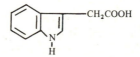

Another mechanism for the control of IAA levels within cells is the formation of conjugates. When IAA is fed to *Colchicum* leaves, it is found to produce a conjugate with glucose. Conjugates with aspartic acid are known from a variety of plants. In *Zea mays*, at least 16 different conjugates of IAA have been identified, including an ester of IAA with myo-inositol. It has been suggested from this work that the conjugates represent inactive forms of IAA, and that environmental stimuli to growth may be expressed as an increase in the activity of enzymes which degrade the conjugate and release free IAA. Conversely, growth inhibition could be mediated by an increase in the rate of formation of conjugates. The existence of this type of control emphasises the difficulty of interpreting results from experiments with externally applied hormones. Not only are there uncertainties concerning the possible uptake of materials in such situations, but it is certain that the relationship between the amount of an applied growth substance in the medium and the amount within any given cell will not be simple.

Whilst it is generally accepted that IAA is the natural auxin in most plant tissues, there is evidence to suggest that other types of compound may serve the same function in certain cases. In *Coleus*, strong auxin activity has been found associated with non-indolic substances which behave chromatographically as steroids. In citrus fruits, an unknown substance with auxin activity has been isolated and termed the citrus auxin. This compound seems to be limited to the reproductive tissues, since a single auxin identified as IAA was found in elongating shoots, bark and branch wood of the same species. It has also been found that in some plants the amino acid phenylalanine can be converted to phenylacetic acid, and this material also shows some auxin activity.

Gibberellins

The next group of hormones or growth substances to be recognised was the gibberellins. The 'bakanae' or 'foolish seedling' disease of rice has as one of its symptoms an abnormal degree of elongation of the young plant stem, which eventually causes it, foolishly, to fall over. This disease is produced by a fungus (*Gibberella fujikuroi*) and, in 1926, Japanese workers were able to reproduce the elongation caused during infection by the application of a sterilised filtrate of medium in which the fungus had been growing. This suggested that a soluble component of the medium was the direct cause of elongation, and that the fungus was not required to be present. More than 10 years later, this substance was crystallised and given the name gibberellin. It was not until 1954 that the structure was finally elucidated; 'gibberellin' was in fact a mixture of compounds, and the first one to be analysed fully was called gibberellic acid (GA). It is now known that the gibberellins are a large family of related compounds, and the gibberellic acid of the 1950s is now called GA_3 (Fig. 5.3).

The chemical complexity of the gibberellins is immense, and its study is beyond the scope of the present discussion. Over 50 different related compounds have been structurally identified. These are named conventionally GA_1, GA_2, and so on. More than 40 GAs have been structurally identified from higher plants; the remainder occur in *Gibberella fujikuroi*. The elucidation of the molecular structure of a gibberellin requires highly sophisticated techniques, including mass spectrometry and nuclear magnetic resonance spectroscopy (techniques which are beyond the reach of most plant physiologists). For this reason the number of species in which structural elucidation has been carried out is small. However, it is clear that gibberellin-like activity in specific bioassays is a widespread property of plant extracts and, for this reason, there is no doubt that the occurrence of these compounds is similarly general. Vegetative parts of plants tend to contain lower amounts of GAs than reproductive tissues; for example, typical levels might be of the order of a few micrograms GA per kilogram of fresh weight for leaves, but 10–100 mg GA per kilogram fresh weight of immature seeds. The most striking effects of GAs on plant growth is the elongation of internodes; it is found that genetically dwarf plants can be induced to assume normal dimensions by the application of GAs. This is the basis of many of the bioassays used for the detection of GA activity.

Fig. 5.3. Gibberellic acid (GA_3).

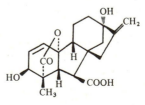

The biochemistry of gibberellins

The biosynthesis of gibberellins proceeds from mevalonic acid. It has been studied in fungi and in cell-free extracts of plants. Attempts to incorporate added precursors into gibberellins in whole plants have so far not been successful. The synthesis proceeds on a complex pathway via kaurene, and it has been shown that the enzymes on the first part of the pathway from mevalonic acid to kaurene are soluble. Those from kaurene to gibberellin are particle-bound, and possibly associated with the endoplasmic reticulum or plasma membrane. Most of the cell-free studies have been carried out with extracts from seeds, but there is evidence that gibberellin synthesis also occurs in root and shoot tips. It has been shown that enzymes in both parts of the pathway are sensitive to inhibition by growth retardants used commercially to restrict the size of crop plants and ornamentals. For example Amo-1618 inhibits one of the soluble enzymes, whereas the oxidation of kaurene is inhibited by the substituted pyrimidine ancymidol.

The 50 or so designated GAs are structurally different. In view of the difficulty in elucidating the molecular structure of such compounds, it is as yet unclear what relationships (if any) exist between different species of plants and the GAs which they can produce. It seems likely however that some GAs are not present in particular species because of a lack of the relevant enzymes. The large number of different GAs also raises the interesting possibility of some specificity in their action. This is not thought likely to occur with auxins, since the number of naturally produced variants in structure is much smaller. There is good reason to believe that the differences which appear between normal and genetically dwarfed plants are related to GA metabolism. Extracts of dwarf plants are less active in producing GAs from added precursors. However, when extracts are made of dark- and light-grown plants (which differ in their rates of stem elongation) no correlation is found between elongation rates and the ability to synthesise GAs.

Clearly, in a situation where many subtle variations exist in molecular structure of a complex chemical such as a gibberellin, it is very difficult indeed to study the control of levels of active forms within cells. It is not surprising that progress in the field of GA metabolism has been slow. Nonetheless, a few reactions have been identified which lead to inactive forms of GA and so could in principle represent control mechanisms. 2β-hydroxylation of GAs is one such reaction, and it has been shown to occur in peas. Another is the formation of glucoside conjugates of GAs. In maturing seeds, the level of free GA falls and at the same time the level of glucoside conjugates rises. This has lead to the speculation that conjugation may represent a way of inactivating GA in the developing seed, and holding it ready in store for hydrolysis to the active form when the seed germinates.

Cytokinins

The culture of plant cells proceeded from an initial false start to the culture of whole organs such as excised roots. This can now be rationalised by the knowledge that cultured plant organs are much less demanding in terms of their nutritional requirements than are isolated cells. Indeed, excised organs often require no added growth factors at all for their continued growth. The culture of undifferentiated callus tissues was not achieved until media containing auxin were utilised. Even then, it was found that certain types of tissue grew very poorly; it was during experiments with one of these – tobacco (*Nicotiana tabacum*) pith – that the cytokinins were discovered.

Whole tobacco pith, when cultured on a medium which contains auxin, shows good growth and proliferation of cells. However, if dissected pieces of pith which do not contain vascular tissue are cultured on the same medium, cell enlargement takes place, but no cell division. It was found that extracts of vascular tissue, or coconut milk, or malt extract could stimulate the dissected pith into cell division. This suggested that a soluble substance was present in these extracts which promoted cell division. The base adenine was found to be effective in this way, and this lead to the discovery that degraded DNA produced an even more marked stimulation of cell proliferation. The active substance was identified as 6-furfurylaminopurine, and given the name kinetin (Fig. 5.4). Substituted purines of this type are much simpler chemically than either auxins or gibberellins, and it is perhaps a measure of this simplicity that it was only 3 days after the discovery of kinetin that the first synthetic analogue, 6-benzylaminopurine was found to be effective as a cytokinin. The first natural cytokinin to be isolated and fully characterised was 6-(4-hydroxy-3-methylbut-trans-2-enylaminopurine), known as zeatin, (Fig. 5.4). 1 mg of this substance was obtained in crystalline form from 60 kg of *Zea mays* kernels. The great majority of known naturally occurring cytokinins which have been structurally analysed are fairly closely related to zeatin. However, cytokinin activity is widespread amongst substituted purines.

The biochemistry of cytokinins

The biosynthesis of cytokinins is likely to proceed from adenine or adenosine monophosphate. Study of the biosynthesis is extremely difficult because of the central importance of these probable precursors in cell metabolism, and particularly in RNA metabolism. This means that in experiments where exogenously applied adenine is used, it will inevitably be diverted into many pathways other than cytokinin synthesis. The chemical identification of likely intermediates is made more difficult by the very low levels of cytokinins found within cells. The site of synthesis of cytokinins in the plant seems to be the roots. This has been shown by experiments with rooted cuttings. The appearance of roots at the base of a cutting is correlated with appearance of cytokinins at the apex. If roots do not develop, the levels of cytokinins at the apex remain low. In this situation, it is clear that the cytokinin must be transported from the roots to the tip of the shoot. In radish (*Raphanus sativus*) seedlings it is the riboside of zeatin which is the form in which the cytokinin is transported. In *Phaseolus vulgaris*, the leaves accumulate dihydro-zeatin*O*-β-D-glucoside. This does not seem to be the transported form, however, since it is not found in roots or root exudates.

The appearance of conjugates with sugars is one complication of cytokinin metabolism; another is the presence of cytokinins in RNA. When transfer RNAs from plants, micro-organisms and animals are hydrolysed, it is found that the hydrolysates contain cytokinin activity. This does not happen when purified ribosomal RNA is hydrolysed. These findings are of considerable interest for two reasons. First, it is possible that tRNAs could contribute cytokinins to the cell by their hydrolysis. The extent of the importance of this potential source of cytokinins is as yet unclear. Second, it is possible that cytokinins within a cell could be incorporated into tRNAs and hence modify the properties of those RNAs. This would be a very attractive explanation of the hormonal action of cytokinins. It is known that modifications to the bases in tRNA at the positions at which cytokinins are found can affect protein synthesis, but as yet there is no clear evidence of a relationship between the level of free cytokinin within a cell and the amount of its incorporation into tRNA. These uncertainties reflect a general difficulty in the study of plant hormones; their levels in cells are often extremely low and, in any case, they may exist in a multiplicity of chemical forms which are interconvertible and have differing biological activities.

Ethylene

The knowledge that certain types of air pollution could modify plant development has been available for a very long time to commercial growers. Pineapple growers, for example, traditionally lit bonfires near their plantations in order to stimulate and synchronise flower development. Flower growers observed that traces of illuminating gas within their glasshouses could cause premature fading and closure of blossoms, and it was also noticed in the last century that citrus fruit transported in railway waggons heated by paraffin stoves ripened more rapidly than if steam-heated waggons were used. All these effects are now known to be due to the presence of small amounts of the gas ethylene. Its concentration in the air is very low – of the order of 10–50 parts per billion (ppb) in city areas – but, in smoke, it may be much higher. Exhaust fumes contain up to 400 parts

Fig. 5.4. *a*, Kinetin (6-furfurylaminopurine); *b*, zeatin (6-(4-hydroxy-3-methylbut-2-enyl) aminopurine).

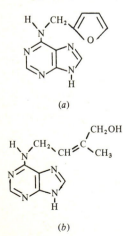

(a)

(b)

per million (ppm) of ethylene, and coal gas contains 10 ppm of ethylene.

The difficulties of using the word hormone when considering plants have already been discussed. Sites of synthesis may be known in general terms (as is the case with the formation of cytokinins by roots) and, similarly, sites of action may be identifiable in terms of whole organs. However, no recognisable subpopulation of cells can be confidently described as being specialised in the production of hormones or in their interception. These problems are acute enough in the case of soluble molecules moving about in the liquid phase. At least in principle in such cases, they are retained within the plant and therefore are constantly subject to modifications which may alter their effectiveness. In the case of a gas, no such restrictions apply. This is well illustrated in the case of ripening fruits, where it is a common observation that the onset of ripening of one fruit within an enclosed space containing others will induce those others to ripen more rapidly. In the intact plant it appears most likely that ethylene exerts its action close to its site of production. Paradoxically, one of the few examples of identifiable cell populations responsive to a change in hormone level are those in the abscission zone (p. 126), which respond to ethylene.

The biochemistry of ethylene

For some time it was doubted that ethylene had any role to play in the normal life of the plant. It is now accepted that plants can synthesise ethylene and, furthermore, that this is the event which explains at least some of the actions of high concentrations of auxin. The biosynthesis of ethylene uses methionine as its starting material. The pathway proceeds through S-adenosylmethionine and 1-aminocyclopropane-1-carboxylic acid (ACC). ACC is efficiently converted to ethylene when supplied to apple tissues. Labelled ACC is not diluted in its conversion to ethylene by the presence of applied unlabelled methionine, but labelled methionine does not give rise to labelled ethylene in the presence of unlabelled ACC. The inhibitor aminoethoxyvinylglycine is active in the early part of the pathway, and so inhibits the formation of ethylene from methionine, but not from ACC. These observations have lead to the formulation of the scheme illustrated in Fig. 5.5. The effect of IAA in greatly stimulating the production of ethylene after a lag period is thought to be due to the activation

of the synthesis of the enzyme which produces ACC. In vegetative tissues this step appears to be rate limiting. By contrast, in the ripening of fruit, where there is a massive rise in the production of ethylene, it is likely that the conversion of ACC to ethylene is the rate-limiting step. In the case of fruits with internal cavities, the concentration of ethylene within the enclosed space may become quite appreciable. In apples (*Malus*) it may reach 2.5 ml l^{-1}.

By analogy with other growth substances, it might be thought that metabolism of ethylene could play a part in the limitation of its concentration within cells. It is known, for example, that externally applied ethylene can be converted by pea epicotyl tissue into ethylene glycol via the formation of a glucoside. This is however thought not to be of physiological importance, since it is a very inefficient process. Furthermore, inhibition of this reaction does not lead to an increase in the concentration of ethylene in the

Fig. 5.5. Suggested pathway for the synthesis of ethylene from methionine via 1-aminocyclopropane-1-carboxylic acid (ACC). AD, adenosine.

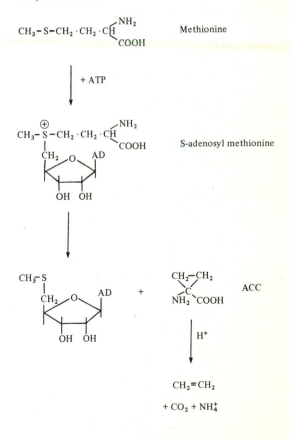

tissue. It is possible that the oxidation reaction which has been detected in fact represents the mode of action of ethylene at a chemical level. At present these aspects of the biochemistry of ethylene remain mysterious.

Abscisic acid

Auxins, gibberellins and cytokinins were all first identified on the basis of their growth-stimulating properties. Abscisic acid, on the other hand, was identified as an acidic substance in extracts of young cotton fruits which could accelerate the process of abscission in a bioassay; originally this substance was termed abscisin II. Another material, this time named 'dormin', was found to be an acidic constituent of leaves which was able to induce dormancy in buds. Chemical elucidation of these substances showed that they were in fact identical, and they are now known by the name abscisic acid (ABA; Fig. 5.6). It should be mentioned that, although the discovery of this substance was made using bioassays involving abscission or dormancy, there is no very strong evidence to suggest that abscisic acid has a physiological role in either of these processes in intact plants. The best demonstration of a function for ABA *in vivo* is its effect in causing stomatal closure during water stress.

All naturally isolated examples of ABA consist entirely of the (+)-enantiomer and, apart from metabolites and conjugates, it appears that (+)-ABA is the only member of its class of growth substances. When the requirement for the correct optical isomer is examined, it is usually found that both the (+) and (−) forms of ABA are potent inhibitors in bioassays, although sometimes the (+) isomer is noticeably more active. In the case of the closure of stomata by ABA, it has been found that the naturally occurring form is the only one which is active (at least in barley leaves).

The biochemistry of abscisic acid

ABA is synthesised in the plant using mevalonic acid as the starting point, and the pathway proceeds, in all probability, through the same initial stages as the synthesis of gibberellins. There is considerable evidence to suggest that ABA is synthesised within the plastids. Several observations point to this conclusion. Externally applied mevalonic acid is poorly incorporated into ABA. This is because mevalonic acid penetrates poorly into plastids. Isolated preparations of lysed chloroplasts do incorporate mevalonic acid into ABA, however. In normally growing spinach, furthermore, it is found that most of the ABA present in whole leaves can be accounted for in terms of that which is present within chloroplasts. Chloroplasts are not the only type of plastid which can synthesise ABA; in avocado fruits, the colourless tissue yields plastids which are just as capable of incorporating mevalonic acid into ABA as are the chloroplasts from green tissues.

The synthesis of the ABA within the plastid seems to be controlled by a feedback mechanism. During water stress, ABA leaks out of the chloroplasts and acts to effect the closure of the stomata. The lowering of the concentration of the ABA within the chloroplast results in the restarting of its synthesis. This produces an impressive increase in the levels of ABA within leaves during wilting. A turgid leaf contains typically 10–20 mg kg^{-1} of ABA. On wilting, this concentration may be increased by a factor of 40. Under the same conditions, the concentration of ABA in stems only rises by a factor of 2, suggesting that the leaves are the most important organs in the economy of ABA within the plant.

ABA can be metabolised within cells to a variety of other compounds including a glucose ester and phaseic acid. It is not clear whether these pathways represent a physiological mechanism for the control of hormone levels, though this would seem likely in view of the decreased activity of the derivatives in bioassays. The levels of the glucose ester vary during water stress experiments, but it is not certain that a fall in the level of the conjugate necessarily means that it has been reconverted into an active form. Studies with birch trees suggest that the formation of a conjugate may be irreversible. At the onset of dormancy in this situation a disaccharide ester of ABA is formed, but it is thereafter stored and not remobilised during subsequent bud break.

Hormones and the differentiation of isolated cells

The five groups of plant growth substances which have been briefly introduced have profound

Fig. 5.6. (+)-2-*cis*-Abscisic acid (ABA).

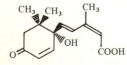

effects on the growth of intact plants. In a small number of cases, there is good evidence to support the view that they may act physiologically within intact plants to mediate certain growth processes. For the time being, however, we shall ignore both of these types of actions of the growth substances, in order to return to the idea stated at the beginning of the present chapter. Is it possible with isolated cells to induce differentiation using environmental factors, of which plant growth substances are likely to be the most potent? Do any general rules emerge from the study of isolated cells which might be applicable to intact plants and normal physiology? The answer to both these questions is a qualified 'yes' – qualified, as will be seen, by the enormous difficulty in achieving control over the experimental systems which are used (and therefore in the precise interpretation of results). The experimental control of cell development *in vitro* remains an area in which the rational basis for the design of experiments is very limited, and progress is as likely to follow from chance observations as it is from existing knowledge. In order to appreciate those aspects of the work which are now to be considered, it is first necessary to understand the basic methods of tissue culture with plant materials.

Tissue culture

The starting point of a tissue culture of a higher plant is a sterile piece of tissue, whether this be a stem segment or a piece of storage organ (such as a potato tuber). Such a piece of tissue is called the explant. The explant is placed on to a sterile culture medium which contains salts, usually a carbon source such as sucrose, and a variety of other organic materials such as vitamins and growth substances. Auxin may be used at this stage as the only growth substance, or a combination of an auxin and a cytokinin may be required.

The excision of the explant inevitably causes wounding in the surface layers. Plants respond to wounding by the stimulation of cell division and the production of a protective layer of wound tissue. Sometimes this is accompanied by callus formation. It is a fairly common observation in horticulture, for example, that cuttings may produce a formless callus tissue before any roots are formed. Usually the extent of this callus formation is limited. However, when an explant is placed on a medium containing growth substances, cell enlargement and division are both stimulated and extensive callus production results (Fig. 5.7).

Fig. 5.7. Diagram to illustrate the steps in producing a callus culture. *a*, The explant (here a piece of stem) is cultured on solidified medium which contains growth substances. Cell proliferation occurs at the tip of the explant, and this callus growth can be removed and maintained on a similar medium. *b*, To produce a liquid suspension culture, pieces of solid callus are agitated in liquid medium. This causes single cells and small cell clumps to slough off into the liquid.

(*a*)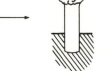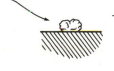

(*b*)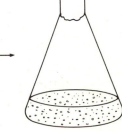

After a certain time, usually a period of a few weeks, this piece of callus tissue can be cut away from the original explant and placed on to fresh medium. Provided that the constitution of the medium is suitable, the callus will then continue to grow as a solid mass of formless tissue (Fig. 5.8). This can be repeatedly divided up and the pieces subcultured. In the growth of solid callus tissues in this way it is customary to use media which have been solidified with agar, both for the initial growth from the explant and for the continued culture of the callus. Liquid suspension cultures can be obtained from solid callus tissues by placing a piece of the tissue into liquid medium without agar and agitating it continuously. Cells at the edge of the solid tissue piece will be mechanically sloughed off by this treatment and, provided once again that the medium is a suitable one, they will continue to grow and divide, producing small clumps of cells which remain suspended in the liquid

(Fig. 5.9). Subculture of liquid suspensions of this type is carried out by decantation or dilution. Sometimes mechanical separation of different sized clumps is also practised at the time of subculture, either by filtering the suspension through a fine mesh, or by allowing the larger clumps to settle and by subculturing only the smaller clumps which remain in suspension.

This description of the induction of callus growth from an explant is generally applicable to a wide range of dicotyledonous species. Monocotyledons are more difficult to induce into growth as undifferentiated callus. They may require very high auxin levels and are largely irresponsive to cytokinins. A few instances are also known of callus induction by GA in conjunction with auxin or cytokinin, and cotton ovules produce a callus in liquid medium when exposed to ethylene.

In developmental terms, the most obvious point of interest concerning the induction of an explant is that the cells within it do not retain their existing state of differentiation. If an explant is made of the secondary phloem of a carrot, it does not grow by producing secondary phloem cells. Instead it undergoes a process of 'dedifferentiation' to a more generalised state.

Fig. 5.8. Tissue culture of *Freesia*. The solid callus on the left of the picture can be induced to form shoots as on the right by a change in the ratio of cytokinin to auxin in the medium.

This is true of all explants which produce callus, and it is the reason why the origin of a particular callus cannot be determined by its appearance. In fact, even after 'dedifferentiation' has occurred, certain characteristics of the original explant may be retained. For example, callus tissues from the juvenile and adult phases of growth of the ivy (*Hedera* sp.) show consistent differences in growth rate over many generations; bud formation by callus tissues of certain trees can only be achieved when young material is used as the explant. Callus cultures are frequently referred to as being undifferentiated. This is an unfortunate term insofar as it implies that all the cells within a callus are in a similar state. This may not be the case (see the following section). The process of dedifferentiation involves synthesis of cytoplasmic components and the stimulation of cell division. It is a process which has its counterpart in normal physiology, when surface wounding produces the response of limited cell proliferation and the formation of a wound periderm.

Structure of tissue culture cells

The most striking structural characteristic of tissue culture cells is their uniformity of appearance, regardless of source. They are usually similar in appearance to colourless parenchyma cells in the intact plant, except insofar as they may show a variety of shapes, from small and nearly spherical in rapidly dividing populations, to large and elongated in less actively growing conditions. The cells are almost always highly vacuolated and possess thin cell walls. The nucleus may be located in the thin layer of peripheral cytoplasm or it may appear suspended in a central position connected to the peripheral cytoplasm by strands which cross the vacuole (Fig. 5.10). Cytoplasmic streaming is often easily visible along these strands and in the peripheral cytoplasm. In terms of organelle complement, the cells usually contain all the normal structures found within a meristematic cell, with the exception that plastids

Fig. 5.9. A liquid suspension culture of cells of the soya bean (*Glycine max*). The culture consists of a mixture of single cells and clumps of various sizes. Bar = 100 μm.

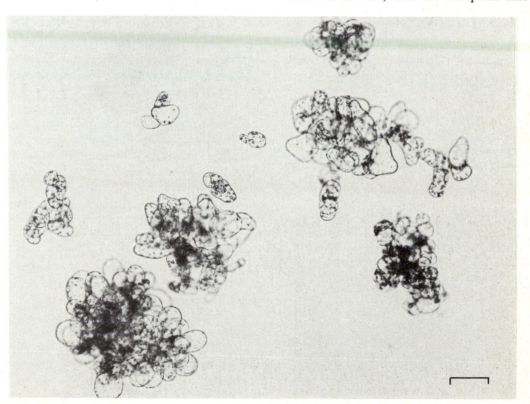

often contain large quantities of starch. A clear difference between a callus tissue and normal organised cellular tissue within the plant body is that the callus cells show very few intercellular connections in the form of plasmodesmata. This can be shown directly by counting plasmodesmata in the electron microscope. It may also be demonstrated physiologically; the electrical resistance between neighbouring cells in a callus tissue is often very high, and the diffusion of dyes which are known to pass through plasmodesmata is very slow. This lack of communication between cells is no doubt of considerable significance in permitting disorganised cell proliferation. There is usually no sign of differentiation in a culture; that is to say, all the cells look much the same, and they may be compared to a vacuolate meristematic cell such as is found in the cambium (p. 123).

However, this appearance is probably an oversimplification, based partly on only a limited number of observations of cultured cells, and partly on wishful thinking. It must be remembered that if a complex tissue was used as the original explant, then the callus which results from it may have arisen from different populations of cells. Although a tissue such as a root contains many cells which would not normally divide again in the intact plant, nonetheless, the act of excision together with the exposure to growth substances in the medium may induce such cells to

multiply and so contribute to the callus. This has been clearly shown in the case of pea roots. When the explant of the root is cultured on a medium containing the synthetic auxin 2,4-D together with yeast extract, the callus which is produced contains cells which are diploid, tetraploid and octaploid. If the yeast extract is omitted from the medium, it is found that all the cells of the proliferating callus are diploid. This reflects a difference in the cells of the root in their sensitivity to the stimulus of growth regulating substances, since the yeast extract would certainly contain cytokinin activity. As a general rule, the more complex the inducing medium, the more likely it is that the resulting callus will be derived from more than one cell type in the explant. The nature of the explant also has some effect on the constitution of the callus. Cultures derived from storage parenchyma, mesophyll or vascular cambium are inherently more likely to be uniform than those which arise from say a stem segment or a root (where very many different cell types are found).

Leaving aside these differences due to the induction of the initial callus from the explant, there are further difficulties to be encountered in terms of the uniformity of cells in culture. At the time of subculture, it is possible that unconscious selection of cells may take place. For example, as has already been mentioned, a liquid suspension culture commonly exists as a heterogenous mixture of cell clumps of different sizes. Decantation or filtering such a suspension at subculture inevitably tends to select those cells which form the smallest clumps. In a solid callus, growth may occur preferentially at one position in the callus – for example, next to the surface of the medium. If this region is selectively subcultured, this too can lead to a slow drift in the nature of the cells being preserved. Long-term drifts in the behaviour of cultured cells are commonly observed, and this type of selection pressure can be one basis for that behaviour. Another possibility is that the use of high levels of growth substances (particularly synthetic auxins such as 2,4-D), may cause chromosome damage.

These preliminary comments are designed to emphasise one fact concerning plant tissue culture. Although its methods and procedures resemble those adopted by microbiologists, there is little reason to assume that a suspension of plant cells is as uniform as say a culture of a yeast. This limitation must be borne in mind during the following sections.

Fig. 5.10. A cell within a liquid suspension culture from soya bean, showing the nucleus and its attendant cytoplasmic strands. The circular structure within the nucleus is the nucleolus. Bar = 20 μm.

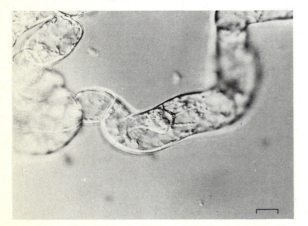

Vascular differentiation in callus cultures

The relatively uniform and undifferentiated cells which comprise a callus culture are capable of continuous subculture for many generations without apparent change. For the reasons described above, it is often found that after several generations of subculture, the growth characteristics of a particular culture may be modified. For example, rates of cell divisions may decrease, or they may increase, or the physical properties of the callus may alter. Generally speaking, however, there is no development of organised cell structure.

The first demonstration of a chemical induction of vascular differentiation was made using a callus isolated from *Syringa*. The callus was maintained on a medium containing a low concentration (0.05–0.1 mg l^{-1}) of the auxin 1-naphthoxyacetic acid (NAA). This callus grew with no visible differentiation, and it was found that movement of auxin through the callus was by simple diffusion in all directions. This emphasises the lack of organisation since, in an intact plant, the movement of auxin is highly polar (p. 199).

Two types of experiment were performed. In the first, blocks of the solid callus were cut into cubes and placed on the surface of a solidified agar medium containing IAA and sucrose. Leaf buds of *Syringa* were then fitted into a groove cut into the flat upper surface of the callus block. Within 10 days the bud had burst and had started to grow. The block of callus was then examined for signs of differentiation within it. It was found that nodules of actively dividing cells were positioned along the lower flanks of the graft and that, as time progressed, these acquired differentiated xylem elements. At a level of 0.5 mm below the graft, other groups of dividing cells appeared. In old cultures, after 54 days of growth in the presence of the graft, these nodules too showed differentiation into xylem cells. Other groups of xylem cells were found scattered throughout the callus block at this stage (Fig. 5.11).

In the second series of experiments, small agar blocks which contained various media were used in place of the grafted buds of *Syringa*. It was found that, when the agar blocks contained auxin and sucrose, rings of vascular nodules were differentiated in the callus tissue at a distance of 1.5–1.75 mm below the position of the agar block. In the initial experiments with grafted buds, only xylem differentiation was reported; however, when agar blocks containing 0.1 mg l^{-1} NAA and 3% sucrose were used in the second experimental series, it was found that both xylem and phloem differentiation took place. Furthermore, within the ring of differentiating cells, xylem formed towards the centre of the ring with phloem on the

Fig. 5.11. The induction of vascular differentiation in callus blocks. *a*, A leaf bud of *Syringa* is fitted into a block of callus tissue and induces the formation of xylem cells within the callus block; *b*, a block of agar which contains auxin and sucrose is placed on the surface of a callus block and induces the formation of a ring of nodules which contains both xylem and phloem cells, below the position of the agar.

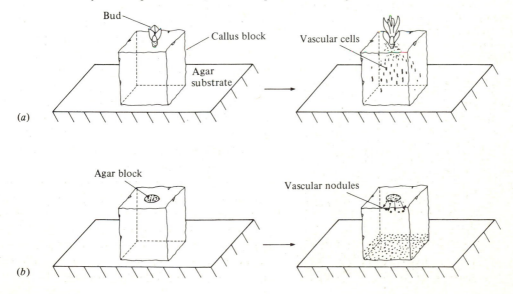

outside. The two regions of differentiating cells were separated by a region of actively dividing cells which resembled a cambium. Rather more haphazard differentiation also occurred in the lower part of the callus block where it was in contact with the medium. This type of more random differentiation also occurred when callus blocks were placed on media containing 0.5 mg l^{-1} NAA and 3% sucrose, but without the agar block on the top surface.

·The importance of concentrations

These experiments suggested strongly that the existence of a gradient of concentration of auxin and sucrose in an apparently homogenous block of tissue was sufficient to induce differentiation in the cells. More remarkable than that, the patterns of nodulation and differentiation were non-random. At a fixed concentration of sucrose, an increase in the concentration of auxin in the agar block at the top of the callus caused an increase in the diameter of the ring of nodules. At a fixed concentration of auxin, increasing the concentration of sucrose caused an increase in the amount of phloem tissue which was formed. In a further development of the second type of experiment, a sterile pipette was inserted into the top of the callus block and filled with a solution which contained auxin and sucrose. Under these conditions, complete rings of vascular tissue could form in the callus around the tip of the pipette. The reasonable conclusion was drawn from these observations that induction of nodulation is a result of the nutritional status of cells, and not of their particular position within the block or of their history. It appeared that any cell which was so positioned in the gradient of nutrients as to receive the stimulation to differentiate, would proceed to do so.

These experiments were later repeated with callus derived from *Phaseolus vulgaris*, and it was shown that differentiation could be expressed in terms of chemical parameters. Thus in blocks which had been induced to differentiate, the xylose/arabinose ratio found in the walls when the whole block was hydrolysed was higher than in control callus. Similarly, the amount of lignin in the differentiated blocks was higher than in the controls. This emphasised that differentiation is accompanied by secondary wall formation and lignification, as we have already seen (p. 101). It was also found from this study that an equimolar ratio of fructose and glucose could

not replace the requirement for sucrose in the formation of organised nodules and phloem tissue. Glucose alone at twice the concentration (2%) produced only xylem differentiation. The importance of the sequence of applications of the inducing substances was also investigated. When IAA alone was used in the treatment, only xylem differentiation occurred. If a 1 week treatment with IAA were followed by a 2 week treatment with sucrose, organised nodules developed, but if a 1 week treatment with sucrose preceded a 2 week treatment with IAA, only xylem differentiation resulted. These findings were interpreted as showing a requirement for cell division as a prerequisite for differentiation. It was reasoned that an initial treatment with IAA would stimulate cell division, and that subsequent exposure to sucrose would, together with residual IAA, allow differentiation to proceed.

These experiments produced three conclusions of possible relevance to development. The first is that the observed differentiation appears to be triggered by particular concentrations of IAA and perhaps sucrose; nodules appear at non-random positions within a diffusion gradient. The second conclusion is that not only the occurrence of differentiation, but also its patterns, may be controlled by chemical gradients. This is shown by the effect of concentration on the radius of the circle of nodules and also by the fact that, just as in an intact stem, xylem is formed towards the centre with phloem on the outside. The third conclusion is that cell division may be a general prerequisite for a change in developmental status.

Differentiation in isolated cells

The first of these conclusions leads to the expectation that even isolated cells or cell clumps growing in liquid culture could be induced to differentiate simply by presenting them with an appropriate mixture of growth substances and perhaps sucrose. In fact, this is not so. There are many instances of the formation of xylem elements in particular within liquid cultures, but their occurrence is always at a very low percentage of the whole cell population. Tracheary element formation may be stimulated by auxin, or cytokinin or gibberellin, depending on the species of the cells in the culture. This immediately points to a possible reason for the failure of wholesale differentiation in such cultures. It is this: whilst the concentration of the applied growth factors may be regulated within close limits, there is no control at all over the fate of these factors once they enter cells (nor over the amount of other growth substances which may be produced by the cells). In other words, two cultures which are subjected to apparently identical experimental treatments may in fact have quite different concentrations of growth factors within the cells themselves. Furthermore, when it is recalled that even within a single liquid suspension culture there may not be very great uniformity between individual cells, the reason for the failure of such experiments becomes clear. In the context of this failure, the induction of organised vascular differentiation within solid callus blocks is more remarkable, and points perhaps to a greater importance for the role of gradients of growth substances rather than for their absolute concentrations.

The failure of isolated tissue culture cells to differentiate may well reflect a general difficulty of tissue culture itself, rather than of the premise that cells may be triggered by specific concentrations of growth substances. Evidence for this view comes from a study of isolated mesophyll cells of the plant *Zinnia elegans*. If young leaves of *Zinnia* are ground in a mortar, instead of the usual debris of broken cells and chloroplasts which is produced by such rough treat-ment, it is found that the tissue breaks down into intact cells, together with fragments of epidermis and existing vascular tissue. The cells may be freed from these fragments by filtration and, if subsequently cultured in a medium which contains auxin and cytokinin, up to 50% of the cells differentiate into xylem elements within a few days (Fig. 5.12). This is a very high figure, and it is also found that 60% of the cells which differentiate do not undergo a preliminary cell division step.

This comparatively recent discovery holds considerable promise for the future in terms of the controls of development at the individual cell level. It is as yet quite uncertain whether the cells which differentiate in culture to produce xylem elements would have followed such a path in the intact plant; it is, however, very unlikely that this is the case. This implies that it is the culture conditions which are triggering development. If this work can be extended to other species, it may form a good example of the control of differentiation by the absolute concentrations of growth factors; clearly in the body of a liquid medium there is little possibility of gradient formation. Furthermore, this isolated cell system provides good evidence that cell division may not be required to redirect the course of differentiation.

Fig. 5.12. An isolated cell from a leaf of *Zinnia elegans* cultured for 5 days in the presence of 0.1 mg l^{-1} NAA and 1 mg l^{-1} benzylaminopurine. The cell has developed the thickened wall characteristic of a xylem element, and its contents have autolysed. Up to 50% of the cells behave in this way. Bar = 20 μm.

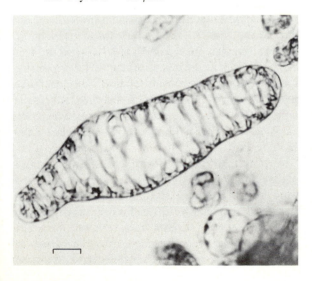

Differentiation following wounding

Evidence concerning the importance of gradients of hormones has been found by the study of vascular differentiation following wounding. When the stem of a plant of *Coleus* is cut in such a way as to sever a vascular strand, it is possible to observe the re-establishment of vascular connections during a subsequent recovery period. The cells of the pith differentiate to reform a continuous link past the region of the cut. Their differentiation proceeds from the upper surface downwards, and it has been shown to require the presence of leaves above the cut. This requirement can be satisfied by externally applied auxin. It is known that auxin is transported basipetally in this tissue (that is, in the same direction as that followed by differentiation after wounding). These observations therefore support the idea of a gradient of hormones being important in the control of differentiation.

Naturally, in a stem segment there remains considerable organisation even after a wound has been made; this, no doubt, accounts for the ability of such a tissue to reform an extended and functioning vascular system. This requires interaction between cells, and the controlled breakdown of specific regions of cell walls, as we have seen already (p. 103). This type of cooperative differentiation is not generally observed in callus tissues (or, at least, not until recognisable organ formation has occurred in response to experimental treatment). The problems of tissue and cell polarity in vascular differentiation are discussed in more detail in Chapter 7.

Hormones and organogenesis in culture

So far we have considered the formation of specialised cells by tissue cultures, in particular cells characteristic of xylem and phloem. This type of development may in some cases be relatively simply triggered by the presence of a growth substance at the appropriate concentration. It can be visualised as a single cell event, both experimentally in the case of *Zinnia* mesophyll, and conceptually in other cases. What, however, of the more complex process of organ formation? Here the result of the stimulus is not a change which affects only one cell comparatively directly, but one which gives rise to extensive organised development from an initial starting point. This type of behaviour seems to result from the interplay between different groups of growth substances.

The interaction of different growth substances was dramatically demonstrated in a series of experiments with tobacco stem pith tissue, which eventually led to the discovery of cytokinins. Initial experiments showed that stem segments of tobacco which were placed horizontally on the surface of a simple medium without growth factors produced a small amount of callus at the basal end, together with buds. The addition of adenine sulphate to the medium at rather high concentrations (40–80 mg l^{-1}) led to a marked increase in the number of buds which were formed. This increase could be antagonised by very small amounts of IAA (5–10 μg l^{-1}). It was found that bud formation was dependent not on the absolute concentrations of the growth factors, but on their relative amounts. The suppression of bud formation by increased IAA levels could be overcome by an increase in the level of the adenine. The inhibitor of auxin transport, triiodobenzoic acid (p. 202) was found to decrease the number of buds formed, but to allow their formation over the entire surface of the segment rather than just at the basal end.

It was during the course of these experiments that it was noticed that tobacco pith required the presence of vascular tissue for its successful division in the presence of auxin; this led to the discovery of kinetin as a highly active cytokinin (p. 133). Kinetin was found to be capable of stimulating bud formation not only with segments of stems, but with callus tissue prepared from stem explants. Furthermore, it was discovered that kinetin is active in this way at very low concentrations, and shows the same pattern of interaction with IAA as did adenine in the case of the stem segments. For example, bud formation was stimulated in callus tissue in the presence of 2 mg l^{-1} IAA by as little as 0.5–1 mg l^{-1} of kinetin. This meant that kinetin and IAA had only to be present in roughly equimolar proportions for bud formation to occur. Earlier calculations had suggested that one molecule of IAA was capable of antagonising the bud-promoting effect of 15 000 molecules of adenine.

These experiments pointed the way to a general technique which has been shown to be widely, but not universally, applicable. In order to induce bud formation by very many species of callus tissue, it is only necessary to transfer the callus from its maintenance medium (usually characterised by a high ratio of auxin to cytokinin) to a medium in which the ratio of cytokinin to auxin is greater than one (Fig. 5.8). Only exceptionally does the absolute concentration of the two growth factors have a determining effect on organ development. For example, in tomato (*Lycopersicon esculentum*) callus cultures, a medium containing 2 mg l^{-1} IAA and 2 mg l^{-1} kinetin induces root formation, whereas 4 mg l^{-1} IAA and 4 mg l^{-1} kinetin induces shoots. In such a case it is of course perfectly possible that the endogenous levels of hormones could be quite different to those which are applied, and so the general rule of the importance of the ratio of growth substances to organ formation may not be violated by this example.

It is unfortunately true, however, that this experimental formula does not always produce the same effect. Some of the exceptions to it are of themselves instructive. The callus produced by alfalfa (*Medicago sativa*) presents such a case. Here the requirement for organ formation is not the presence of growth factors within the medium, but rather their absence. Where the callus is maintained on a medium which contains both auxins and cytokinins, it does not develop organised structure. However, if it is subsequently transferred to a medium which lacks growth factors, organogenesis can occur. The nature of the organs formed is determined by the previous history of the callus. If it is transferred from a medium which contains a high concentration of auxin and a low concentration of cytokinin, shoot formation is the result. By contrast, roots are formed when the transfer is from a medium with high levels of cytokinin and low levels of auxin. This type of behaviour is explained by the proposition that organ formation takes place first by a process of further dedifferentiation to a meristematic primordium, which subsequently develops according to the nutritional environment. Thus, if a primordium can develop in the same conditions as were used to induce it, visible organ formation will be the result. However, if the inducing conditions are antagonistic to the further development of the primordium, nothing will appear to happen until the conditions are changed. The nature of the change itself could influence not only the fact that the primordium develops at all, but also the type of organ which it produces.

This idea of generalised primordia which can be induced to form different organs by chemical switching has further support from other studies than *Medicago* callus. Cultured roots of *Convolvulus* for example can be induced to form buds or roots from pre-existing primordia, simply by variations in the chemical stimulus given to them. In whole-plant physiology, similar effects are observed; for example, in grape vines the primordia which precede flower formation are produced in the autumn. In the spring they may develop into flowers, or alternatively, they may produce tendrils.

Other factors in organogenesis

Not every callus will respond to a switch in the auxin/cytokinin ratio by producing buds or roots. Many variations exist in responses to growth substances, even when they apparently belong to the same family. Hypocotyl segments of the aubergine *Solanum melongea*, for example, produce shoots in the presence of IAA, but 1-naphthyl acetic acid, a synthetic auxin, stimulates root formation. Naphthoxy-acetic acid promotes callus growth without organ formation at all, whereas 2,4-D has no effect even on growth. These variations of response to a single class of substance – in this case, auxins – serve to point to the great complexity of interpreting the behaviour of tissues or cells as a result of chemical treatments.

Organogenesis is not only induced by treatments with auxins and cytokinins. In potato cultures, abscisic acid induces shoot formation, whereas kinetin does not. Gibberellins are able to induce shoot formation in cultures of *Chrysanthemum*, but are antagonistic to shoot formation in *Nicotiana* (tobacco). In addition to hormonal influences, it may also be necessary to control the environment of the callus tissue in order to promote the formation of organs. Light, temperature, and even the inorganic constitution of the medium have all been shown to exert effects on the development of organised structure within tissue cultures. All these observations, and many others besides, serve to demonstrate the near impossibility of predicting the response of a particular callus or culture to a given set of experimental conditions. The situation has been summarised as follows by H. E. Street, a well known worker in the field of tissue culture: 'all cases of experimental morphogenesis in culture, despite their growing number, are in a sense special cases, cases in which we have succeeded without knowing why.'

This is nowhere more clearly illustrated than in the division between callus cultures from dicotyledons and those from monocotyledons. As we have seen, it is generally true that the ratio of auxin to cytokinin determines organogenesis in dicotyledons. Monocotyledons are by contrast more difficult to grow in culture at all and, usually, show very little response to cytokinins. In our present state of knowledge it is simply not possible to say whether this type of behavioural difference is due to intrinsic differences in the way in which morphogenesis proceeds, or whether it merely reflects a lack of understanding of the uptake and metabolism of applied growth substances.

Plant development following organogenesis

A callus tissue may produce shoots or roots, depending upon the precise environmental conditions. As we have seen, a high cytokinin/auxin ratio favours shoot formation, whereas rooting is often dependent upon a high auxin/cytokinin ratio. Rooting is, in fact, a more common pattern of behaviour than shoot formation and is often observed in freshly isolated callus tissues without any special treatment having been given (Fig. 5.13). This is thought to represent the retention of meristematic primordia during callus induction. The ability to produce roots spontaneously often declines after several cycles of growth and subculture. This has been explained in terms of the dilution of morphogenetic factors of unknown nature which were present in the explant, but may equally be due to drift in the cell population following selective subculture methods, or to loss of morphogenetic potential due to chromosome abnormalities.

The formation of roots versus shoots has an interesting consequence for the further development of organisation. Once a shoot has formed, it is usually a relatively easy matter to regain an intact plant. The process may be regarded as akin to making a cutting in horticulture. The presence of the shoot creates the axis of growth, and very probably creates the correct nutritional and hormonal conditions for subsequent root formation. Hence once a shoot has been initiated, it is often sufficient to remove it from the medium used to induce it and transplant it onto sterile maintenance medium, or even sterile compost. Rooting will then follow, and a normal plant is regained. By contrast, the formation of roots within a callus is very often its final state of development. A callus with roots will almost never form shoots and, unless shoots are formed, there is no prospect of plant production. These considerations are of course of particular interest for commercial uses of tissue culture. This is one of the reasons for the general failure of tissue culture with monocotyledons such as cereals.

The ability to form whole plants from callus tissues via the pathway of shoots and roots is a general confirmation of the important principle of totipotency in plant cells. This means that even the differentiated cells which contributed to the explant and the original callus retain all the genetic material needed to programme the formation of an entire plant. It implies that differentiation does not result in or arise from the gradual loss of genetic material, but rather from its sequential expression. This is a theme which will be discussed in more detail in Chapter 8.

Embryogenesis in culture

The development of organised regions within a callus mass depends on the manipulation of the environment, both in its hormonal aspects and, to some extent, in its physical aspects also. It is also true

Fig. 5.13. Root formation by two cultivars of *Dianthus deltoides*. These roots have formed in agitated liquid medium from cells produced by mechanical grinding of leaves. Medium contains coconut milk, NAA and kinetin.

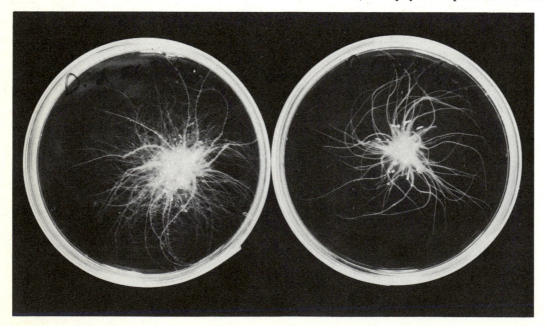

to say that however successful these manipulations are in producing the desired response, the target cells cannot be identified and constitute a very low proportion of the population as a whole. Thus the development of a shoot which may subsequently be rooted and give rise to a normal plant expresses the idea of totipotency of plant cells in a general way, but not rigorously. In the strictest terms, totipotency must mean that a cell has the ability to reproduce all the structures in a normal plant. In seed plants, this means that development should proceed not by a route involving shoot and root formation, but via the embryo.

The most successful tissue culture by far in the production of somatic embryos has been that of the carrot (*Daucus carota*). Explants of carrots consisting of secondary phloem were used as the starting material, and when these were agitated in a medium containing coconut milk, proliferation of the cells occurred; this led to the formation of a liquid suspension culture. These cultures comprised a mixture of units from single cells up to small cell clumps. It was found that if the entire population were subcultured, then growth could be maintained indefinitely. One of the most striking characteristics of the carrot culture is that within the population of cells and cell clumps appear units which resemble various stages in the normal embryonic development of the plant. Thus both globular and torpedo-shaped embryos can be found and, at earlier stages, aggregates of cells which appear to be dividing in the organised sequences which form part of the earliest development of the zygote (p. 155; Figs 5.14, 5.15).

Much of the work with embryogenesis emphasises that it is not simply under hormonal control. The need for a rapidly proliferating cell culture is paramount and, in this respect, a medium containing coconut milk has been repeatedly shown to be particularly effective. Embryo formation is stimulated by low levels of auxins and can be antagonised by concentrations as low as 1 μM 2,4-D in the

Fig. 5.14. Diagram to illustrate the cycle of culture of carrot cells and the recovery of plants via embryogenesis.

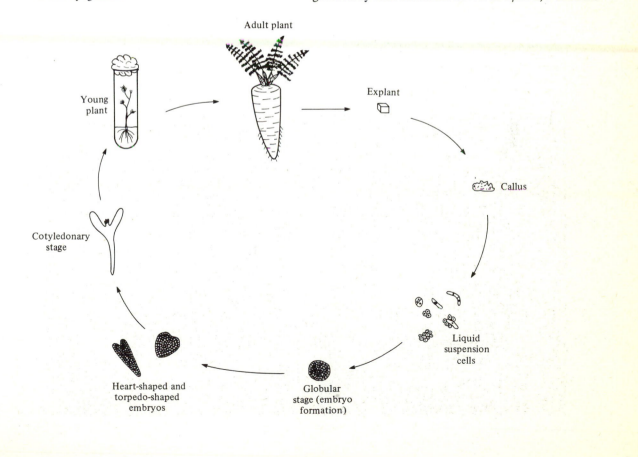

Adult plant

Young plant

Explant

Callus

Cotyledonary stage

Liquid suspension cells

Heart-shaped and torpedo-shaped embryos

Globular stage (embryo formation)

medium. Growth of embryos may be stimulated by darkness; further development may be arrested by increasing the osmotic pressure of the medium. Such 'poised' embryos can be stimulated into regrowth by a reversal of the increase in osmotic pressure, and under these conditions, growth is better if the poised embryos were kept in darkness. Embryogenesis has also been achieved from isolated protoplasts of carrot cultures, and an experiment carried out on a joint USA/USSR space mission showed that embryogenesis can occur under conditions of zero gravity.

The research with carrot cultures has shown that not only may a variety of environmental stimuli be effective in inducing development, but their sequence of application is also important. The explant must be made to proliferate quickly by the presence of a complex medium, typified by coconut milk. The cells and cell clumps are maintained on a medium with coconut milk and IAA. In the presence of an adequate nitrogen supply from the medium, the cells will continue to proliferate; if nitrogen is lacking, root formation occurs. Withdrawal of the auxin favours the development of embryos, and the further progress of these is dependent upon osmotic conditions and light/dark regimens. It has been pointed out that normal embryogenesis and indeed cell development in the intact plant occur against a similar background of changing environmental and nutritional stimuli.

Anther and pollen culture

In the context of embryogenesis, a further novel system is of particular interest. This is the behaviour of cultured anthers and pollen grains.

Angiosperm pollen is produced within the anther by a process involving meiosis to give groups of four microspores which initially each contain a single nucleus. This nucleus contains half the quantity of genetic material of the originating mother cells. In a diploid species, the microspores are haploid. A highly asymmetric division of the microspore nucleus (p. 120) gives rise to a generative cell which comes to lie free in the cytoplasm of the young pollen grain. This cytoplasm also contains the vegetative nucleus. The generative cell divides again to produce two sperms; this may occur within the pollen grain before it is shed, or it may occur after germination of the pollen tube. The normal behaviour of the pollen grain is to germinate to produce a long tube which delivers the sperms to the ovary for fertilisation to take place.

This normal pattern of behaviour can be modi-

Fig. 5.15. Scanning electron micrograph of a heart-shaped embryo which has formed in a liquid suspension culture of carrot cells. Bar = 0.5 μm. (Picture supplied by Dr Clive Lloyd.)

Fig. 5.16. Embryo formation by pollen grains: this picture shows a pollen grain from tobacco which has split open as a result of the division and growth of the vegetative cell. Bar = 10 μm.

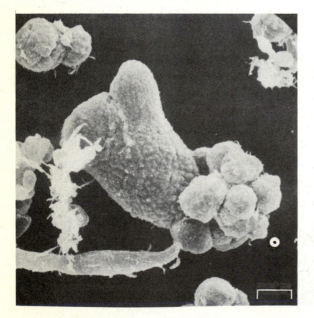

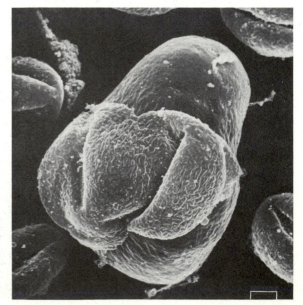

fied by excising the anthers at an appropriate stage and culturing them on suitable media. In some cases (such as *Datura* and *Nicotiana*) it is necessary to provide only a mixture of salts, sucrose and vitamins in order to induce the production of embryos directly from the immature pollen grains. This process occurs at a low frequency, and takes several weeks. Other plants require the presence of growth regulators (usually auxins) for the induction of embryos (*Brassica, Asparagus*). In some cases the shed pollen may be cultured outside the anther – exemplified by *Datura*. However, in many species the inducibility of the pollen only occurs during a short period of its development, and removal of immature pollen grains leads to their death. In such species the whole anther must be cultured.

The interest in this system lies in the fact that the response of the cells at the earliest stages of embryo induction can be observed. The most common behaviour is for the vegetative nucleus to divide repeatedly. In normal pollen development of course this nucleus does not divide, but exists to sustain the growth of the pollen tube. A series of divisions of the vegetative nucleus gives rise to as many as 20–30 cells within the internal space of the pollen grain. Finally the pollen grain wall ruptures and the proembryo is liberated (Fig. 5.16). The generative cell in such developing pollen grains either degenerates, or divides a few times before being lost as the embryo breaks out of the pollen grain. An embryo formed in this way will be haploid. In one species (*Hyoscyamus niger*) it has been shown that the generative cell itself can sometimes form an embryo whilst the vegetative nucleus remains inert. Finally, in *Datura*, fusion can occur between the generative and vegetative nuclei after one or more rounds of DNA replication. In this case the resultant embryo is not haploid.

The production of whole plants from the vegetative cell of the developing pollen grain is a good example of expressed totipotency in a highly specialised starting material. Of great interest is the fact that totipotency is shown in this system by many cereals, such as barley (*Hordeum vulgare*) – species which are difficult to culture by more traditional methods. In the case of barley the pollen within the anther produces a callus which can be regenerated into whole plants by the procedure of shoot induction followed by rooting (Fig. 5.17). Embryos are not formed directly by this species, a fact which is possibly related to the obligatory use of high concentrations of auxin. It is also interesting that in those species which do produce embryos (such as *Nicotiana tabacum*), it is clear that the initial divisions of the vegetative nucleus are not particularly well ordered, in spatial terms. This introduces the idea that organised structures do not inevitably have to be produced by following rigidly geometric patterns of cell division. This theme will be returned to in the following chapter (p. 155).

Fig. 5.17. Recovery of plants of barley (*Hordeum vulgare*) from solid callus tissue which has been produced from cultured anthers. Conventional callus cultures of monocotyledons are far less easy to regenerate back to plants.

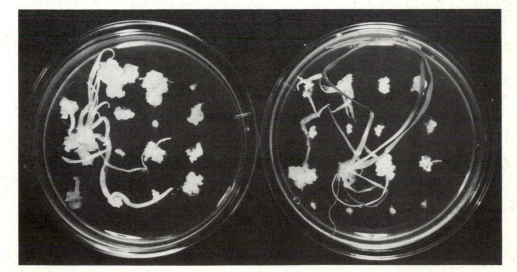

Totipotency

One of the most important general conclusions to be drawn from the work with embryogenesis in culture is that of the totipotency of plant cells. This is a concept which is basic to any further discussion of possible physiological controls of development (Chapter 8). It is clear that embryogenesis from callus cultures is fairly uncommon and more an intrinsic property of particular species than something which can be imposed from outside. Its existence is however an important proof of the suspicion arising from other work (notably that with organ formation) that plant cells retain their full developmental ability even when they are themselves in a differentiated state within the plant. The carrot cultures which have been described were derived from secondary phloem of the root (clearly, a highly differentiated structure by any definition). Even in the case of plant development from callus via the pathway of shoot formation, totipotency is finally expressed, since such plants behave quite normally and can undergo complete cycles of sexual reproduction in subsequent generations. This must imply that the starting material (from whatever source) contained cells which were in full possession of the total genetic requirement for complete development. The experiments with carrot embryogenesis represent a refinement on this scheme insofar as the proportion of cells initially undergoing normal development can be extremely high whereas, in the case of bud formation from a callus, it is usually extremely low.

The concept of totipotency of cells, however, carries implications other than that each cell carries a full genetic complement. The problem can be expressed simply in the following terms. A callus culture may be in a general way totipotent; that is, it may be able to be regenerated back to a normal plant, given certain manipulations of the medium and the environment. However, taken individually, the cells which comprise that callus in fact can only directly contribute to a very limited series of events. They may, for example, divide to give rise to a meristematic nodule. This may subsequently give rise to vascular differentiation or it may form a primordium capable of giving rise to a shoot or a root. The original cells may differentiate directly into a mature cell type characteristic of the vascular system – this is most commonly a xylem element. Rarely, a cell may produce an embryo.

These types of behaviour represent a very limited series of responses at the cell level. In this respect the cells of the callus tissue may be said to resemble closely the cells from the intact plant; whilst retaining the full potential in genetic terms to generate all the structures found throughout the entire life cycle, each individual cell can in fact only carry out a certain restricted number of these functions.

Morphogenesis in thin cell layers

The possibility of direct programming of particular cell types to new forms of developmental pathways has been investigated using thin cell layers. Under these circumstances, the explant is a small piece of tissue removed from the plant and consisting of a small number of identifiable cell types. For example, an explant from a surface position would contain only epidermal and subepidermal cells. The explant is cultured on a medium which contains salts, vitamins, sugars and hormones in the same general way as is used to obtain proliferating callus (p. 137). However, it is found that variations in the constituents of the medium can produce not only callus proliferation, but also other identifiable patterns of development.

In an experiment with surface thin layers taken from floral parts of tobacco, four separate types of development could be induced by manipulation of the hormonal conditions. With equal quantities of auxin and kinetin in the presence of glucose, floral bud formation occurred on the surface of the thin layer without intermediate callus formation. An excess of kinetin over auxin in the presence or absence of glucose gave rise to vegetative buds. Roots were formed in the presence of a high auxin to kinetin ratio, together with sucrose in the medium, whereas callus could form if the kinetin concentration were very low, and in the presence of auxin and glucose. In the case of the epidermal layers removed from the main vein of leaves of *Begonia*, it was found that roots or unicellular hairs could be induced to form simply by manipulating the absolute concentration of auxin in the medium.

Microscopic examination of the thin layers from tobacco showed that both the callus formation and the development of different organ types all took place from the subepidermal cells of the explant (Fig. 5.18).

This is a clear example of chemical switching of development in a very simple cell system.

Certain other observations were made during the course of this work. The nature of the explant affected its potential for development. For example, floral development was not obtained from explants of non-flowering plants. This indicates an overall switch in the onset of flowering, which will be discussed in more detail later (p. 209). It was also found that when careful dissection of cell types was carried out, developmental potential could be limited. Excised cambium could be induced to form only a small amount of callus and vegetative buds whereas, if it were left in contact with subepidermal cells, flower formation could be induced. These and similar results may show the importance in the intact plant of intercellular interactions. It must be emphasised however that as yet there is no explanation for this behaviour; one

cautionary factor which has to be borne in mind is the possibly very serious effects of wounding which dissection of very small pieces of tissue inevitably involves.

The ability to switch development by external factors in a thin layer of cells opens the possibility of examining the structural changes which appear during the operation of the supposed trigger stimulus. This work is as yet at a very early stage of its development. So far it seems that floral development from subepidermal cells is characterised by the formation of prominent starch grains in the cytoplasm, whereas an early sign of the development of a callus is the stimulation of incorporation of material into the cell wall. The whole field of structural modifications associated with hormonal treatments is fraught with difficulties, but these simplified systems seem to hold great promise for the future.

Effects of growth substances on cell structure

We have already shown that progress along many developmental pathways is accompanied by noticeable structural change within cells (Chapter 4).

Fig. 5.18. Scanning electron micrograph of the surface of a stem segment of tobacco (*Nicotiana tabacum*) cv. Xanthi, after 10 days' growth in the presence of equimolar IAA and kinetin. Proliferation of cells below the epidermis (left of picture) eventually leads to the production of small floral primordia (right of picture). Bar = 200 μm.

The present discussion indicates that growth substances may have a part to play in the initiation of differentiation in isolated cells and tissues. Is there any relationship therefore between the presence of growth substances and cell structure?

In framing this question, it is essential to be aware of inevitable restrictions on the type of answer which may be obtained to it. Electron microscopy is an excellent tool for the analysis of certain types of subcellular change; for example, the differentiation of individual organelles, or the specific positioning of subcellular structures in space. By careful analysis it is even possible to obtain quantitative estimates of change in the numbers of organelles and even in the amount of membrane present within the cytoplasm. However, impressive though its capabilities are, they are limited severely in certain other areas. The detailed substructure of membranes can only be analysed by freeze–fracture, and here the identification of structural components is difficult. The structure of the wall can be described morphologically, but not chemically. The complexity of the DNA in the nucleus is such that very little idea can be gained of possible changes at this site.

These are serious limitations in the context of early effects of growth substances. Since the development of organised wall thickenings, for example, is a feature of xylem element differentiation, and since this can be induced by applied growth substances, there is in some sense a connection between the two. However, the main interest in structural change induced by growth substances must rest in its very earliest stages, and these are likely to be associated with membranes, the wall and the nucleoplasm – precisely those sites where electron microscopy is least fitted to analyse the situation.

With these limitations in mind, it is nonetheless possible to find a few convincing examples of effects on ultrastructure of cells. One of the best documented is the increase in endoplasmic reticulum which follows the treatment of cells of the aleurone layer of barley with GA. GA is known to be important in the germination process and appears to act by stimulating the production of the enzyme α-amylase which is responsible for the mobilisation of storage starch within the germinating seed. Within 2 h of exposure to GA, the cells show greatly increased amounts of endoplasmic reticulum (ER) which forms stacks of lamellae in the cytoplasm. Biochemical studies in this

system seem to indicate that this effect is not due to net synthesis of membrane components, but rather their rearrangement into ER lamellae from other sources within the cell. A similar conclusion has been drawn from the effect of ethylene in stimulating the rapid expansion of the pea epicotyl. Here, long sheets of rough ER are found within the cells after 6–12 h of exposure to the growth substance. There is some evidence also that ethylene may induce the formation of ER in abscission zones of tobacco flower pedicels. In all these situations, the physiological effect of the growth substance is an increase in synthetic activity and secretory activity in the target cell.

The dictyosomes have frequently been cited as organelles which show a rapid response to various growth substances. There is an increase in the number of dictyosomes in the cells of hypocotyl tissue of lettuce (*Lactuca sativa*) following growth stimulation by GA. Similarly, the auxin-induced growth shown by Jerusalem artichoke (*Helianthus tuberosus*) tissue is also correlated with increased numbers of dictyosomes within the cells. In both these cases it may be inferred that the dictyosomes were in some way contributing to the expansion of the cell wall which accompanies growth.

Perhaps the best illustration of the difficulty associated with chemical effects on ultrastructure is given by studies with microtubules. Many of the effects of growth substances relate to general growth phenomena; an increase in elongation or perhaps in diameter of a tissue such as a stem segment. In *Vigna angularis*, for example, IAA stimulates elongation whereas kinetin inhibits elongation and stimulates lateral expansion of the epicotyl. GA promotes elongation and inhibits lateral expansion. In a study of the orientation of microtubules in epidermal cells of this tissue it was found that treatment with kinetin produced a predominant arrangement of microtubules parallel to the axis of the cell whereas, following GA treatment, the preferred orientation was transverse to the cell axis. This is an interesting result; it was concluded that a link had been demonstrated between the effects of growth substances on differential cell expansion and on the ultrastructure of the cytoplasm. This may be an over-interpretation however, since considerable doubt remains concerning the role of microtubules in the orientation of cell wall components.

It has to be concluded that whilst there are many

examples of a change in ultrastructure following as a fairly rapid response to a chemical treatment, their character is such that they are almost certainly not a primary response. A tissue which is programmed to behave in a certain way to a given signal will inevitably show structural changes as a result of the signal being received. However, the same response will not be elicited in other tissues or cell types. In other words, if the primary effect of say GA were to increase the amount of endoplasmic reticulum, then it would be reasonable to expect to observe this in any cell. This is not in fact the case. The most plausible explanation of this is that the reaction of the endoplasmic reticulum is not the primary response.

Summary

The field of tissue culture is an extensive one, and the contents of this chapter survey only a tiny part of the work which has been carried out. The experiments which have been described have been selected on the basis of their general interest for development. They are a deliberately limited series of observations, since it is unfortunately true with this field that the more facts are gained the less able are the existing hypotheses to accommodate them. Tissue culture has certainly not given rise to predictable and widely reproducible procedures for the control of development in isolated cells. This may be regarded purely in a negative way, as a failure of the technique. On the other hand, it should encourage the seeking of explanations.

Much of the inconsistency in the behaviour of callus tissues no doubt stems from the physiology of the cells themselves. Callus tissues are widely described as being 'undifferentiated', but such a term implies a degree of uniformity in the population which is not likely to be present. One important conclusion from this is that, within an organised tissue, there must be a considerable degree of regulation exerted between cells. Within a callus tissue or a liquid suspension, cells are much less in communication with each other and so the possibility of organisation is much reduced.

There is no doubt that applied growth substances have profound effects on the growth and even on the development of callus tissues. Unfortunately it is not possible to formulate a general scheme for the actions of growth substances on the basis of tissue culture work since, as we have seen, the effects produced by a given treatment may be quite different in different species and even, in some cases, in different cultures from the same species. However, tissue culture clearly suggests that within a more regulated environment such as an intact plant tissue, the known growth substances are likely to be powerful effectors in matters of development. Of course, in much tissue culture work, the regulators which have been used are completely synthetic and are often present, in the medium at least, at high concentrations compared to the known levels of the equivalent molecules in intact tissues.

The idea of hormonal switching of development is, in any case, probably naive in terms of whole-plant development. The plant cell in its normal environment is not subject to constant conditions of light, temperature, and supply of inorganic nutrients, awaiting only a hormonal signal to develop in a particular way. From the moment of its formation, for example in a meristem, the plant cell is normally experiencing a constantly changing physical and chemical stimulus which may be presumed to include such parameters as nutrient supply, light levels, gradients of growth substances, and even such factors as physical constraint from its neighbours. All these factors can be reproduced in tissue culture work, and they have all been shown to affect development. Viewed in this way, it is perhaps not surprising to witness a lack of uniformity in response to variation in only one class of stimulus.

Tissue culture studies then, confirm convincingly the idea of totipotency in plant cells, and suggest strongly that normal development is controlled on the basis of organisation at a level higher than the individual cell. In order to appreciate that organisation, it is necessary to consider development in the whole plant.

6

Patterns and organisation in the whole plant

Introduction

We have so far considered the cell as an individual unit within the plant. Its development can be described in terms of a pathway of change from a starting point to its conclusion. The patterns of change, which progress along this pathway involves, can be defined in structural and biochemical terms. They result in the formation of a functioning mature cell. In principle these pathways might be elicited in isolated cells provided the correct stimulus were given and, as the last chapter showed, this goal can be achieved in a limited and somewhat unpredictable way with various isolated cell systems and fragments of tissue.

The concept of cell development occurring as a sequence in isolation is of course a gross oversimplification in terms of normal plant growth. That this is so we have already deduced from the facts concerning the behaviour of cultured cells. Organisation into populations of cells is one of the basic properties of all multicellular plants; there is no doubt that the development of a particular cell within such a complex, whether it be a filament of a fern gametophyte or the flower of an orchid, depends on its relationship to other cells in its vicinity. In order to grasp this idea in more detail, it is necessary now to consider patterns of real organisation in whole plants. To begin with, we shall consider what is known of the development of cell populations from a study of the embryo and the growing apices of higher plants. We shall then consider how the actions of these populations may be orchestrated within the plant to give rise to its normal growth behaviour.

The growth of the embryo

The fertilised zygote is a highly vacuolated cell suspended in a structure called the embryo sac. It is formed by the fusion of the egg cell with one of the sperms which is carried to the embryo sac within the pollen tube. The other sperm fuses with the two haploid nuclei of the central cell of the embryo sac to give rise to the primary triploid endosperm nucleus. This nucleus then divides and produces the endosperm, which is a nutritive tissue, and which forms the immediate environment of the developing embryo. Some endosperms are comprised of cells (cellular endosperm) whereas, in others, partitioning cell walls are not formed. Coconut milk is a well known example of a liquid endosperm of this second type, and its role in the nutrition of the embryo is the rationale behind its use in much tissue culture work. Following fertilisation, there is usually a delay before further development occurs. This may last a few hours (as in *Crepis capillaris*) or it may last several months (as in *Colchicum autumnale*). Usually, but not invariably, the endosperm develops in advance of the zygote.

The classic species for the study of embryo development has been the shepherd's purse, *Capsella bursa-pastoris* (Fig. 6.1). The first division of the zygote is asymmetric and expresses the polarity which will remain throughout the life of the plant (Chapter 7). The two daughter cells of the first division are different in size and developmental potential. The larger basal cell will produce the suspensor, whilst the smaller terminal cell will produce the embryo proper.

The suspensor consists of a short column of cells formed by repeated transverse divisions. These cells have many plasmodesmata across their transverse walls, and the most basal cell develops wall ingrowths in the manner of a transfer cell (p. 104). The function of the suspensor is probably to act as an efficient pathway of nutrient transfer from the endosperm into the developing embryo.

The terminal cell undergoes an orderly series of transverse and longitudinal divisions to give rise to a globular structure which eventually reaches 64 cells in number. This process involves progressive reductions in the size of the cells and an increase in the density of their cytoplasm. There is no visible ultrastructural differentiation in the globular embryo up to the 32 cell stage. This implies that differences which surely exist in the cells at this stage are expressed at a molecular level and relate to the position of the cell rather than to its specialised structure. Differentiation of the two cotyledons produces a heart-shaped embryo, and the cotyledons elongate and curve round to fill the space occupied by the endosperm. Chloroplast differentiation occurs as the cotyledons become elevated. The main root–shoot axis of the embryo is located centrally within the embryo, with the shoot tip between the two cotyledons. At maturity, the root tip has a differentiated root cap.

Organisation in the embryo

This description of embryo development represents only one example of a process which has a multitude of variations. The mature embryo does not always consist of a simple axis. In grasses, for example, both the coleoptile and the first leaf are preformed at the embryo stage. The attainment of such organised structure might be thought to depend critically on fixed patterns of division during growth; this does indeed appear to be the case at the early stages of formation of the globular embryo in *Capsella*. However, it is not always so. In cotton (*Gossypium hirsutum*) for example, there seems to be little discernible pattern in the initial divisions. It is also known that in *Eranthis hiemalis*, killing the developing embryo causes the formation of one or more new embryos from the remaining cells of the suspensor. In gymnosperms, the zygote divides repeatedly without cytokinesis, so that the initial stages of embryogenesis consist of the formation of a free nuclear embryo in which up to 256 nuclei can co-exist in a single cytoplasm. Partitioning of these nuclei at a later stage gives rise to a mass of cells, each one containing a single nucleus. Further development involves cell divisions with normal cytokinesis; these proceed more rapidly in the terminal region of the embryo, giving rise to a marked polarity in the distribution of cell sizes. The small cells at the terminal end of the embryo give rise to the shoot and root apices and the cotyledons. Even in angiosperms, there is evidence from carrot (*Daucus carota*) and cotton that different embryos may develop along different pathways. In lower plants the early cell divisions are more symmetrical, giving rise to a globular mass of cells in which the shoot apex, the root, the foot and the first leaf are differentiated. This differentiation is classically regarded as occurring in each of the four quadrants of the embryo, although such a simple analysis may not be widely applicable. Many attempts have been made to classify embryos according to cell division patterns during their early development.

None of these is entirely satisfactory and, indeed, it might be said that such an idea represents the imposition of a concept of order which is not justified when compared with the known facts.

At maturity the embryo is a highly polar structure with differentiated apices, a procambium which will form the vascular system, and a protoderm which will form the epidermis. Subsequent development of the plant rests on the activities of the meristematic tissues.

Apical organisation: the root

The apex of the root is apparently a fairly simple structure, with the meristem enclosed within developing tissues – the root cap at its apex and the vascular

Fig. 6.1. Diagram of the course of embryo development in *Capsella bursa-pastoris*. The initial asymmetric division of the zygote defines the polarity which persists throughout the development of the embryo. The embryo is derived from the small dense product of this first division, whilst the large vacuolate cell gives rise to the suspensor.

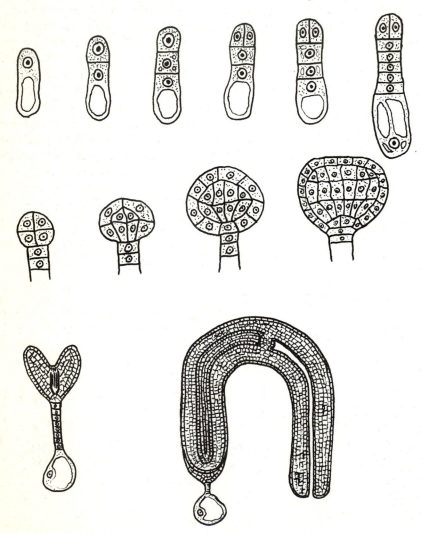

and ground tissues of the root in a basipetal direction (Fig. 6.2). The structure of roots in fact shows considerable diversity between species, and it is difficult to make very many generalisations which can apply universally. In many lower plants all the structures of the root are generated from a single apical cell. This is exemplified by *Azolla* and *Ceratopteris* which will be considered in the following section. In the majority of the higher plants it is not possible to identify with certainty any initial cells.

One of the earliest attempts to rationalise the structure of the root apex was based on the observation that files of cells in the maturing tissues can be traced back, in certain cases, to tiers of meristematic cells. These tiers were thought of as giving rise to particular parts of the root, and were named histogens. Thus in the root of radish, *Raphanus sativa*, three histogens can be identified (Fig. 6.3a). The most distal of these gives rise to the root cap and the

epidermis; the middle tier of cells gives rise to the cortex and the innermost layer generates the vascular system. Variations on this arrangement are possible; for example in the root tip of maize (*Zea mays*) (Fig. 6.3b) a distinct meristem called the calyptrogen donates cells to the root cap. The other three histogens donate cells to the epidermis, cortex and vascular system respectively.

This simple idea is unable to withstand completely two important experimental observations. The first is that quite large parts of the meristematic region of the root can be damaged or removed surgically without destroying the ability of the root to generate all its normal tissues. At the very least this implies that the number of cells within each histogen is not very small, and it strongly suggests that the idea of particular cell populations committed to producing one particular set of mature tissues is an oversimplification. Indeed, it is possible entirely to decapitate a root and to not destroy its ability to regenerate a new meristem. Furthermore, an analysis of the rates of division in root apices has shown that the centre of the meristem, which was formerly regarded as the seat of the activity of the cells giving rise to the mature tissues of the root in the histogen theory is, in fact, mitotically quiescent.

The precise significance of the quiescent centre is unclear at present. It is often absent from the very

Fig. 6.2. A section through the root tip of the radish (*Raphanus sativus*). Bar = 5 μm.

Fig. 6.3. The histogen theory of root apical organisation. *a*, In radish (*Raphanus sativus*) three histogens contribute cells to the root cap and epidermis, the cortex and the vascular cylinder; *b*, in maize (*Zea mays*), the root cap has its own histogen, the calyptrogen.

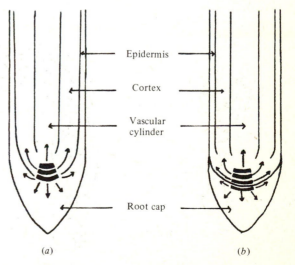

Epidermis

Cortex

Vascular cylinder

Root cap

(a)   (b)

earliest stages of root growth during seedling germination, and also in lateral roots. The quiescent centre is not completely unable to undergo mitosis, but rather is a region where the cell cycle is unusually long. The average duration of the cell cycle in the quiescent centre of maize roots is 174 h, whereas, in the root cap meristem just distal to the quiescent centre, the duration of the cell cycle is 12 h. Removing the root cap induces a more rapid rate of division in the quiescent centre of maize, and leads to the formation of a new meristem; the subsequent reformation of the root cap causes the cell cycle in the quiescent centre to slow down once again. The quiescent centre may comprise as few as 500 cells in a meristem which contains 125 000 cells; the cells of the quiescent centre are recognisable structurally because of the few mitochondria they contain, the relative paucity of endoplasmic reticulum, and the small size of their dictyosomes, nucleus and nucleolus. The cells of the quiescent centre are held in G_1 (p. 26) and, because of this, they are particularly resistant to damage by ionising radiation. Following a treatment which causes damage elsewhere in the meristem, the quiescent centre cells divide and restore the structure of the apex.

The root meristem is not located superficially. Branching of roots does not occur from the epidermal surface, but from the pericycle, which is a layer of cells at the outermost edge of the vascular system. These cells retain their diploid nature and meristematic character; lateral root development from the pericycle involves the formation of a primordium which differentiates into a root apex before emerging through the cortex and epidermis. The mechanism of initiation of lateral roots is of interest, since presumably all the cells of the pericycle are capable of this behaviour, whereas only a few of them will actually initiate laterals during normal growth. Removal of the root tip stimulates lateral root formation, whereas it has been shown in *Pisum* that removal of the cotyledons in sterile-grown seedlings inhibits the process. It appears, therefore, that lateral root initiation may be controlled by a balance of stimuli from the mature tissues and inhibitors from the tip. The possible role of growth substances in this process will be discussed later (p. 171).

The root meristem of *Azolla* and *Ceratopteris*

The above description of the meristem of the

root tip suggests strongly that whilst it might be expected that rather precise patterning of cells would be required to generate a complex functioning tissue such as a root, attempts to discover such patterns are not entirely successful. Surgical experiments suggest that such precision may not even be absolutely necessary. This situation, no doubt, in part reflects the complexity and size of the tissue under consideration (factors which prevent a detailed analysis by electron microscopy, simply in terms of the enormity of the task). However, as a counterbalance to this state of affairs, we shall now consider the structure of the root tip of the water fern *Azolla* (Fig. 6.4). This material is not typical of higher plants, since the root is generated by a single apical cell. However, its small size and regular development have allowed a remark-

Fig. 6.4. A section through the root tip of *Azolla pinnata*, showing the large apical cell. Bar = 10 μm. Picture kindly provided by Professor B. E. S. Gunning. (First published in *Planta*, **143**, 121–44, 1978.)

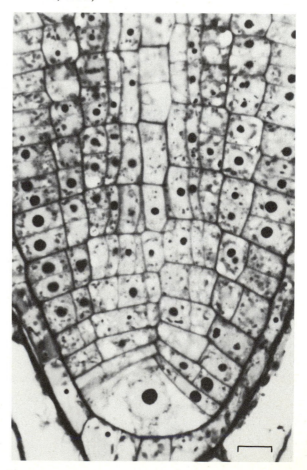

able insight into the production of a multicellular tissue.

The apical cell of *Azolla* is four-sided. The side next to the root cap is curved, and the wall crossed by relatively few plasmodesmata. The other three sides are flatter and approximately triangular, with many plasmodesmata within the walls. In young roots the apical cell is sparsely vacuolated; vacuolation increases as the root ages, although the size of the apical cell does not change. The root cap is generated by a single division of the apical cell parallel to its curved face. The daughter cell of this division then divides once more in the same plane to give a two-cell covering to the curved face of the apical cell. This two-cell layer then undergoes repeated divisions at right angles to the original direction, to generate the entire structure of the root cap which is everywhere only two cells thick.

Following this initial division to produce the root cap, the behaviour of the apical cell changes to a pattern which remains constant throughout its meristematic life. Divisions occur parallel to each of the three flatter faces of the cell in turn, giving rise to a series of successive segments. These segments are known as merophytes. This sequence of divisions clearly has a built in 'handedness'; it may proceed clockwise or counter clockwise. In fact, when this is examined, it is found that the handedness of half the roots is one way, and the opposite way in the other half; this depends on the position at which the root arises from the frond. When the patterns of cell divisions are examined, it emerges that they occur in precise sequences. The immediate derivatives of the apical cell can of course divide again, either parallel to the original direction of their formation or at right angles to this direction. In a study in which every cell in 26 root tips of 2 species of *Azolla* was examined, it was found that the patterns of cell division were predictable and species-specific. This meant that, for each mature cell type, it was found possible to describe the numbers and types of divisions which had preceded its formation. This situation is also true in another fern, *Ceratopteris* (Fig. 6.5).

The *Azolla* root is determinate; that is to say, it

Fig. 6.5. Diagram to illustrate the predictable patterns of cell division during root development in *Ceratopteris thalictroides*. Divisions from only one face of the apical cell are shown. The fate of the cells following seven cycles of division from this face is as described in the final diagram. The precise development of the vascular tissues depends on the originating face from the apical cell, so that the vascular system has a different symmetry from that of the apical cell.

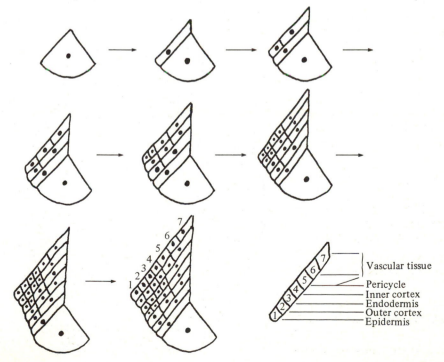

Vascular tissue
Pericycle
Inner cortex
Endodermis
Outer cortex
Epidermis

grows for a certain time and then growth ceases. The ordered structure of the root allows an estimate of the number of times which the apical cell divides. This is about 55. This determination is expressed within the structure of the apical cell, which becomes more vacuolate with age, and within the young walls of its derivatives, which have fewer plasmodesmata as the root ages. It thus appears possible that progressive ageing of the root results in the gradual isolation of the apical cell from its supply of nutrients or stimuli from the shoot, and that this could be a reason underlying the cessation of its meristematic activity. The predictability of the division patterns means that the positioning of the preprophase band of microtubules (p. 119) may be studied in some detail. In every case a positive correlation was found between the positions of the band and the positions of the future cell plate, even in highly asymmetric divisions. Furthermore, in *Azolla* it has been found that when a division involves the migration of a nucleus, the preprophase band of microtubules may occur in its correct position before such migration takes place (Fig. 6.6).

Perhaps the most intriguing aspect of the structure of the *Azolla* root is that although it is formed by sequential divisions on three faces of the apical cell, its mature structure shows a two-fold symmetry. This must mean that each file of cells does not develop as though it were in isolation since, if this were the case, three-fold symmetry would result. This shows that the important and indeed overriding consideration in the developmental sequence is not the history of a particular cell but its position in the tissue and its relationship to its neighbours. In this observation it is

possible to glimpse the reason why in other systems such precise patterns of divisions are not observed. It may be, as we have said, that the complexity of the tissue prevents their observation; on the other hand, it may be that such precise patterns are simply not necessary. What remains clear from *Azolla* is that although developmental fate is determined by the position of the cell in the root, it is also predetermined by the division status of the cell. A cell must usually undergo a certain number of divisions before it can differentiate into say a xylem element, for example. On the other hand, this does not appear to be universally true, since in old roots, root hairs can form progressively nearer the root tip, and indeed directly from the apical cell itself. The concept of determining new patterns of behaviour by cell division processes will be returned to later (p. 231).

Azolla and *Ceratopteris* it must be stressed, represent very unusual and simple cases of organisation. Nonetheless, they point the way to several new approaches to development, and make the process more amenable to direct experimentation. There is no doubt that much will be learned from further study of these elegant roots.

Apical organisation: the shoot

The shoot apex is considerably more complex to describe than the root apex. This complexity arises from the wide variety of structures which the shoot apical meristem may originate. In contrast to the root, the shoot apical meristem is superficial, and branching proceeds from the tip. The most obvious product of the growth of the shoot apical meristem are the leaves and stem of the plant; however, the same region initiates the formation of axillary buds, thorns, tendrils and reproductive organs. Some of these structures are determinate in their growth (such as the leaves) whereas axillary buds, for example, are inde-

Fig. 6.6. The relationship between the position of the nucleus and the position of the preprophase band of microtubules. In certain asymmetric divisions in the *Azolla* root tip, the preprophase band appears to define nuclear migration.

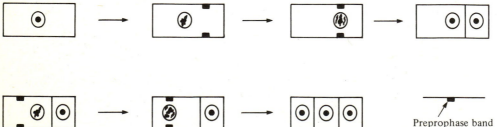

Preprophase band
of microtubules

terminate. In many flowering plants the onset of reproductive development causes profound changes in the structure of the apical meristem which signal the end of the growth of the shoot.

As we have already seen with root structure, overall order within a living complex of cells does not necessarily require a mechanically precise arrangement of cells in the meristem to generate that complex. Many attempts have been made to describe the structure of the shoot apex in precise terms, and most of these attempts could fairly be said to represent impositions of theoretical concepts which do not withstand experimental examination. There is considerable variation in the structure of apical meristems of shoots when different families of plants are considered; there is also variation within the life of a single plant. Lower plants, for example, may have an easily recognisable single apical cell. This is the case with *Osmunda* and *Equisetum* (Fig. 6.7*a*). Gymnosperms, of which *Gingko biloba* is a well studied example, may have generalised zones of cells which are recognisable in terms of their staining properties and which appear to donate cells to different parts of the growing tissue, rather in the manner of the histogens on the root apex (Fig. 6.7*b*).

The most common description of the angiosperm shoot apex is called the tunica–corpus theory (Fig. 6.7*c*). This states that there are two principal populations of cells which are defined in terms of their predominant plane of division. In the surface layers,

the tunica, the predominant plane of division is at right angles to the surface. This results in the formation of increasing numbers of cells in the surface plane. Within the tissue, the corpus, the planes of cell division are in all directions, and their result is to increase the bulk of the central tissue in all directions. More than half the dicotyledon species which have been examined have a two-layered tunica, with numbers of layers up to five reported.

The independence of the tunica and corpus layers has been investigated by the use of chimaeras. Chimaeras have already been introduced in a discussion of plastid variation (p. 65). It is possible to induce polyploidy in one or more of the layers of the shoot apex by the use of colchicine. This drug, by interfering with the formation of the mitotic spindle, produces cells in which the amount of DNA within the nucleus may be doubled or quadrupled. This effect can be recognised by the staining properties of the nucleus, and by using this technique, it has been possible to show that the outermost layer of the tunica for example gives rise to the epidermis, whereas inner layers may donate cells to internal tissues. This emphasises an essential difference between the tunica–corpus theory and the histogen theory of apical organisation; namely that, in the former, there is no predetermination of the eventual fate of the cells produced by division (groups of initial cells do not give rise to specific tissue types).

There is some evidence to suggest that just as the root apex has a quiescent centre, so too does the shoot apex have a region which has a lower rate of mitotic cycling than the surrounding tissue. This region is known as the meristème d'attente (the waiting meristem), since the idea of its existence was propounded by French workers. The concept of the meristème d'attente is less well established than is that of the quiescent centre in the root tip. In particu-

Fig. 6.7. Diagrammatic representation of types of shoot organisation. *a*, A single apical cell (A) exemplified by *Equisetum*; *b*, an apex with a central zone (CZ) of cells which donate progeny to the bulk of the tissue, exemplified by *Gingko biloba*; *c*, the angiosperm apex with a two-cell tunica overlying the corpus in which cell divisions occur in all planes.

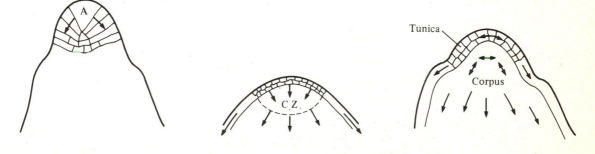

(a)   (b)   (c)

lar, it is reasoned that if the quiescent region of the shoot were donating cells to the underlying tissues, and these cells were themselves dividing, then a low frequency of mitosis in the quiescent region would not reflect a special significance, but merely the fact that sufficient cells can be generated by a low mitotic cycling rate. Proponents of the meristème d'attente theory however have held the view that this region is waiting for the onset of reproductive development, and does not contribute cells to the vegetative growth of the shoot. It appears likely that the truth may lie between these two extremes; what is not in doubt in any case is that unlike the quiescent centre of the root,

th meristème d'attente does show considerable division activity at the onset of reproductive development.

The formation of leaves

The first sign of a developing leaf is a swelling on the side of the shoot at some distance from the tip (Fig. 6.8). This swelling results from increased cell division in one or more layers of the tunica and the corpus. In some cases, only the outer tunica layer contributes to the primordium – this is common in monocotyledons. In dicotyledons, the tunica continues to divide only anticlinally and the primordium is formed as a result of periclinal division in a cell of the second or third layer. Other neighbouring cells also divide to give rise to a raised primordium. The young leaf grows at its apex for a certain time, apical growth then ceases. This expresses the determinacy of the leaf; apical growth may cease very early on, before the primordium has attained a length of 1 mm. Further growth modifies the circular symmetry of the primordium into the lamellar form of the final leaf blade. The internal cells of the leaf are produced by the division of dispersed initial cells at the margin of the developing lamella.

Cells of the different tissues of the leaf may cease to divide at different stages of its development. For example, in tobacco (*Nicotiana tabacum*), the epidermal cells cease to divide when the leaf is about one fifth of its final size. Further growth of the leaf is accompanied by expansion of these cells. The palisade cells continue to divide until the leaf is mature, but spongy mesophyll cells cease to divide and grow at an earlier stage. This results in a continuous layer of expanded epidermal cells overlying the palisade tissue with small air spaces; the mesophyll, by contrast, is characterised by very large air spaces (Fig. 6.9). The vascular system of the leaf is initially developed from the vascular procambium of the shoot itself. The midvein is formed first, with protophloem elements preceding the protoxylem. The phloem seems to develop in continuity with the

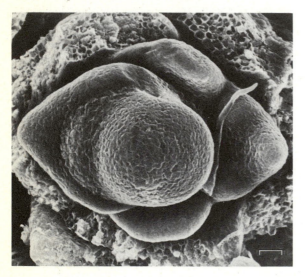

Fig. 6.8. Scanning electron micrograph of a dissected vegetative apex of the pea (*Pisum sativum*). Leaves arise as swellings on the sides of the apical dome. Bar = 100 μm.

Fig. 6.9. Simplified diagram to illustrate the structure of a leaf from a plant of tobacco. A continuous epidermis overlies a palisade layer of cells which are in close contact. The mesophyll by contrast has many air spaces and the cells are not simply elongate in shape.

Epidermis

Palisade parenchyma

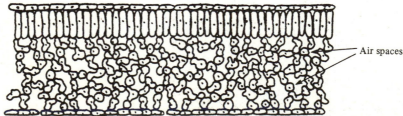

Air spaces

Spongy mesophyll

Epidermis

vascular system of the shoot, but the formation of xylem is discontinuous; the first-formed xylem elements extend both acropetally and basipetally to meet up with the mature vascular system of the shoot.

The pattern of leaves around the stem of a plant is clearly highly ordered. Leaves may be alternate, or opposite, or helically arranged in a spiral. The patterning of leaves is known as phyllotaxis, and it has been the subject of considerable mathematical analysis. In the very common arrangement of the so called genetic spiral for example, primordia are separated from each other in an angular arc which approaches 137° about the centre of the shoot apex (Fig. 6.10). The position of each leaf can be expressed as the cross over point between two sets of opposing logarithmic spirals drawn about the centre of the apex. These spirals are known as parastichies. This predictability of the positioning of new primordia has allowed experimentation to be carried out to analyse its cause. It is found in *Lupinus*, for example, that if a surgical incision is made between a primordium and the apex of the shoot, no effect is observed on the positioning of the next primordium, but that the one after is dis-

placed towards the position of the primordium which was isolated by the cut. This has lead to the proposition that a zone of inhibition surrounds each new primordium. If a similar inhibitory zone is assumed to occur at the apex of the shoot itself, this goes some way to explain why the meristem as a whole does not become a mass of disorganised primordia. The nature of the inhibition is however entirely unknown; it could conceivably represent a nutritional effect or be due to the presence of a growth substance.

The development of some leaves does not seem to be inevitably and irreversibly programmed from the earliest stages of primordial development at the apex. In an experiment with *Osmunda* (the royal fern), successive primordia were dissected out and cultured. It was found that the oldest primordia all developed into leaves, but that younger primordia sometimes gave rise to shoots, this proportion tending to rise as the age of the excised primordium was decreased. This suggests that the final form of the primordium is not predetermined, something which was also inferred from studies of organogenesis with tissue cultures (p. 144). In the case of a normally growing apex, such variation does not occur and all the primordia would have formed leaves if they had not been excised and cultured. Surgical techniques *in vivo* showed that isolation of a primordium from the apex by an incision increased the chance that a 'leaf' primordium would develop into a shoot. Even more

Fig. 6.10. A plant of *Sempervivum rubin*. The diagram on the right is a tracing of the photograph. Angles AB and BC are 137° and quite accurately define the position of adjacent leaves in the spiral.

remarkable, it was found with *Osmunda* that if older primordia are cut into several pieces, then each one produced the whole leaf structure. This type of experiment has been taken to show that the determination of a leaf is a process involving organisation at a level higher than the individual cell, although it must be admitted that such plasticity of behaviour is not widespread.

Other products of the shoot apex

So far the shoot has been regarded only as the source of leaves. Even this simplification is not without its complications since, in some families of plants, leaf form can vary throughout the life cycle. The production of different types of leaves at different stages is known as heterophylly. Sometimes this is an age-related effect, (heteroblasty; Fig. 6.11), with juvenile plants having a different leaf form from adults; in others it may be a direct result of an environmental circumstance. This is most strikingly shown by plants which may have terrestrial and aquatic forms. *Ranunculus heterophyllus*, for example, has highly dissected leaves when it is submerged, but expanded lobed leaves when it is growing as a land plant (Fig. 6.12). The fern *Marsilea drummondii* has simple leaves initially, whilst older plants produce leaves with first two, then four lobes. By aseptic culture of the sporelings it was shown that this change in leaf form could be induced by the presence of glucose in the culture medium. This effect could be antagonised by applied GA. It is not clear whether changes in leaf form in general are the result of nutrient status or are controlled by levels of growth substances.

Apart from leaves, the shoot may produce other shoots, and thus give rise to branching of the stem. In lower plants this is frequently a result of an equal or unequal division of the shoot apex itself. It is known that if the apex is divided surgically, regions on either side of the cut can regenerate normal shoots; branching of the apex would seem to be the normal physiological equivalent of this drastic procedure. The mechanism of such terminal branching is far from clear in its details. The new shoots arise on either side of the old apex, and the apical cell, if present, may be

Fig. 6.11. Heteroblasty in ivy (*Hedera helix*). The 'typical' ivy leaves are formed during the juvenile phase of growth; the adult stage leaf is not lobed and has a different venation pattern.

induced to divide and form a cluster of smaller cells at the original centre of the old apex. Whether this change precedes the formation of the two side shoots or is a result of it is not clear.

In higher plants, buds are commonly found in the leaf axils. They may arise close to the apex, or they may not appear until a later stage in development, from so called detached meristems (small regions of meristematic cells which have been separated from the main apex by the growth of intervening tissues). The bud develops to a certain stage and then becomes arrested by the phenomenon known as apical dominance. The formation of buds is not a universal consequence of the development of a leaf; sometimes only certain leaves have axillary buds associated with them. In *Nymphea* (water lily), buds and leaves are formed alternately at the shoot apex. Lateral buds do not always undergo indeterminate growth as side shoots; they may develop into thorns, for example, which are determinate. The thorn of *Gleditsia* originates with a normal domed apex like that of an axillary shoot, but rapid cell divisions at the tip produce a narrow and elongated structure which differentiates schlerenchyma to become very hard and rigid. The sharp point of the thorn is produced by the retention of apical growth to a late stage in its development. This contrasts with the growth of the leaf, where

apical divisions cease very early on in development.

The relationship of reproductive tissues to the shoot varies with the evolutionary status of the plant. In ferns, the leaves bear the reproductive tissues and (in many cases) the spore-bearing leaves and the vegetative fronds are very similar in morphology. In horsetails and club mosses, terminal cones are produced which bring to an end the vegetative growth of the shoot on which they develop. In gymnosperms and in some angiosperms also the entire shoot apex may be the source of a single cone or flower but, in many other cases, an inflorescence which consists of many flowers is produced from one shoot. Flowering usually signals a change from unlimited to determinate growth, but this is not always the case.

In the production of a single flower, each of the floral parts is formed in rapid succession as small appendages at the surface of the apex. Thus, in *Aquilegia* for example (Fig. 6.13), 5 sepals are formed in rapid succession, followed by 5 petals. These are established simultaneously in positions which alternate with the sepals. The 40–5 stamens are then produced in simultaneous whorls of 5, followed by 2 whorls of petaloid stamenodia. The 5 carpels are then produced at the summit of the apex. Thus in the development of the flower the zonation of the apex which characterises the vegetative state is replaced by a complex of appendages which are all determinate. In particular the 'waiting meristem' of apparently quiescent cells in the central zone is lost as its mitotic activity increases to match that in the rest of the meristem.

Fig. 6.12. Drawing of *Ranunculus heterophyllus*, showing the difference in leaf form between aerial and submerged leaves.

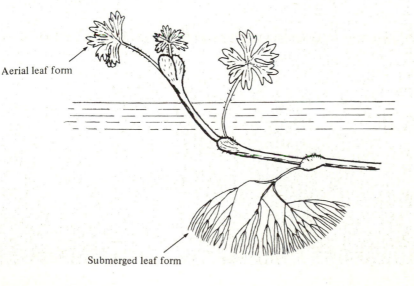

Aerial leaf form

Submerged leaf form

Cells and the organisation of tissues

This brief and very superficial survey of the development of embryos and the behaviour of apices is intended as a background to several important questions of developmental significance. Most of these structures, whilst they can be described in a general way, do not conform rigidly to predictable patterns. The structure of any particular embryo or apical meristem may be fitted into a general pattern in a more or less convincing manner, but it cannot be deduced from first principles through knowledge of its behaviour. Conversely, its behaviour cannot be deduced by a study of its structure. Moreover, this does not seem to represent merely a failure of knowledge; the more widely the techniques of microscopy are applied to problems related to the structure of apices, the more diverse the patterns become. In addition, electron microscopy has so far failed to discern differences in cells of apical meristems which are related to their future development. This is, of course, precisely why the meristematic cell is such a useful standard of the undifferentiated state. However, it does mean that there is a limit to the usefulness of anatomical and structural studies in terms of the organisation of the growing regions of the plant. Cells within meristems are clearly capable of a degree of organisation which is not necessarily matched by discernible patterns of predictable cell division planes or growth directions.

Clearly, although patterns of cells within embryonic or meristematic tissues do not appear to us to be as regular or as well organised as we would like, a very high degree of organisation does in fact exist in these circumstances. This implies that processes of self regulation are active in such tissues at all times. There are many examples of the presence of such regulation. For example, it has already been seen (p. 146) that isolated cells from mature tissues, when grown in culture, may express the ability to form embryos under certain conditions. They do not do so within the plant. A shoot has the potential to produce a wide range of structures, some determinate, some of unlimited growth potential. In the plant this wealth of diversity is expressed in strict sequences in space and in time. Furthermore, fragmented meristems and even fragmented primordia may generate whole replacements in some cases. This indicates that regulation is not a feature which necessarily involves

Fig. 6.13. The shoot apex of the columbine, *Aquilegia vulgaris*. *a*, A vegetative apex showing leaf formation from the flanks of the apical dome. By contrast, flower formation *b*, involves sequential development of many different types of appendage, and results in modification of the entire apex. Bar = 100 μm.

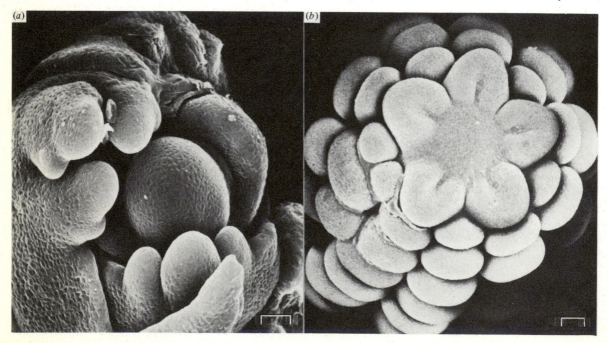

the entire structure of a meristem or an organ, but which probably acts as a level higher than that of the single cell.

Aside from these local considerations of the structure and activities of the growing points of the plant, it is clear also that regulation extends to the development of the plant body as a whole. This is a simple logical requirement, and is easily illustrated. Newly formed organs such as leaves have to be supplied with vascular tissue in order for them to obtain nutrients and water. The development of absorbing surfaces such as roots has to keep pace with the growth of aerial parts in order that the plant may maintain itself in a state in which it may function. The onset of flowering has to be controlled so that the plant is able to produce its progeny in sufficient numbers and at a time of the season which is appropriate. The developing embryo has to be protected from premature germination. Perennial plants have leaves with a limited lifespan; these leaves have to be shed in a regulated manner. In deciduous species, abscission of leaves is more or less simultaneous; in evergreens it may be more continuous but, nevertheless, under the control of the plant. Growth itself may show regulated seasonal variations controlled by dormancy mechanisms. Plants must be able to respond to environmental stresses; drought, for example, elicits wilting in order to conserve water supplies, and shading will cause a plant to accelerate its elongation rate in order to reach the light.

The role of plant growth substances

Some of these processes of the regulation of growth and development involve a spatial separation between the perception of the stimulus (for example, light level or quality) and the operation of the control. This suggests that the mechanisms of regulation in the plant might be expected to result from a transmissible product of a stimulus. We have already seen (Chapter 5) that plant growth substances can exert profound effects not only on growth and cell division, but on differentiation and organogenesis in cell cultures. It is a reasonable assumption therefore that the same substances might represent at least part of the mechanism of control of growth and development in the whole plant. It is the examination of this concept which will take up the rest of the present chapter.

In order to be satisfactorily implicated as an element in a control process, it is essential that a plant growth substance should conform to certain criteria. It should be present in the tissue, and at a level which is appropriate for the observed response. If possible, changes in the response which it is supposed to control should be correlated with changes in the level of the controlling substance. These apparently undemanding criteria are in fact very difficult to meet with certainty in real experimental situations. The reasons for this are several. Plant growth substances are present in cells at extremely low concentrations and, furthermore, they may be localised in particular compartments of the cell – ABA within plastids, for example. Growth substances may occur in different chemical forms with different activities. Finally, as we have already seen, there is considerable evidence to suggest that growth substances act in combinations to produce a range of effects on a variety of cell types. All these factors serve to complicate considerably the interpretation of experimental observations. The difficulties are increased when it is borne in mind that the detection of the presence of a growth substance may be made not by an analytical method, but by the use of a bioassay. In such circumstances what is detected is not a particular molecular species, but simply an 'activity' which again may be the result of the presence of a combination of molecules.

The presence of plant growth substances in roots

The presence of IAA has been demonstrated conclusively in roots of *Zea mays*, *Ricinus communis*, and *Lupinus luteus*. Gas chromatogaphy coupled to mass spectroscopy have confirmed without a doubt that IAA is found in normal tissues. Studies on the distribution of IAA in *Zea* roots have shown that the stelar tissue contains more IAA on a fresh weight basis than does cortical tissue, and that the root cap contains the highest concentration of all. The actual amounts of IAA present in the tissue are minute – a figure of 356 μg kg^{-1} fresh weight was obtained for root cap tissue, with around 100 μg kg^{-1} fresh weight in the stelar tissue. This represents a concentration in the *Zea* root as a whole of 5×10^{-7} M. The source of the IAA in roots is a matter for some debate. If auxins are applied to shoots, they are rapidly transported to the root tip in an intact plant. Segments of roots are able to transport auxin in a polar fashion from the base towards the tip. The rate of this transport is about 5 mm h^{-1}. It would thus appear likely that, in the intact plant, IAA in the roots is provided by the shoot. However, experiments with excised roots cast doubt on this idea. Excised roots in culture do not require an external supply of IAA for growth, which implies that if IAA has any role in root growth, roots must be able to synthesise their own supply. It has been suggested that the mature tissues may act as the source of IAA and, in particular, that IAA is produced as a by-product of the autolysis which accompanies differentiation of xylem elements.

The evidence concerning the presence of gibberellins in roots comes from the use of bioassays. These techniques have shown that cereal roots contain GA activity in amounts comparable to those found in shoots, and that cultured excised tomato roots also contain GA activity. Gibberellins have also been identified in the bleeding sap produced when plants are decapitated. This represents the contents of the xylem vessels, expelled by root pressure. It has been shown that the amount of GA activity in such exudates varies with the physiological status of the roots. For example, during flooding of the root system of the tomato (*Lycopersicon esculentum*) the level of GA activity in the sap was found to decrease, and this correlated with a decrease in the rate of elongation of the stem. Chemical analysis of the sap of *Acer* and *Ulmus* suggests that the gibberellins are conjugated and exist as glucosides in this situation. Whether GA is synthesised in roots is not clear; application of growth retardants, which are inhibitors of GA synthesis, has been shown to reduce the GA activity of root exudates of *Pisum*. In a chemical study of the bean plant, *Phaseolus vulgaris*, it was found that the major GA of intact plants was GA_1. Removal of the root tips lead to an accumulation of GA_{19}. This result suggests that root tips may not be the site of *de novo* synthesis of GAs, but of their interconversion. GA_{19} was thought to be synthesised in the shoot, then transported to the root where it was converted to GA_1. This implies a role for the root tip in modifying the activities of GAs, and also the need for transport mechanisms which act in different directions for different forms of GA. In segments of *Zea* roots transport of applied GA_3 is from the tip to the base – the opposite direction for IAA transport in the same tissue.

Cytokinins are known to occur in roots, and chemical identification of several different active molecules has been made. In *Pisum* for example, zeatin and its riboside are present, together with 2ip – 6-(3-methyl-2butenylamino) purine – and its riboside. It was found that the terminal millimetre of the root tip contained more than 40 times as much cytokinin as the next 4 mm. Substituted purines are found as minor components of tRNAs (p. 134). In pea (*Pisum sativum*) roots, the amount of cytokinin present in tRNA was 3.7% of that of free cytokinin. In a study of cytokinins in *Zea* roots, it was found that the detachable root cap contained only 20% of the cytokinin present in the terminal millimetre of the tip. Thus the highest concentration of cytokinins in root tips seems to be localised in the meristem of the root proper. Levels of cytokinins in xylem sap from roots show variations which correlate with applied stress. For example, flooding of the roots of sunflowers for 72 h produces a reversible decline in the amount of cytokinin in xylem exudate. The result of such a stress in the whole plant is the senescence of the lower leaves (Fig. 6.14). The biosynthesis of cytokinins by roots seems likely in view of these results, but is difficult to demonstrate due to the central role of adenine in metabolism. (p. 134).

ABA has been reported to occur in root extracts of *Pisum*, *Zea* and *Lens culinaris*. It has also been shown to be present in xylem exudates, but there seems little evidence to suggest that roots are an important synthetic source of this growth inhibitor. ABA may be responsible for the inhibition of growth which is the mediating mechanism of geotropism (p. 170). It is known to be present in the root cap; however, a second growth inhibitor, probably xanthoxin, is also present in root tips.

Ethylene is produced by root tips of *Pisum*, and this production is greatly stimulated by applied IAA. There is general agreement that many of the inhibitory effects shown by high levels of IAA are probably due to a stimulation of ethylene production. An interesting function for ethylene in roots is suggested by the existence of a tomato mutant which lacks this auxin-stimulated ethylene production. Roots of intact plants of this mutant fail to form lateral roots, and this abnormality can be overcome by spraying seedling leaves with IAA or by exposing the roots to ethylene. However, excised roots do not form laterals even when exposed to ethylene levels up to 1000 times those required for normal branching in the intact plant. These experiments seem to indicate a synergistic role of IAA and ethylene in the control of lateral root formation. They also serve as a clear illustration of the dangers of comparing behaviour patterns between whole plants and tissue segments or excised fragments of plants.

Control of root growth

It is thus fairly well established that all the major classes of growth substance are present in some form in roots, and it is not unreasonable to assume that roots may be a major source of production of cytokinins. What evidence is there for the effects of these endogenous growth factors on the growth and development of roots? There are three obvious processes which are important for the plant in respect of its roots. First they have to proliferate and form a continuous vascular system in order to supply water

Fig. 6.14. The effect of flooding on sunflowers (*Helianthus annuus*). The plant on the left has been watered normally; the plant on the right had its roots submerged in water for 5 days prior to the photograph being taken. The two lowest pairs of leaves have senesced almost completely, and the uppermost pair is beginning to lose chlorophyll at the leaf margins.

and minerals to the aerial parts of the plant. Second, they should show positive geotropism in order to seek out water in the soil, provide anchorage and indeed, in the case of roots initiated above ground, seek out the soil surface. Finally roots should be able to branch and produce root hairs in order to increase their absorbing surface as much as possible.

There is little good evidence for a role of endogenous growth factors in the stimulation of growth of roots. The highest concentrations of IAA and cytokinins occur at the root tip, and not in the region back from the tip where elongation is at its greatest. On the other hand, experiments which reduce the rate of polar IAA transport into the root do slow down its elongation rate. The sensitivity to externally applied auxins is very high; concentrations above about 10^{-8} M are often inhibitory. Auxin is known to be transported mainly within the stele of roots, but removal of the stele from segments of roots of *Zea* does not lead to any marked change in the elongation rate. It has been suggested that the function of the auxin in the stele is to maintain the pericycle in a meristematic state in order that lateral roots may be formed. GA has been shown to stimulate the elongation rate of root segments, but there is little evidence for an effect on intact roots. Similarly cytokinins are not thought to be involved in the elongation of roots, due to the lack of demonstrable stimulation effects. There is some limited evidence showing a correlation between the levels of endogenous growth inhibitors and root elongation rates. Light inhibits the elongation of roots in *Pisum*, and this has been shown to correlate with an increase in endogenous ABA levels. In *Lepidium sativum* (cress) light inhibition of elongation is correlated with an increase in ethylene production.

Geotropism

Geotropism is an example of the control of elongation in a rather special sense. When a root is placed in a horizontal position, it will turn downwards due to differential growth rates between the upper and lower parts of the root. The structure of the cells responsible for the perception of the gravitational field has been described previously (p. 110). It was originally thought that the curvature was the result of an asymmetric distribution of auxin in the root, which caused the upper part of the horizontal axis of the root to elongate faster than the lower part, and hence produce a downwards curvature. A series of ingenious and elegant surgical experiments with the roots of *Zea* has recently cast doubt on this idea.

Zea roots are characterised by the presence of a root cap which is easily removed (Fig. 6.15). The cap must be present for the geotropic response to occur. If half the cap is removed, the root will always curve in the direction of the remaining half. This is so even if a mica strip is placed between the growing part of the root and the cap region, provided that it is on the decapitated side. If such a mica strip is placed between the remaining half of the root cap and the elongating region of the root, then the curvature is away from the side with the root cap. It is possible to induce curvature in vertical roots by replacing their caps with caps taken from roots which have been growing horizontally for a few hours. When this is

done, the vertically growing root will turn in the direction corresponding to 'down' for the added root cap. If a horizontal root is bisected by a piece of mica which penetrates the root cap, then its curvature will be greater than if the root cap and the underlying region of the root itself are bisected.

These results are consistent with the idea that the root cap produces an inhibitor of elongation which moves back into the root. The inhibitor is asymmetrically distributed, being more concentrated in the lower part of the horizontal axis. This asymmetry is thought to arise both from greater production in the lower part of the root cap and also from lateral transport as the inhibitor moves back into the root. It has been found that certain varieties of *Zea* require light for the geotropic reaction. Such roots will not curve downwards when placed horizontally unless

Fig. 6.15. Geotropism in maize roots. *a*, The normal behaviour of intact roots. *b*, The response is prevented by removal of the root cap. Removal of half the root cap (*c*) results in curvature towards the remaining half. Insertion of a mica strip behind the apex does not alter this behaviour (*d*) unless it is on the root cap side (*e*). Removal of a prestimulated root cap and its replacement on a vertical root (*f*) causes that root to curve as though it were horizontal. Bisecting the root cap (*g*) with a mica strip does not abolish the curvature, but it is reduced if the mica strip extends back into the root proper (*h*). Broken lines represent the directions and magnitudes of movement of a growth inhibitor.

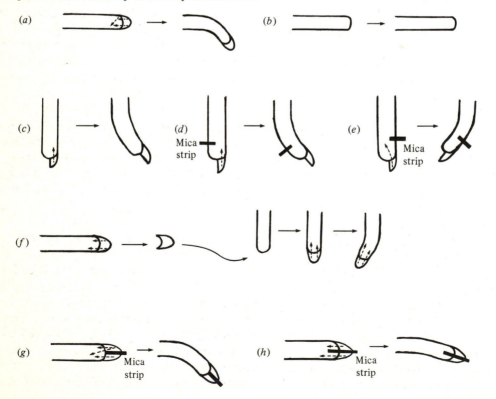

they are first illuminated. However, they will curve if the root cap is replaced by one from a horizontally stimulated root which has no light requirement. This suggests that the root cap itself is the site of the light perception, and this has been confirmed by local irradiation of small areas of the root. Geotropism is stimulated when only the root cap is illuminated, but if only the elongating part of the root is exposed to light, no geotropism is observed. Pretreating vertical roots with light before turning them into a horizontal position allows them to show curvature, which suggests that a preformed substance is used in the reaction. The decay of this substance can be measured in terms of how long the light period is 'remembered' during an intervening dark period before the root is turned horizontal.

The nature of the inhibitor is not known for certain, but ABA is present in the root cap of *Zea*, and its level is increased by a factor of about 50% in the light-requiring roots when they are placed in a horizontal position in the light as compared to the levels in the dark. Furthermore, agar blocks containing ABA in the concentration range of 10^{-8}–10^{-4} M produced an increasing series of curvatures when applied to the decapitated root tips of either normal or light-requiring varieties. In a recent analytical study, it was found that a transient difference in the level of ABA in the upper and lower parts of horizontal *Zea* roots could be measured, amounting to about a 28% excess of the inhibitor in the lower part of the root. All these facts suggest that ABA may be the endogenous growth inhibitor responsible for mediating the geotropic response. However, they do not eliminate the possibility that other inhibitors may be present; it is also feasible that the response is the result of a combination of inhibition and stimulation caused by more than one class of growth factor. Even highly sensitive and specific analytical techniques such as have been used to measure ABA levels in these roots give no idea of the cellular distribution of the molecules which they detect, or of their potency in particular tissues. It is also remarkable that although applied ABA shows an effect over four orders of magnitude of concentration, the measured differences in ABA in intact roots is only 28%.

Lateral root formation

Lateral roots arise from the pericycle at the periphery of the stele. Their position is not random; in herbaceous plants they may be formed opposite the vascular bundles, and they are also rather regularly spaced along the long axis of the root. In a growing root the first lateral will normally appear before the region which contains the mature protophloem elements. The distribution of lateral roots along a root axis itself suggests that the process of their formation may be under the influence of gradients of inhibitors (laterals are not formed close to the apex) or promoters. Experiments with excised roots growing in liquid culture have shown that auxin markedly promotes the formation of laterals; in root segments of *Haplopappus ravenii* up to 60 lateral roots per centimeter were formed in response to the presence of α-naphthalene acetic acid in the medium. This stimulation of lateral formation could be antagonised by the presence of the cytokinin benzylaminopurine. GA was without effect in this test system. Root tips are thought to be a source of cytokinin in the plant, and thus it is certainly possible that endogenous cytokinins from the tip represent the inhibitory influence which prevents lateral formation in this position. ABA in the root cap would also be expected to have a similar effect. Experimental results with ABA are in fact contradictory; some reports claim that it is inhibitory to lateral formation, whereas others suggest that it may stimulate the process. A role for ethylene in lateral root formation is suggested by the existence of the mutant of tomato which does not normally form laterals in the presence of small amounts of externally applied ethylene (p. 168).

Very little convincing information has come from all these studies in terms of whole-plant control of lateral root formation. The confusion no doubt rests in large part on the use of excised roots as a test system. The unsatisfactory nature of using only parts of plants for investigating normal control processes was clearly illustrated above (p. 168).

Hormones and leaf development

The hormonal physiology and status of the stem apex is largely unknown at present, although dramatic effects can be produced within apices by the external application of hormones. Most of our present knowledge concerning endogenous levels of hormones is related to the shoot as a whole. Rigorous identification of IAA, several GAs, ABA and ethylene have been reported from shoots of various species; cytokinin activity has been reported in diffusates and extracts of apical buds of *Dahlia*. Thus it appears likely that all the major classes of hormones are present in the shoot.

Leaves are known to contain auxins; if a leaf is detached from the plant and its diffusate into agar is tested for auxin activity by bioassay, it is positive. Interestingly, the ability to obtain auxin activity by diffusion from young leaves is light dependent. Tips from dark-grown plants yield little or no diffusible auxin, although extraction of the tissue allows the presence of auxin activity to be demonstrated. When the distribution of diffusible auxin is examined in a maturing leaf, it is found to decrease at the tip of the leaf first. This corresponds to the fact that in a maturing leaf, growth ceases first at the tip (p. 162). There is also a general decline in the amount of auxin which diffuses out of a leaf with age. This means that in woody perennial plants, for example, there is a surge of auxin production as the buds break and the leaf primordia begin to grow in the spring, and this parallels the rise in cambial activity in the stem. The stem of a plant commonly shows a gradient of diffusable auxin, with a maximum towards the base, although the best sites for production of auxin are the young leaves at the tip. This is due to the operation of polar transport of the auxin, which will be discussed in the next chapter (p. 199). There are a few reports of correlations between growth rates of the shoot and the presence of growth substances. In *Gingko biloba*, for example, there is a peak of auxin activity which precedes the period of maximum elongation of the shoot. In rice (*Oryza sativa*), there is a good correlation between growth rates throughout the life cycle and the presence of GA activity in extracts. Correlations between extractable cytokinin activity and growth rate have not been found.

There is evidence that leaf development may be influenced by combinations of different growth substances. In the sweet potato (*Ipomoea batatas*), for example, it was found that both IAA and GA cause elongation of young leaf primordia when applied separately; if they are applied together, however, the effect obtained is much greater, even though each is supplied at its optimal concentration. In cultured tobacco, the ratio of cytokinin to GA is critical in the development of normal leaf shape. When GA is present in great excess, the leaves formed are long and thin. Addition of cytokinin to the medium causes an increase in cell expansion in the blade and inhibited elongation, so that the leaf assumes normal morphology as a result. Of some interest in this context is the observation that if roots of tomato plants are flooded, the GA content of the shoot decreases. This would seem to indicate that there is a relationship between the cytokinins produced by the root (and decreased in amount by flooding) and the synthesis of GA which is thought to take place in the shoot.

GA has an effect on the type of leaves produced by plants which show heteroblasty. Unfortunately it seems to act in different ways in different species. Thus in ivy (*Hedera helix*), adult leaves are induced in juvenile plants by the application of small amounts of exogenous GAs. Root formation on the stem is also inhibited. In *Ipomoea*, the reverse is true; application of GA to adult plants induces the production of juvenile leaves.

As we have already seen (p. 136) ABA is probably synthesised mainly in the chloroplasts of leaves. It accumulates in this particular compartment, and thus is not readily available in diffusion experiments. However, wilting on the leaf will cause leakage of the ABA from the chloroplasts and this gives rise to a rapid and dramatic increase in the amount of ABA which can be extracted by diffusion. This mobilisation of ABA is thought to be the mediating response in stomatal closure. If excised leaves are dipped into solutions of ABA, rapid stomatal closure results.

These results, and many others, serve to emphasise the complexity of the effects of growth substances within the developing shoot. On the one hand, experiments with excised tissues and cultured apices suggest strongly that applied growth factors can exert profound effects on the morphogenesis of structures at the shoot apex. On the other hand, very little hard information is available concerning the endogenous levels of such factors and the nature of their interactions *in vivo*.

Apical dominance

The development of leaves by the shoot apex is highly ordered. As we have seen, surgical experiments with apices suggest that each new primoridium is surrounded by a zone of inhibition. The nature of this inhibitory process is unknown. The inhibition may be overcome by treatment of the apex with externally applied growth substances; for example, treating the apex of *Lupinus albus* with IAA in lanolin causes fusion of primordia and the development of ring fasciations. This should come as no surprise at this stage; equally it must be admitted that such observations cast little if any light on the causes of normally operating controls at the shoot apex.

Rather more amenable to analysis in terms of endogenous controls are various growth correlations in the plant, as opposed to actual morphogenetic events taking place in tiny volumes of apices. One of the best studied of these is the phenomenon of apical dominance. The formation of lateral buds in leaf axils proceeds to various extents in different plants. Lateral buds may be easily visible, or they may be very much reduced. They all have in common the fact that, at some stage, their development will be arrested and will not proceed until the growth of the stem results in displacement of the bud well away from the apex, or until the apex is itself removed or physically damaged. Thus, in a typical plant stem there is a gradient not only in the age of leaves with the youngest at the top nearest the growing point, but also in the development of lateral branches. This phenomenon of the priority of apical over lateral growth, is called apical dominance. Because of its gradual nature, it is reasonable to suppose that it is controlled by gradients in the stem.

Initially it was thought that gradients of nutrients controlled the development of lateral buds. The apex was regarded as an active sink of nutrients which deprived the lateral buds of their means to grow. There is some evidence to suggest that this is, in certain conditions, part of the explanation. For example, plants of flax (*Linum*) show very strong apical dominance if grown under conditions of low nitrogen supply, but will branch quite freely when nitrogen is readily available to the roots. *Coleus* plants branch more readily when grown in strong light than in poor light. Both these environmental responses mean that the plant limits its growth effectively to match the conditions under which it is grown. However, early experiments cast doubt on the general validity of the nutrient theory. If a minor bud on a potato (*Solanum tuberosum*) tuber is isolated surgically from the main terminal bud, which is normally dominant, then the minor bud is stimulated into growth (Fig. 6.16). This, and similar observations, suggested that it might be the production of an inhibitory substance by the dominant bud which prevented the growth of lateral buds. Thus when the dominant bud or shoot tip is removed, the source of the inhibition is also taken away and lateral branching results.

The role of auxin

When auxin was discovered, it was quickly demonstrated that its application to the cut stem after removal of the shoot tip prevented the formation of lateral branches (Fig. 6.17). This suggested strongly that it is the auxin which is produced in the developing shoot apex which is the endogenous inhibitor controlling lateral bud development. Normally the dominance of the apex is so strong that the presence of leaves along the stem is not very important. However, it was found in *Syringa* that the number of lateral buds which developed following simple decapitation of the shoot tip was increased further by the removal of the leaves. It was also found that if auxin was applied to the cut petiole remaining on the stem where a leaf had been removed, the lateral bud in that axil remained dormant (Fig. 6.18). Thus it appears that lateral buds are held in check not only by the apex itself, but by the leaf with which they are associated. This may be presumed to have considerable survival value in plants which are subject to grazing. Damage to the main axis or to the foliage will produce replacement shoots in appropriate positions.

One of the interesting features of apical dominance is that in some plants removal of the apex of a shoot will result in the most apical lateral buds being stimulated into growth, whereas in others the most basal buds develop. This behaviour may vary within a single species depending upon the conditions under which it is grown. In *Sambucus*, for example, the most basal buds develop at a temperature of around 15 °C, whereas if the temperature is around 25 °C, it is the most apical buds which grow. It can be rationalised that this behaviour might be a result of relative rates of auxin transport and production. If auxin is produced high on the stem and transported in a polar fashion towards the base, clearly its actual concentration at any point in the stem would be a function of both the rate of production and the rate of transport. Under conditions of low production and high transport rates, the upper part of the stem would be more depleted of auxin than under the reverse conditions. The importance of transport phenomenon is also indicated in special cases. In horizontal stems of *Cordyline*, buds on the lower part of the stem develop into rhizomes, whereas those on the upper part become leafy shoots. In apple trees, buds on the lower sides of horizontal stems tend to produce flowers, whereas those on the upper sides give rise to vegetative shoots. These patterns of behaviour may well reflect lateral as well as axial transport of auxin in the stem.

Auxin is, of course, not normally regarded as an inhibitor of growth. Its action in the inhibition of lateral bud development has been studied using isolated buds attached to a piece of internode from etiolated pea seedlings. It was found that, at a concentration of only 10 ppm, IAA was almost 100% effective in preventing the development of the lateral bud. Synthetic auxins such as NAA and 2,4-D were even more potent in this assay system, and retained their effectiveness for a longer period. This reflects

the ability of the tissue to metabolise IAA. It was found that if the application of IAA were delayed for 24 h following the excision of the segment from the plant, the bud was able to grow and this growth was not completely arrested by IAA. Using radioactively labelled auxin it was found that the degree of inhibition which any bud demonstrated was closely correlated with the amount of IAA it had accumulated. This type of experiment also showed that IAA applied to cut stem segments could find its way into the lateral bud.

Superficially, all these results add up to a fairly convincing case for transported auxin being a signal which prevents the development of lateral buds. However, the situation is not so simple as it first appears. The essential feature of the model is that IAA can find its way into the bud and there accumulate to produce a concentration which is superoptimal for growth. Whilst this seems to have been

Fig. 6.16. Apical dominance in potato (*Solanum tuberosum*). On the shoot in the photograph (left), one subsidiary bud has been isolated from the main shoot by means of a cut (arrow). This isolated bud has, as a result, grown out whereas a similar bud left undisturbed is still under the dominant influence of the main bud. The drawing (right) is a tracing of the photograph and shows the position of the buds and the cut (arrow).

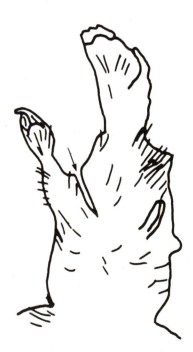

demonstrated by the work on excised stem segments which has just been described, other findings have contradicted the model. Direct determination of the amounts of auxin in lateral buds of *Syringa vulgaris* and *Lupinus* have shown that the amount of auxin present is not superoptimal. It is also now known that the application of small concentrations of auxin to cut stems can actually increase the growth of lateral buds, and that the inhibition which was originally observed is a function of high concentration.

Other factors in apical dominance

Several other lines of work have pointed to a role in bud break for cytokinins. For example, in the work with cytokinins and bud formation in callus cultures (Chapter 5) it is frequently observed that buds may be induced very close together on the surface of a callus, and appear to suffer no inhibition from the proximity of neighbouring buds. In the isolated pea internode system, it was found that kinetin could relieve the inhibition produced by application of IAA, and that this effect depended upon the relative concentration of the two growth substances. If a derooted (but otherwise intact) plant is placed in a

solution containing kinetin, the apical dominance which is normally present may be overcome. Finally, direct application of cytokinins to inhibited buds can cause relief of their inhibition. All these observations suggest that cytokinins arising in the root system of the plant may have a role to play in the degree of inhibition which is found in any particular bud.

There is no evidence to suggest that either GA or ethylene have any role to play in apical dominance. The latter fact is of particular interest in view of the possibility of ethylene being produced in experiments using high concentrations of auxin. However, it has been found that in *Xanthium* plants grown under long-day conditions, apical dominance is increased by the addition of a short period of far-red illumination at the end of the light period. This was correlated with a large increase in the amount of ABA in the inhibited buds, an increase which was dependent on the presence of the youngest leaves.

A possible morphological mechanism for the operation of apical dominance has been suggested. The growth of a bud clearly requires that it should be

Fig. 6.17. Effect of auxin in apical dominance. Removal of the top of the stem (*a*) causes lateral buds to break (*b*). However, if the top of the stem is replaced with an agar block containing auxin – see arrow – (*c*) the lateral buds remain dormant.

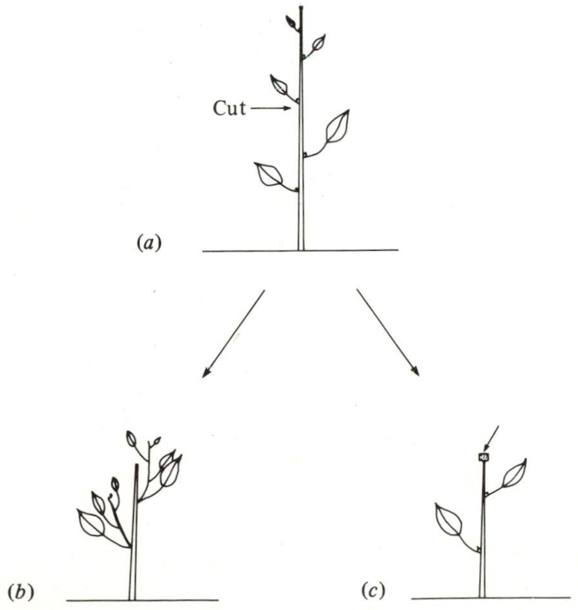

Fig. 6.18. Effect of leaves on apical dominance. Removal of the top of the stem and the leaves encourages the breaking of all the lateral buds in the leaf axils (*a–b*). However, if some of the leaves are replaced with agar blocks containing auxin (arrows), the lateral buds in those leaf axils remain dormant (*a–c*).

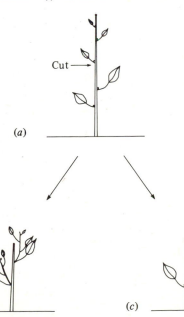

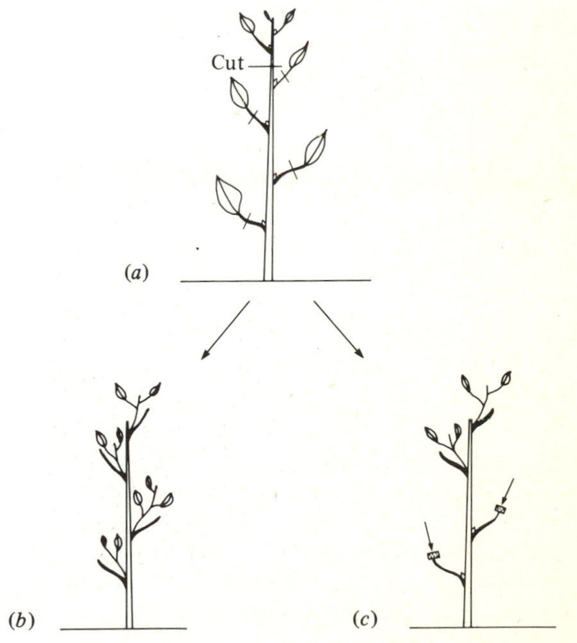

supplied with nutrients and water, and this means that it must be connected to the vascular system of the whole plant. Normally this occurs by a process of xylem differentiation in both acropetal and basipetal directions, so that preformed vascular tissue in the base of the bud becomes connected to the system in the stem. In the experiments with isolated pea internodes, it was found that buds which had been released from inhibition were in possession of a well developed vascular system, whereas those which were inhibited by auxin were not. In experiments using decapitated stems, it was found that this connection could be controlled by external application of auxin. If the amount of auxin applied to the cut end of the stem exceeded that applied to the branch bud, connection between the two vascular systems was not made. The reverse was true; auxin applied in greater concentration to the branch allowed vascular connections to be made.

This system is clearly fairly far removed from normal physiology, but it does offer some explanation for the auxin effect. It has some counterpart in studies with wound vessel formation (Chapter 7). In the intact plant it can be envisaged that the relative amounts of auxin in the stem (controlled by production and transport from the tip) and in the bud might affect the degree of vascularisation to the bud. Drastic reduction of the supply of auxin in the stem, caused by decapitation, would then change the relative concentrations in the two situations, and hence allow lateral bud development. Unfortunately for this simple and elegant idea, there are many examples known where growth of the bud precedes the development of a good vascular system, and furthermore, there are also examples of inhibited buds which appear to have perfectly adequate vascular connections to the stem.

Leaf senescence

So far our discussion of the organisation of the plant body has been concerned with growth processes: the formation of new structures or their enlargement. The need for the regulation of development in such processes is quite obvious for, without it, the plant would quickly develop into a uncoordinated mass of different tissues and organs, quite unsuited to the efficient exploitation of its environment. The same considerations apply with equal force to processes of decline and death. This can be illustrated by an examination of leaf senescence.

The most striking manifestation of leaf senescence is the simultaneous change which takes place in the autumn in the leaves of deciduous trees. Senescence is, however, a natural fate of any mature leaf after a certain period of photosynthetic function. The reasons for the shedding of leaves are not hard to find. In a growing plant the formation of new leaves at the tip of the shoot will increasingly result in the shading of older leaves in lower parts of the stem. This will have the effect of reducing the efficiency of the lower leaves as net producers of photosynthetic products so that, eventually, they may become a burden to the economy of the plant. Again, transpiration from leaves is an essential mechanism for the retrieval of nutrients from the soil and is not normally detrimental to the plant, provided that the uptake from the roots can match the loss of water from the leaves. However, during cold seasons, when root action is curtailed, massive continuing transpiration would lead to net loss of water. Hence, a deciduous tree will shed its leaves at the start of winter in order to avoid this possibility. In both these situations, senescence precedes shedding, and represents a mechanism for the partial recovery of useful nutrients from the leaf before it is lost to the plant. Senescence is not a random process, then, but one which is controlled in both space and time.

There are two ways of studying senescence. Using intact plants, it is possible to detect degenerative changes as they occur in basal leaves (Fig. 6.19). The first sign of incipient senescence in leaves of *Phaseolus vulgaris* is the formation within chloroplasts of increased numbers of lipid droplets amongst the thylakoid membranes. At a gross level this corresponds to a change in colour of the leaf from green towards the eventual yellow of the fully senescent organ. In cells of these leaves it was found that the plasma membrane appeared intact, and that both chloroplasts and mitochondria could still be recognised, although in a much altered form. The chloroplasts were filled with densely stained lipid droplets, and assumed an overall spherical appearance. Thylakoid membranes were absent. Mitochondria contained swollen and distorted cristae. The nucleus was also still present in fully senescent cells, together with its membrane, but the contents of the nucleus were less densely stained than in control leaves. These findings suggest that not only at a physiological level within the whole plant, but also at a cellular level, the process of senescence is quite closely regulated. Wholesale release of hydrolytic enzymes, for example, would be expected to produce far more degenerative changes than are in fact found. In this regard, the senescent cell should be contrasted with the fully differentiated xylem element (a product of a developmental process which involves the total loss of all cytoplasmic contents).

In detached leaves it is possible to study the biochemical events in more detail. On detaching a leaf from the plant, the first sign of senescence is a rapid fall in the amount of protein present; this may take place during the first few hours after detachment. This fall is accompanied by a rise in the quantities of free amino acids in the leaf, and these changes accelerate in time. The chlorophyll content of detached leaves is stable for the first day if the leaves are kept in the dark, but then it too begins to fall at an increasing rate.

One of the first effects of a growth substance on senescence to be discovered was that produced by cytokinins. If a drop of a solution which contains kinetin is placed on the surface of a detached leaf, the area covered by the solution will remain green long after the rest of the leaf has senesced (Fig. 6.20). This is found to correspond to a maintenance of the rate of protein synthesis in the treated part of the leaf. Protein synthesis in a detached leaf usually continues for up to 40 h after excision, after which time it stops. The effect of kinetin is to delay this decline.

Cytokinins are thought to be produced in the root. When root exudates were examined for their effects on senescence it was found that as well as containing substances which inhibited the process, they also contain a promoter of senescence. This has been identified as the amino acid L-serine. As a further complication it was found that the promoting effect of L-serine could be antagonised by L-arginine, although L-arginine had no effect on senescence when applied to a detached leaf in the absence of serine. This represents a complex balance between inhibitors and promoters of senescence, particularly when it is remembered that the early stages in the process involve the release of amino acids from proteins in the leaf. When the effect of the protein

Fig. 6.19. Part of a senescent leaf of celery (*Apium graveolens*). The leaf itself was quite yellow when fixed, and the abnormal structure of the chloroplasts reflects this. Nonetheless, plastids and mitochondria are still clearly recognisable. Bar = 1 μm.

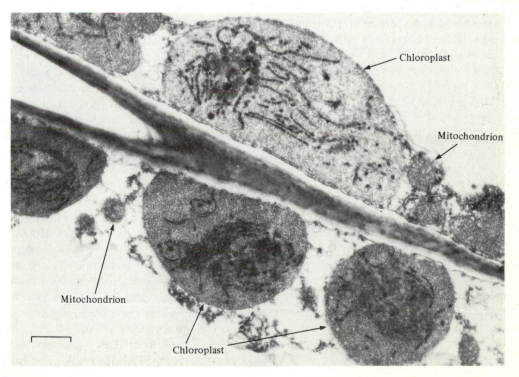

synthesis inhibitor cycloheximide was studied, it was found that its presence at a concentration sufficient to inhibit protein synthesis also inhibited senescence. In other words, protein synthesis is necessary for senescence to proceed. This may point to the need for the synthesis of hydrolytic enzymes *de novo* involved in bringing about the degenerative changes which occur within the cells. When the levels of proteolytic enzymes in detached leaves are measured, a rise is observed during the first 4 days of senescence. This rise is enhanced by the presence of L-serine.

An interesting effect emerges when the cellular respiration rate of detached leaves is examined. In untreated leaves the rate rises rapidly after the first day, and may reach a level of two or three times its initial value on the third day. This rise is not seen in leaves which are treated with kinetin. When the sensitivity of the respiration to the uncoupling agent dinitrophenol is examined it is found that the increased rate in control leaves is not affected whereas

Fig. 6.20. Effect of kinetin on leaf senescence. The picture shows a sunflower leaf which has almost totally lost its green colour following detachment from the plant. The dark circular areas in the upper part of the leaf are islands of green tissue remaining where drops of a solution of kinetin in water were allowed to dry on the leaf surface at the time of its detachment from the plant.

the lower rate of respiration present in kinetin treated leaves is. This leads to the suggestion that one of the primary triggers for senescence is the uncoupling of respiration from energy production. This would result in a change in the balance between synthesis and breakdown of polymeric materials such as proteins and RNA. This is an appealing idea, but one which does not withstand some observations; in particular that the respiration rates in senescing attached leaves do not always show a rapid rise during the early stages of the process. It is as likely, for example, that net loss of protein would result from the rapid export of amino acids from the leaf as it begins to senesce. This in fact does occur with leaves which are attached to the plant, although of course it is not observed in experiments with detached leaves.

There is some evidence to suggest that the process of senescence at the cellular level is under the control of the nucleus. For example, in an experiment in which mesophyll cells of *Elodea densa* were plasmolysed and observed over a period of 8 days in culture, it was found that those fragments of cytoplasm which contained a nucleus senesced more rapidly than enucleate fragments. Isolated chloroplasts often senesce more slowly than do chloroplasts which are left in intact detached leaves. The behaviour of a mutant of *Festuca pratensis* also sheds light on the role of nuclear genes in senescence. This mutant is characterised by a failure of the leaves to turn yellow during senescence – due to retention of chlorophyll. The characteristic is inherited in a Mendelian fashion, and hence is probably due to a nuclear gene.

These studies shed little light on the likely control processes operating in whole plants. Certainly the presence of endogenous cytokinin may prevent senescence – flooding of roots, for example, is often followed by senescence of basal leaves. In addition, senescence can in some cases be reversed; for example, removal of the shoot of mature tobacco plants will reverse the senescence of the lower leaves, as will spraying the leaves with solutions of ammonium nitrate. In such circumstances the dilemma of the plant physiologist is made abundantly clear. One experiment (removing the shoot) might be expected to alter endogenous levels of controlling substances, but the other (spraying with a nitrogen source) is much more likely to have a direct metabolic effect on the cells and their nutritional status.

Abscission of leaves

The senescence of leaves of dicotyledons is usually followed by leaf fall, or abscission. Abscission is not an invariable successor to senescence however; grasses do not in general show it, as evidenced by their turning brown in late summer without loss of leaves. Ferns are also very slow to abscind – the brown leaves of bracken, for example, are a common winter sight. Finally, abscission is not necessarily a seasonal phenomenon; the leaves of evergreens may survive for several years before senescence and abscission take place.

In contrast to senescence, which takes place in all the cells of a leaf, the cells active in the abscission process are located in a small region, the abscission zone (p. 126). The control of the process of abscission seems to be fairly clearly related to the production of ethylene. Abscission does not normally occur until the leaf has senesced. This is of advantage to the plant as a whole since it allows a recovery of the amino acids and other nutrients which are produced during senescence. It appears highly probable that it is ethylene produced during the terminal stages of senescence that induces the changes which lead to abscission. Ethylene is known to be a potent promoter of abscission, and may show this effect at concentrations as low as 100 ppm.

Using isolated tissue pieces with abscission zones, it has been found that auxin can prevent abscission, but only up to a certain stage in the progress of the event. Thereafter, it will promote abscission. This is taken to mean that auxin may prevent the onset of abscission, but that once the process has begun, the effect of auxin is to increase the rate of ethylene production and hence accelerate abscission itself. In whole-plant terms, it could therefore reasonably be supposed that the auxin which is exported from the young leaf (in amounts which decline as the leaf reaches maturity) acts to prevent premature abscission of the leaf before its useful life is over. The control of abscission therefore seems to be relatively well explained in terms of signals given by growth substances. In this regard, it perhaps should be emphasised that although abscisic acid was discovered in part as a result of work on the abscission of cotton fruits, it does not seem to have a general role in the process. Indeed, in certain situations, such as abscission in *Begonia*, ABA antagonises the effects of IAA in promoting abscission. This is a complete reversal of the usual pattern of inhibition and promotion by these two substances.

The process of abscission itself, in chemical and biological terms, involves the ethylene-stimulated synthesis and secretion of hydrolytic enzymes which weaken the walls of the cell in the abscission zone. This, coupled with cell expansion and division on the stem side of the abscission, produces the separation effect (p. 127).

Plant growth substances and organisation

In this chapter and in Chapter 5, we have seen that the plant growth substances are very important in the life of plants. Applied to isolated cells and tissues they can induce new patterns of behaviour and, applied to whole plants, they may affect styles of growth. In addition, they are present in the major organs of the plant and, in a few cases, it appears that changes in their level may be correlated with changes in behaviour of tissues and cells. What can be said of their role in the normal plant?

There are two types of answer to this question. The first has been the more generally accepted, and that is that the plant growth substances work in the manner of hormones. The widespread continuing use of the term 'plant hormones' attests to this approach. Before accepting this idea however, it is as well to consider its implications in more detail. Some of these have been summarised previously (p. 130), but will now be restated before considering an alternative view of the importance of growth regulators.

In animals, hormones are products of particular cell types. They are transported in the blood stream and have their effect on distant cell types. This effect is mediated by changes in concentration of the hormone. Much research on plant growth substances has been directed on the assumption that a similar state of affairs may exist in plants. It is clear however, that the analogy between animal hormones and plant growth substances is not a very good one. Localised synthesis of plant growth substances does not seem to occur. They are present in the vascular system and may move long distances by this route, but target cells for their action have not been clearly identified. Furthermore, very many of the attempts to correlate changes in endogenous levels of growth substances with changes in growth rate or behaviour have failed.

Many of these difficulties of the hormonal theory of plant growth regulation may fairly be ascribed to technical inadequacies. The level of growth substances within tissues is usually very low, and this has meant that chemical identification of specific molecules is difficult. As a result, early work, and much current research, relies on the use of bioassay to detect levels of growth substance activity. This means that living material (for example, the oat coleoptile, or tobacco pith) is used to assay the changing levels of a growth substance. Clearly, since the test tissue is itself alive, this may disguise the real nature of a chemical change in the substance being tested. Thus if a growth stimulus declines between two tests, this may mean that the concentration of a growth stimulator has decreased. On the other hand, it might also mean that the concentration of a growth inhibitor has increased.

The most modern research relies increasingly on sophisticated chemical techniques for the detection of specific molecules and this is, in some cases, causing a change in the explanation of responses to growth regulators. An example of this may be given in the case of genetically dwarf plants. These can be restored to more or less normal habit by the application of exogenous gibberellin. However, when endogenous levels of gibberellins are measured in a range of normal and dwarf peas, it is found that there is no good correlation between actual concentrations of the growth substance and the ultimate height of the untreated plant. In apical dominance, there is no doubt that applied auxin can duplicate the effect of an absent shoot tip. On the other hand, levels of auxin rise in buds as they break into leaf. Finally, when actual changes in concentration are measured during a response which is supposedly mediated by a growth regulator, it is often found that they are very small. This is true in stomatal closure and also in the lateral redistribution of inhibitors during geotropic curvature of roots.

Proponents of the hormonal theory may rightly point out that these results are obtained in too crude a fashion. Since the mechanism of action of growth regulators is not known, it is possible that their local concentration at their sites of action varies more dramatically than analysis of even small tissue pieces would suggest. This may well be the explanation for many of the observed discrepancies in experimental results which are a common feature of research with plant growth regulation.

Growth substance sensitivity

An alternative to the view described above has been propounded. This is that far from being designed to respond to small changes in the levels of growth substances, plant cells are in fact very insensitive to such changes. The evidence for this is based largely on the argument that in bioassays, test tissues respond to growth substances over a very wide range of concentration – often four or five orders of magnitude. It is proposed that in the intact plant the response of a cell to a growth substance is not finely tuned to the concentration of that substance, but that it is due to an intrinsic sensitivity on the part of the cell. Thus, growth correlations reflect not merely gradients or changes of concentration of growth substances, but also the different levels of sensitivity of the cells within a gradient. This idea predicts that changes in growth rate or pattern will not necessarily be reflected by changes in concentration of growth substances, since what is actually changing may be the sensitivity of the cells themselves. This can be expressed in an oversimplified way as follows. Suppose that the first action of a growth substance is to bind to a receptor. The effect of the growth substance will presumably depend on the level of such binding. Clearly the amount of the growth-substance–receptor complex depends not only on the amount of the growth substance itself, but also of the receptor. Thus a cell with more receptor sites would show more of an effect to a given amount of growth substance than a cell in the same environment with fewer receptors.

This novel idea is far from being accepted by the majority of workers in the field of plant growth regulation (not least because it implies that much of the research in this area is being conducted in the wrong way). The idea itself suffers seriously from the defect that until receptors for plant growth substances can be identified, it cannot be directly tested. The question also arises as to why plants need growth regulators at all if individual cells can determine their own sensitivity. The most plausible explanation for this seems to be that the growth regulators may represent a more or less constant environmental feature of each cell in an organism where most environmental parameters are highly variable. Examples of variable parameters include temperature, light level, and the nutrient status of the cellular environment. Against such a chaotic background of variables, it is argued that plant growth regulators might act as a stable reference point to growth and development. The hormonal theory, in complete contrast, requires that cell growth is limited and determined by changing plant growth substance concentrations.

Organisation and electrical fields

In order to place these general remarks in some perspective, it is instructive briefly to consider the situation relating to growth effects and electric fields in plants. The idea that electrical fields may be important in plant growth has interested a minority of plant physiologists for many years. Early work consisted of direct measurement of potentials in various plant organs and even whole plants, and these produced a wealth of data, some of which is very interesting. For example, in a series of measurements on intact trees of the Douglas fir (*Pseudotsuga menziesii*), it was found that the main shoot axis was at a positive potential compared to subsidiary, lateral shoot tips, and that these were themselves at positive potentials compared to lower parts of the stem. In this situation, the positive potential of the tissues correlates exactly with the growth dominance. In intact roots of the

onion, *Allium cepa*, it was found that the region of the root immediately behind the tip was the site of the most positive potential. This is the region of most rapid elongation of the root. Subsidiary positive peaks of potential were found along the root at points of lateral root outgrowth.

Much of this type of early work can be and has been dismissed on technical grounds. However, recent advances have put the measurement of very small potential differences on a much firmer footing. In particular, the invention of the vibrating probe technique means that nowadays it is possible to measure the distribution of ionic currents around the surface of a single cell. Briefly, the apparatus for such measurements consists of a very small platinum black ball (30 μm in diameter), which is vibrated at high frequency over a distance of a few tens of micrometers by means of a piezoelectric crystal. Any steady potential difference which exists between the two extreme positions of the vibration will be translated into an alternating voltage of the same frequency as the mechanical vibration of the probe. Because of this, and the possibility of filtering out random noise, the vibrating probe can reliably measure very small potential differences indeed, of the order of 10^{-9} V. One of the best known applications of the technique has been to study current patterns around the outside surface of intact eggs of the *Fucales* (p. 197). Here it is found that a 100 pA current of density approximately 1 μA cm^{-2} enters the rhizoid pole and leaves the cell at the thallus pole. Other systems which have been examined using this technique include germinating pollen grains and root hairs (Fig. 6.21). In each case, a small current is detected which enters the tissue at the growing point and leaves it at a basal point. The

Fig. 6.21. Patterns of current distribution around intact cells. *a*, In the germinating *Pelvetia* zygote, current enters at the rhizoid pole; *b*, in a germinating pollen tube, current enters at the tip of the tube; *c*, in a root hair of *Hordeum*, current enters at the tip of the root hair.

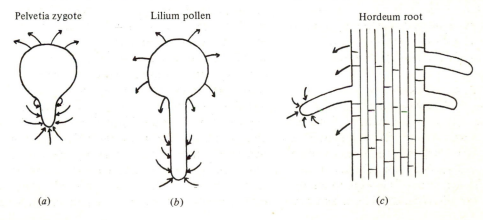

Pelvetia zygote   Lilium pollen   Hordeum root

(a)   (b)   (c)

presence of such currents lends some authority to the earlier observations of a grosser nature made with less reliable equipment.

For such electric effects to have any importance in normal growth, two further sets of circumstances must be satisfied. First, the patterns of current distribution should be liable to change by environmental factors which affect growth. In the early work, this was shown by means of changing parameters such as oxygen tension, temperature or light levels. Variations in these conditions lead to variations in the magnitude of the potentials present in intact plants. The modern equivalent of these experiments may be illustrated by reference to the behaviour of roots of the plant *Lepidium sativum*. When these roots are grown in a vertical position, they show a reproducible pattern of current distribution (Fig. 6.22*a*). Current enters the root at its tip, with a maximum density of about 1 μA cm^{-2} in the region 500–1000 μm from the tip of the cap. The pattern of this entry is symmetrical around the axis of the root. Turning the root to a horizontal position changes the pattern of current distribution in a way which is consistent with its leaving the tip of the root on the upper surface and entering it at the lower surface (Fig. 6.22*b*). These changes in current distribution could be detected as early as 3 min after changing the position of the root. Curvature was not visible until 7 min had elapsed with the root in the horizontal position. This change in the current distribution therefore precedes the change in growth pattern.

The second necessary condition for these

Fig. 6.22. Patterns of current distribution around the root tip of *Lepidium sativum*. In the vertical root the patterns of current distribution are symmetrical around the root axis. In the root which is turned into a horizontal position, current appears to enter the lower part of the root and exit from the upper part.

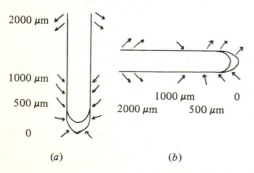

(a)   (b)

effects to be of likely physiological importance is that applied voltages should be able to influence the growth patterns of cells or tissues. Early work with onion (*Allium cepa*) roots suggested that growth could be accelerated by applied voltages which were in the same direction as the naturally occurring field, and growth could be retarded by applied voltages which opposed the natural field. The modern equivalent of such experiments is the determination of polarity in *Fucus* eggs and moss protoplasts which will be described in the next chapter. These experiments show without doubt that externally applied electric fields can control the behaviour of intact single cells.

Taken together the observations present an intriguing comparison with the hormonal theory of growth regulation. Electrical fields are present in intact growing tissues. The magnitude of the electrical fields varies with growth conditions, and externally applied electrical fields may control growth patterns. Each of these statements could be repeated with the words 'growth substances' in place of the words 'electrical field'. What then, is the connection between the two, if any? This is a highly speculative area at present, but it might be suggested for example that the distribution of any receptors for plant growth substances could be influenced by currents flowing through a tissue. As we shall see in the next chapter, there is some evidence to suggest that polarity in single cells, and the phenomenon of polar auxin transport, can both be described in terms of asymmetries in the distribution of proteins within the fluid mosaic membrane. It is not an unreasonable suggestion that proteins could be constrained to move within the membrane under the influence of small electrical currents. In order to assess the reality of this situation it will be necessary to identify receptors for growth substances and discover their properties.

Summary

This brief survey of a few selected types of physiological process in the whole plant is far from exhaustive. However, it does raise a number of points concerning the mechanisms of regulation of growth and development. In the first part of the chapter we were able to conclude that patterns within meristems were in many cases cryptic, and even apparently random, as in the case of embryo development in such species as cotton. In a simple system such as the *Azolla* root tip, an apparently highly ordered sequence of events nonetheless gives rise to a structure with a symmetry which is different from the originating apical cell. It appeared from these considerations and others that organisation within a multicellular tissue was not simply arranged on a mechanical or geometric basis. This contrasts sharply with the general appearance of many plants, which are often highly ordered in terms of their geometry, and also with facts presented in the first four chapters of this book, where we have seen that the differentiation of an individual cell is also highly ordered in space.

A few examples of controlled processes have been considered at a physiological level. In all of these it is possible to argue a case for the importance of growth substances acting as hormones; that is, signal molecules which, when present in a particular tissue at a particular level (possibly in combination with other such substances), can elicit a specific response. There is no doubt at all that the known growth substances are very effective in their actions on plant cells; the whole body of tissue culture attests to this, as do many current practices in horticulture and agriculture.

However, it is equally clear that in nearly all cases our understanding of the mechanisms for control processes *in vivo* is limited. In a general way, for example, it is possible to class growth substances as either promoters or inhibitors of an action. Thus gibberellins promote stem elongation, and ethylene promotes abscission. Auxin inhibits lateral bud development, and cytokinins inhibit senescence. The very concepts of promotion and inhibition imply a reference to a normal state. Inhibition of a process means that it takes place less actively than normal, whilst promotion means that it occurs more actively. It is very common in considering the actions of growth substances in the whole plant that the control of the normal state is not recognisable. Thus, whilst it is fairly clear that cytokinins will inhibit the senescence of leaves, it is much less clear that any one process can be described as promoting the process and hence providing the balance of 'normal' behaviour. Similarly the inhibitory influence of auxin in apical dominance

is quite well documented, but precisely how a lateral bud is stimulated to grow in the whole plant is far from established.

There are two approaches to this dilemma. The first is to question the whole basis of the hormonal theory of the control of plant development. This is an extreme position and one which is probably not justified in terms of the facts of the case. The known growth substances are physiologically active, occur in plant tissues and (in a few cases) appear to correlate well with behaviour of those tissues. It is less drastic to question the implied assumption of the hormonal theory, that variations in levels of hormones represent the major or sole mechanism for transmitting a stimulus. The second answer to the dilemma is related to this criticism. In our present state of knowledge there is insufficient information concerning endogenous levels of hormones, the variations in those levels, and the variation in the chemical composition and cellular compartmentalisation of hormones for a unified hypothesis of their action to be formulated. What is clear from a consideration of whole-plant physiology is that the behaviour of a cell or group of cells is determined by a multitude of influences, not merely by chemical signalling.

One of the interesting facts concerning the actions of growth substances and their study is the way in which theories of action change with time. For example, before the discovery of growth inhibitors, a convincing case was made for the role of auxin in root geotropism. At present, this role is assigned to abscisic acid. It would be a rash conclusion indeed to assume that the present-day status of research represents anything approaching a final analysis. It is as though we are at present attempting to reconstruct a symphony by listening to only one or two sections of the orchestra. A musician might make an intelligent guess at the other parts, but he could be quite misled. In the analogy of the whole plant, there is no doubt that the orchestra is represented not merely by our presently known growth substances but, possibly, by others and certainly by direct environmental influences, nutrition and (probably most important of all) by the genes.

Polarity and development

Introduction

The development of the plant body proceeds by a coordinated series of events comprising division, cell enlargement and differentiation. In simple systems such as filamentous organisms or even within a developing line in a complex linear organ such as root, these processes can be described quite completely in terms of the planes of cell division and the directions of elongation. It is principally within the tissues of meristems that the simple mechanical analysis of events breaks down; growth does not appear here to be geometrically organised. Nonetheless, clearly organisation is present, since its result is a structure which is easily and specifically recognisable. The shape of one leaf can often be sufficient to allow the identification of the species of a plant; even the bare outline of a leafless tree viewed at a distance is sufficiently characteristic to allow a trained eye to recognise the type of plant it represents.

Overall form of this type can be regarded as the development of different symmetries around axial structures. The most obvious example of this is the pattern of the arrangement of leaves around a stem, or the pattern of petals of a flower, but it can equally apply to the distribution of tissues within a single leaf or petal. Clearly the operation of axial growth and symmetry is not simple; if it were plants would present a rather uniform and geometric appearance to the world, which in most cases they do not. Nevertheless, the idea of regarding growth as occurring along axes with lateral symmetries is a reasonably sound one.

It is a matter of common observation that axes of growth are polar. This means that they differ from one end to the other. A tree shows polarity insofar as at one end of its growth axis it develops leafy shoots, whereas at the other it develops roots. This polarity is not merely something which concerns the overall structure of the plant body, however. Within a particular tissue, such as a root, there are polar axes in terms of developmental processes. The apical end of these axes is the meristematic cell; the basal end is the mature cell. Within a given cell line there is a polar gradient of structure along the axis, as we have seen in the case of the developing vascular system. Even with a single cell, polarity exists and may be obvious. Examples of this type of polarity which have already been encountered include the secretion of the root cap slime across the outer wall of the outer cap cells; and the development of wall ingrowths on only one side of certain transfer cells associated with the vascular system. This cellular polarity points to why it is of fundamental importance to understand polarity in as much detail as possible. Polarity is an expression of the spatial component of the development process. At a genetic level, development is thought to be regulated by temporal sequences of gene expression. However, in very many cases, the products of gene expression become localised within particular places in the cell. This is most obvious in the development of the various wall specialisations which were discussed in Chapters 3 and 4, but it is more generally true. Cells are not simply packages of genes and enzymes, and their functions are highly organised in space. Polarity expresses one general aspect of this organisation.

Polarity is of importance in another more specific sense also. Without stable axes of growth, plants would exist as formless masses of cells, rather in the manner of callus tissues in culture. On the other hand, if axes were irrevocably fixed from an early stage of growth, then plants could not produce the diversity of structure which in fact they do. Many important developmental changes in plants involve the creation of new axes of growth. This is most easily seen in the transition from filamentous growth to two dimensional growth in a fern gametophyte, but it is equally important in the formation of say a lateral root, a root hair, or the development of the stomatal complex. These examples of development involve an obvious and predictable change in planes of cell division and growth axes. It is perfectly reasonable to assume, although more difficult to demonstrate, that morphological changes leading to, for example, leaf initiation in the shoot apex also involve changes in the direction of cell axes.

Thus there are several important questions concerning polarity which reflect deeply on the whole process of cell development from the molecular level upwards. What is the nature of polarity? Under what circumstances is it fixed and how can it be changed? Is it inevitably conserved from one generation to the next? These are the questions which this chapter will consider. Polarity will be examined first in higher plants, and then in several lower plant systems which have experimental advantages over more complex organisms. Finally we shall consider the operation of polarity at a physiological level.

Polarity in cuttings

The fact of polarity in cuttings of higher plants has been known for a long time. The classic experiments were carried out with stem cuttings of willow (*Salix*), which will regenerate shoots and roots when suspended in a moist atmosphere. A stem cutting of willow will always produce new shoots at the morphological tip and roots at the morphological base, no matter how long or short the cutting is, and whether it is suspended in the correct orientation, or horizontally or upside down (Fig. 7.1). Furthermore if a ring of bark is removed from an intact stem, then shoots will form at the lower cut surface and roots at the upper

Fig. 7.1. Polarity in cuttings of willow. Shoots regenerate at the tip end of a stem cutting, even if it is suspended upside down.

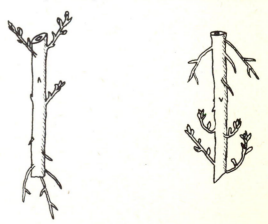

cut surface, just as though the stem had itself been completely severed. This is true wherever the position of the ring on the stem; it shows that the ability to produce roots or shoots is not an intrinsic property of different portions of the stem, but of the relationship of the regenerating tissue to the rest of the stem. These simple experiments were interpreted as showing that polarity is an irreversible characteristic of the plant, uninfluenced by gravity, and that it is also a property of the individual cells within a given piece of tissue. Polarity of this type is not limited to stem cuttings; segments of roots of the dandelion (*Taraxacum*) also show the same behaviour. Roots are reformed at the apical end of the segment, and shoots at the basal end. Leaf cuttings do not show such marked polarity and may produce both roots and shoots from the leaf base. Monocotyledons show a tendency to produce roots at nodes, regardless of their position within the segment.

With our present knowledge of growth substances, it is tempting to conclude that the polarity expressed within a cutting of a stem segment for example merely reflects gradients of hormone concentration. Specifically, it would be a reasonable hypothesis that the transport of auxin from the tip of a shoot to its base would give rise to an accumulation of this substance at a cut basal end, and that this would then trigger root formation. Equally, its transport away from a cut apical end would result in the formation of shoots due to a switch in the relative levels of auxin to cytokinin. Taking this view it comes as no surprise to discover that inversion of a cutting in a gravitational field for example does not lead to inversion of the polarity of regeneration.

Fig. 7.2. The retention of polarity in stems of ivy. The original plant (*a*) was induced to root at the tip of the shoot (*b*–*c*). The rooted part of the stem (*d*) was allowed to grow for several seasons. The original stem was retained as a 'leg' to the plant (*e*). When segments of this leg were tested for their ability to act as cuttings, it was found that the original polarity of the stem had not been altered by the intervening period of inverted growth (*f*).

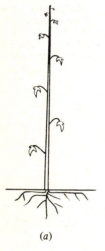

(*a*)

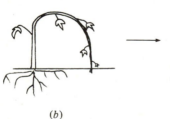

(*b*)

(*c*)

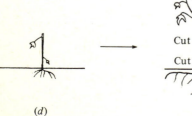

(*d*)

Cut

Cut

(*e*)

Roots

Shoot

(*f*)

There are a few observations and experiments which at first appear to indicate that this may not be the whole explanation. Shoot tips of willow or bramble (*Rubus fruticosus*) will root when they come into contact with the soil for example. This is also true of ivy (*Hedera helix*), and this enabled a very remarkable experiment to be carried out (Fig. 7.2). A plant of ivy was bent over so that its shoot tip could be buried in the soil. Roots formed at this tip, which was then cut off from the rest of the plant. Shoots then developed from dormant lateral buds on the severed piece of stem. This is not particularly remarkable, and indeed is a standard procedure for the propagation of such plants as blackberries. However, the experiment was continued by allowing the plant to grow for a period of years with the original tip-rooted stem as a 'leg' to the base of the plant. This leg was then itself cut into segments and used to make cuttings. It was found that the segments produced roots not at the end which had been closest to the soil in the intervening years of growth as might be expected, but from the end which had been basal in the original plant. In other words, polarity had been remembered and not reversed during the years of growth in the inverted position, despite the fact that presumably during that time the segment of the leg had functioned in a competent manner. This experiment, at first sight highly mysterious in its outcome, in fact confirms the notion of polar gradients of hormones as we shall see later (p. 201).

Polarity in unicellular coenocytes

The degree to which polarity may be influenced by external factors seems to depend in a general way on the complexity of the organism. Thus the situation with higher plant cuttings suggests that polarity is firmly fixed. On the other hand, there are examples from unicellular organisms which indicate that polarity may be rather easily reversed. One of these is provided by the marine alga, *Caulerpa*. This organism is cocnocytic; it consists of a differentiated cell body which is multinucleate but has no dividing walls.

Caulerpa prolifera may grow to a length of many centimeters. Its growth habit is that of a horizontal rhizome which produces lateral organs. On the upper side of the rhizome green blade-like structures are produced, corresponding to 'leaves', whereas on the lower side, the rhizome produces colourless rhizoids.

The plant grows from the tip of the rhizome and rhizoids are induced near the tip in clusters. During normal growth, one new rhizoid cluster is formed every day. The 'leaves' are usually induced between the second and fourth rhizoid cluster from the growing tip (Fig. 7.3*a*). Secondary 'leaves' may be formed from the region of the primary frond.

When this organism is rotated through 180°, it is found that newly formed rhizoids will continue to be induced on the now lower side of the rhizome, and that newly formed leaves will appear on the now upper side (Fig. 7.3*b*). This experiment shows that the polarity of induction of organs is rather easily reversible simply by an inversion of the gravitational field. Analysis of the timing of the events following inversion shows that polarity is only reversed if its expression occurs after inversion. In other words, a pre-

Fig. 7.3. The inversion of polarity in *Caulerpa prolifera*. *a*, In normal growth the horizontal rhizome produces fronds on its upper surface and rhizoids at its lower surface; *b*, inversion of the plant results in the formation of rhizoids from the now lower surface, and fronds from the now upper surface (an inversion of polarity); *c*, if secondary fronds form on an inverted leaf, they grow upwards, corresponding to the new polarity.

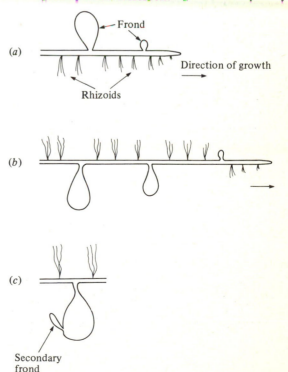

formed leaf will not turn upwards following an inversion, even though a newly formed leaf after inversion grows upwards. This is true even of secondary leaves; if secondary leaves are formed after the inversion of the plant, they will grow upwards, despite the fact that the leaf on which they occur is growing downwards (Fig. 7.3*c*). This seems to indicate that the change in polarity can only be expressed when it is acting upon a newly formed wall/cytoplasm interface. There is no question, either, of rhizoids turning into leaves simply because they find themselves on the upper side of the rhizome following an inversion. The rapidity of the response presumably reflects the movement of gravity sensors within the cytoplasm, just as the geotropic response of roots in higher plant reflects the movement of starch-containing plastids (p. 110). These experiments point to the importance of the cytoplasm in the determination of polarity, and might also be said to suggest that the expression of polarity involves the cytoplasm/wall interface.

The relative roles of the nucleus and the cytoplasm have been examined in the alga *Acetabularia*. This develops from its zygote by forming a rhizoid and a cylindrical stalk. The stalk grows apically for some time, then forms a whorl of hairs and finally a cap. The nucleus remains in one of the branches of the rhizoid at the base of the stalk during this elongation and differentiation stage. The nucleus increases enormously in volume but does not divide. Many interesting experiments can be performed with this cell since it can be surgically manipulated without causing its death (p. 221). From the standpoint of polarity, it has been found that severing the growing stalk from the rhizoid (and hence depriving the stalk of contact with a nucleus) inhibits neither the elongation nor the production of a normal cap. If the nucleus is transferred from its normal position in the rhizoid into the apical end of an enucleate stalk fragment, then rhizoid formation is induced at that position. These observations show that polar gradients can exist within a single cell even in the absence of a nucleus, but that the presence of the nucleus is undoubtedly an important influence on the formation of substances in such gradients.

Polarity in eggs and spores

So far it has been seen that polarity can be expressed in terms of the gross morphology of whole organs and indeed whole plants, but that it may equally be expressed by organisms which are coenocytic, and hence unicellular. In the latter case, polarity seems to be more labile than in the multicellular tissues of a higher plant. The question arises, then, of the relationship between polarity in single cells which form part of the life cycle of multicellular organisms. With higher plants, the zygote is a single cell and, indeed, does show highly polar structure and behaviour. Its first division is often obviously asymmetric (p. 155), and it is not unreasonable to assume that the polarity demonstrated by this division is retained and finds its final expression in the polarity of the embryo and hence of the entire plant. However, experimental investigation into the cause of the polarity within the zygote is difficult, simply because of its physical inaccessibility to the experimenter.

With lower plants, there is a stage in the life cycle when single cells may be released into the environment and thus made more easily available for study. Certain generalisations can be made concerning these cells. The spores of mosses, ferns, and horse tails can be influenced to develop in particular directions by gradients of many environmental parameters. These include light intensity, temperature, gravitational fields, and pH. The ability to influence development by an externally applied gradient suggests that either the cell which is influenced does not have any intrinsic polarity of its own, or that it may be overridden by the environmental influence. It appears likely that, of these two possibilities, the second is the most widespread. In *Cystosira barnata* it has been shown for example that, in the absence of an external stimulus, the rhizoid will emerge from the zygote at the point where the sperm entered it. This would suggest that polarity can by such means be conserved from one generation to the next. It is easy to see how this could be accomplished in higher plants, where the position of entry of the sperm is closely controlled. *Equisetum* spores can be influenced to develop in particular directions by light gradients but, in the absence of such a stimulus, the rhizoid will emerge from a recognisable site in the wall called the rhizoid point. Results of this type suggest that the cell has an intrinsic polarity, but that this can be overridden or redirected by environmental conditions. This idea is reinforced by the finding that *Fucus* eggs and some spores are only susceptible to environmental direction for a relatively brief period.

The situation is slightly complicated by two sorts of observation. The first is that in moss spores, it is possible to determine the direction of the protonema and the rhizoid independently by irradiating the cells at different times from different sides. This argues against the idea of a single axis which is fixed or can be reoriented. Equally, it is found that fertilised *Fucus* eggs will develop rhizoids at opposite sides of the cell if they are grown in polarised light. In this case the axis is the same at both ends, and so not polar.

Certain independent single cells cannot be altered at all in their development by the application of external stimuli. The eggs of *Coccophora* and *Sargassum*, for example, are elongated when they are shed from the mother organism, and the first division is always at right angles to the long axis no matter what type of gradient they are exposed to. In this case polarity is clearly predetermined before shedding and presumably relates to the way in which the cells were formed in the parent tissue. This seems to be the case with many motile algal spores and, interestingly, the polarity of the developing spore is often parallel to that of the parent from which it was derived, or at least bears a predictable relationship to that polarity. Pollen grains of higher plants also develop in a polar fashion, since germination can usually only proceed through pre-existing pores in the pollen wall. This type of polarity is also imposed by the parent tissue, since such pores are formed in response to the position of the pollen grain in the tetrad.

Thus, in early reproductive-type cells, we can say that polarity may be fixed or it may be open to environmental influence for a short while. The susceptibility to environmental influence is likely to be advantageous particularly with non-motile spores which may come to rest on their substrate in completely random positions with respect to normal environmental features such as gravity or light.

Polarity and development of simple systems

Polarity, as we have seen, can be expressed in a variety of ways. In a multicellular higher plant, the most obvious manifestation of polarity is the overall shoot/root polarity of the main axis. Within single cells or more simple organisms, polarity and axial growth can be defined in terms of directions and positions of elongation within walls, or planes of cell

division, or even in terms of the distribution of organelles. These aspects are most clearly seen in the growth and development of filamentous stages. One of the best studied systems is that of the fern gametophyte.

The fern gametophyte is formed by the germination of a spore. The germination process results first in the production of a rhizoid and then in the formation of a filamentous outgrowth. This filament increases in length by repeated divisions of the apical cell alone. A young gametophyte is thus a highly polar organism, with one or more rhizoids at one end of the growth axis, and a filament of cells of decreasing age at the other (Fig. 7.4). Eventually, under appropriate conditions, the gametophyte relinquishes this style of filamentous growth by apical division and instead produces a two dimensional sheet of cells called a prothallus. The production of the prothallus is an important developmental event which involves a reorientation of the polarity of cells within the filament. The transition can sometimes be highly predictable, as in the case of the fern *Dryopteris pseudomas*. In this species, the apical cell divides until a

Fig. 7.4. Initial stages in the growth in white light of the gametophyte of *Dryopteris pseudo-mas*. The spore germinates to produce a rhizoid and a green filament. The filament extends by division of the apical cell only until it is five cells long. At this time the third cell of the filament divides at right angles to the original growth axis. Subsequently, the apical cell may divide, and further divisions in the filament produce the two dimensional prothallus. Rhizoid formation occurs in basal cells of the filament following the transition in the third cell.

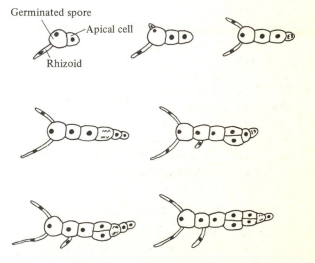

Germinated spore

Apical cell

Rhizoid

filament of five cells has been produced. The next division to take place is then not in the apical cell, but in the third oldest cell within the filament, and its direction is at right angles to the divisions which give rise to filament extension (Fig. 7.4). This reoriented division is then followed by parallel division of the second youngest cell in the filament, and occasionally by further extension of the filament by division of the apical cell. Further divisions in these cells then produce the prothallus.

This very simple system illustrates several points of interest concerning development and the control of polarity within cells. Along the polar axis, the division of all the cells except the apical cell is suppressed until the transition to two dimensional growth occurs. The transition is dependent upon the quality of the light in which the organism is grown. In red light, filamentous growth continues without the formation of a prothallus. In blue or white light, the transition occurs readily; in very high intensity blue or white light, it can occur before the filament has reached the five-cell stage. The site of perception of the light stimulus is the apical cell itself. This is shown by the fact that if the apical cell is fortuitously covered by a piece of spore coat, as sometimes may happen, two dimensional growth of the filament which it produces does not occur even in white light. The apical cell seems to act in an inhibitory manner on the rest of the cells in the filament; if it is damaged by surgery or by ultraviolet (UV) irradiation, then the other cells in the filament can be stimulated to divide. If the whole filament is subjected to a short plasmolysis treatment, this too will stimulate older cells into division, although the apical cell is not killed by this treatment. This suggests that the influence of the apical cell is transmitted symplastically, via plasmodesmata, since these will be at least partially broken by a plasmolysis treatment. These observations show that the simple organism is at least superficially comparable to a multicellular tissue in the operations of its controls. Cells respond to signals along polar gradients in a way which depends upon their position. Alteration to the signal or to the nature of the gradient may cause a modification in their behaviour.

The plasmolysis experiment which has been described has another facet which is relevant to polarity. When the cells of the filament are plasmolysed, they do not contract away from all the surfaces of the wall to the same extent. Cells other than the apical cell contract away from the apical end first, and may remain attached to the basal wall (Fig. 7.5). The basal cell wall in each cell is of course the older of the two end walls. In the apical cell, this pattern is reversed, and the protoplast remains attached to the tip of the filament whilst it contracts away from the first cross wall. This no doubt corresponds to the fact that the apical cell grows at its tip, and it might therefore be expected that the plasma membrane would show an especially close relationship to the wall at this point. Polar plasmolysis of this type, it must be realised, is a measure of relative degrees of attachment, not absolute levels. The existence of polar plasmolysis behaviour shows that not only is the filament as a whole a polarised structure in terms of growth rates and cell ages, but that each cell is itself polarised. Polar plasmolysis is also shown by such cells as side-illuminated developing *Fucus* eggs (p. 195) and by regenerating *Physcomitrella* protoplasts which are growing in electric fields (p. 198). In both these situations, the plasmolysis behaviour expresses the polarity which will emerge later on in development.

The importance of intercellular connections in the suppression of mitosis in the filament has already been seen. There is a further elegant demonstration of their role. Following the transition to two dimensional growth, which involves a transverse division in the third cell of the filament, rhizoids are produced in the basal cells of the filament (Fig. 7.4). Thus it may be said that rhizoid production in these cells is a result of the perception of a blue or white light stimulus in the apical cell. Furthermore, such rhizoids can only be produced by cells which were originally formed in the presence of blue or white light.

This is shown by the following experiment (Fig. 7.6). The spores are germinated and grown in white light until they have developed a filament of three cells. They are then transferred to red light, which allows apical divisions to continue. At the five-cell

Fig. 7.5. Polar plasmolysis in fern gametophyte cells. The apical cell shrinks away from the basal wall, whilst more basal cells shrink away from the apical wall.

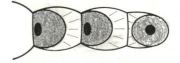

stage the plants are transferred back to white light. The apical cell will continue to divide, and eventually transition to two dimensional growth occurs from the third cell back from the apex. After this, rhizoids begin to form on the three most basal cells of the plant, and also on the most basal cells which were produced after the transfer back to white light. They do not form on the elongated cells which arose by growth and divisions of the apical cell during the red light treatment (Fig. 7.6). This of course does not explain the behaviour of the cells in red light, which in any case differs from those originating in white light in overall morphology and cytoplasmic contents. Nonetheless, it is a clear indication of the passage of a signal down a polar gradient which acts on susceptible cells but can pass through others without eliciting a response. The fitness to respond to the signal is

Fig. 7.6. Transmission of morphogenetic signals through non-target cells in *Dryopteris pseudo-mas*. *a*, The spore is germinated to the three-cell filament stage in blue light. The plant is then transferred to red light. *b*, the cells which grow and divide in red light are elongated. At the five-cell stage the plant is returned to blue light. Further cell division produces isodiametric cells typical of blue-light growth. *c*, Transition occurs in the third oldest of these blue-light cells. *d*, This is followed after further divisions by rhizoid formation in basal cells on either side of those which developed in red light. The red-light cells transmit the signal for rhizoid formation, but do not respond to it.

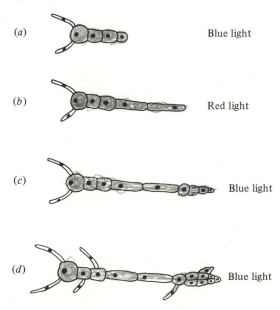

(a)   Blue light

(b)   Red light

(c)   Blue light

(d)   Blue light

determined by light quality in this system; however, it is feasible that other factors could be acting in higher plants.

The transition and polarity

The fern gametophyte illustrates a further aspect of polarity in tissues and cells. Whilst there is a clear gross polarity to the tissue, whether this be the filament of a gametophyte, or the shoot/root axis of a higher plant, it does not follow that all the cells comprising the polarised tissue are themselves polarised in the same direction. In the shoot apex, for example, we have already seen (p. 161) that cells in different layers may have different predominant planes of division. Morphogenesis at the shoot apex involves local changes in directions of division and cell enlargement. In the gametophyte, the development and polarity of individual cells also depends on the state of the whole organism, presumably mediated by some sort of morphogenetic signal which is transported. A filament cell prior to the transition to two dimensional growth is polarised along the axis of the filament. This is shown by its plasmolysis behaviour, which is polar, and its ability to transmit signals from the apex in a polar fashion. However, at the point of transition to two dimensional growth, the polarity of the target cell (the third oldest in *Dryopteris pseudo-mas*) changes; it divides at right angles to the direction of the filament, and its subsequent growth is also at right angles to its original orientation. For these reasons, and because it signals an entirely new phase of development, the transition to two dimensional growth is of special interest.

The most impressive correlation which has been found to explain the change in direction of the polar axis is a simple geometric one. We have already seen that growth in red light results in the formation of cells in the filament which are longer than normal, and that these conditions inhibit the transition to two dimensional growth. Conversely, growth in blue light produces smaller cells which are more isodiametric, and transition is facilitated. White light produces an effect between these two extremes, allowing transition. This has resulted in the idea that red light and blue light have different effects, and that the effect of white light is therefore a combination of the two. When the growth characteristics of the plants are examined in detail, it is found that variations in growth are not caused by the different photosynthetic effects of red, blue or white light. Growth in blue light will always result in early transition, even though its intensity is only just enough to permit continued growth. Similarly, as we have already seen, shading of the apical cell by a piece of the spore coat produces red-light effects in the filament even though those cells are in fact exposed to white light of high intensity. Analysis of the growth of filaments shows that red and blue light do not exert their effects by changing rates of division or elongation; rather they alter the orientation and localisation of expansion. Thus, in blue light, cell expansion occurs over the entire cell surface, giving rise to a more isodiametric outline than occurs in red light, which induces expansion in a polar fashion close to the tip of the apical cell.

The correlation which exists between cell geometry and the plane of cell division is as follows. A cell divides in the plane which produces a new cell wall in the position where it will occupy the least surface area. Thus a long thin cell will divide across its width to produce two shorter cells, whereas a short wide cell will divide at right angles to its axial orientation in the filament. This simple proposition seems to fit the circumstances of the fern gametophyte remarkably well. It explains the behaviour of plants in different qualities of light, and it may also account for other observations, such as that in *Onoclea sensibilis*, applied auxin promotes axial elongation and inhibits the transition to two dimensional growth.

It is clear that although this correlation is a good one, it is a far from complete explanation of the behaviour of the gametophyte; in particular, it says nothing concerning the siting of the division within the filament, or of the mechanism of suppression of division in the intervening and basal cells of the filament. It may not be widely applicable to more complex tissues either; the long and narrow fusiform initials of the cambium divide so as to produce two long and even narrower daughter cells – precisely the opposite behaviour which would be predicted from the hypothesis that the cell plate is positioned so as to occupy the least surface area following division. Extra complications are also indicated by the formation of curved cell plates, as is seen in the development of the stomatal complex (p. 118).

A correlation between two events or circumstances does not of course mean that they are causally related. In this context, a variety of other observations and suggestions have been made concerning the change in polarity which occurs during the transition to two dimensional growth. These include effects on cytoplasmic microtubules, and on patterns of RNA synthesis. At present it is simply not possible to credit any of these changes with the function of the primary control of polarity. Nonetheless, the gametophyte illustrates clearly the importance of cellular polarity within a tissue and the possibility of its regulation by environmental influences, mediated by transported signals.

Polarity and cell structure

The relationship between the polarity expressed by growth and division and the structure of cells has been examined in another filamentous system, the developing protonema of the moss *Funaria hygrometrica*. A germinating spore of this organism gives rise first to a network of green filaments, which are called chloronema. These develop for several days as a single uniform cell type, but, after a while, a second type of tissue with a different cell type is produced. This is the caulonema. The difference between the two cell types is obvious at both a structural and a developmental level. The caulonema cells are characterised by oblique cross walls, rather than the transverse cross walls found in chloronema. Caulonema cells have fewer chloroplasts than chloronema, and a different pattern of proteins shown by electrophoresis. The caulonema cells can give rise to buds as side branches, and these produce the reproductive tissues at a later stage. Thus the whole plant is polar; the original spore inoculum on an agar plate is surrounded by first a ring of chloronema, then a ring of caulonema, with buds at the periphery. As we

Fig. 7.7. Electron micrograph of the tip of the apical cell in caulonema of *Funaria hygrometrica*. The tip region is devoid of all of the larger organelles. Bar = 5 μm. Picture by courtesy of Professor E. Schnepf. (First published in *Planta*, **147**, 405–13, 1980.)

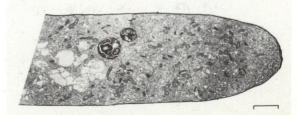

would now expect, this overall polarity is expressed at the level of individual cells, and this aspect of caulonema has been studied in some detail.

Caulonema cells grow from the tip, and have a rather pronounced region of cytoplasm just behind the tip which is devoid of the large cytoplasmic organelles such as the nucleus and plastids (Fig. 7.7). The terminal few micrometers of the cell contain only small vesicles, and behind this is a region which contains dictyosomes and mitochondria. A region of dense cytoplasm which includes the nucleus is found next and, towards the base of the cell, is the vacuole (Fig. 7.8a). Thus the cell is extremely polarised both in its growth characteristics and in the distribution of its cytoplasmic contents. Even within a single class of organelle, the plastids, there is a degree of polarisation. Those ahead of the nucleus are typically filled with starch whereas those in the region of the vacuole are less starch filled and retain more photosynthetic membrane. The mitochondria ahead of the nucleus tend to be small and globular, whereas those behind it are more elongate. The dictyosomes at the tip of the cell have more cisternae on average than those which

are found in the basal part of the cell. During normal growth the nucleus moves forward continuously so that it is a constant distance of about 120 μm from the tip of the cell. When the cell reaches a length of 400 μm, it divides and the daughter nucleus in the new apical cell continues to maintain its correct position.

This represents a highly ordered and polar structure at the level of a single cell. Various experiments have been carried out to see if this polarity can be disturbed by external influences, or by chemicals. The drug colchicine was found to have no effect on the distribution of the organelles, although it did reduce the rate of elongation and produce malformation of the tip. One of the actions of colchicine is to disrupt microtubules. Deuterium oxide (heavy water) acts in the opposite manner on the microtubule equilibrium – it stabilises microtubules. Treatment of the caulonema with heavy water also reduced the growth rate. However, in the presence of heavy water, the nucleus became temporarily displaced towards the basal end of the cell, returning to its normal position after a few hours. It was occasionally observed that whilst the nucleus was in its displaced position, a new tip arose at the basal end of the cell. If this happened, its further growth was slowly towards the centre of the inoculum, that is, in the opposite direction from normal in these cultures.

Physical disruption of the cytoplasm was achieved by centrifuging the cells at a low speed. This caused the starch-containing plastids to come to rest against the basal cell wall, with the nucleus next to

Fig. 7.8. Effect of centrifugation on the apical cell of *Funaria hygrometrica* caulonema. The normally growing cell (*a*) has a pronounced tip region, then a region which contains plastids and mitochondria; the nucleus is situated behind this. The vacuole is at the base of the cell. Following centrifugation (*b*), the vacuole is displaced to the tip of the cell, and the denser organelles lie towards its base.

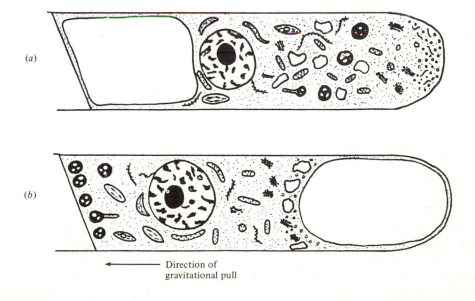

(*a*)

(*b*)

← Direction of
gravitational pull

them. The tip body which lacked large organelles was displaced from the apical region of the cell, which was instead occupied by the vacuole (Fig. 7.8*b*). If the cell was in interphase at the time of the centrifugation (which lasted 20 min), subsequent growth outside the centrifuge restored the normal polarity of the cytoplasm. On the other hand, if the nucleus was approaching division at the time of the centrifugation, it did not migrate back to its normal position. Two types of behaviour were observed instead. In one, the basal portion of the cell produced a new tip which grew in a reversed direction towards the centre of the inoculum, and this was followed by nuclear division donating one daughter nucleus to the new tip cell. In the other, the nucleus divided in its basal position within the original cell, producing a new cross wall. The subapical cell thus formed then regenerated a new tip which grew back towards the centre of the inoculum. Both these types of behaviour represent a reversal of polarity of the original apical cell.

These experiments suggest that the nucleus has, by its position, an overriding influence on the polarity of the cell in this situation. They also argue against the importance of the cytoplasm/wall interface, since at the gravitational fields used, no physical disturbance to the components of this region would be expected.

Side-branch formation

The 'natural' counterpart of the branching of the apical cell which is induced following centrifugation or heavy water treatment is the formation of side branches in the caulonema. In a normally growing plant, this is quite a predictable process. As we have seen, the apical cell divides when it reaches a length of 400 μm. The tip cell after division is about 120 μm long, and the basal cell about 280 μm. This basal cell does not undergo further elongation. Simultaneously with the division of the apical cell, the second oldest cell in the filament (which, after the apical division, is of course cell number 3) begins to produce a side branch at its tip (Fig. 7.9*a*). This side branch elongates and eventually the nucleus in the cell from which it arose divides and contributes one daughter nucleus to the side branch. This establishes a new polar growth axis in the branch, and so side-branch formation in mosses is in a simple way analagous to the transition to two dimensional growth in ferns; a developmental step which involves a change in a polar

axis. In the case of *Funaria*, the new polarity is established before a reoriented nuclear division, and the appearance of the tip of the side branch is a clear manifestation of this. In the transition in *Dryopteris*, it is reasonable to assume that a change in polarity precedes nuclear division also, although there is no easily detectable cellular change which might prove the point.

The division in the subapical cell with its side branch occurs more or less simultaneously with the division in the apical cell. The position of the side branch is predictable with respect to the position of the nucleus. The new tip of the side branch is initiated slightly forward of the nucleus, and on the opposite side of the cell from the nucleus (if the nucleus happens to be laterally displaced). The nucleus moves to the base of the side branch when the branch has attained a length of about 22 μm. When it reaches the base of the side branch, it divides. The side branch grows as chloronema, that is, without a distinct tip region and with well differentiated chloroplasts.

The effects of heavy water and of centrifugation

Fig. 7.9. Side-branch formation in *Funaria hygrometrica*. *a*, Normal development: the side branch is initiated in the subapical cell at about the time when the apical cell divides. *b*, The nucleus of the subapical cell can be displaced by treatment with heavy water; if it divides in its displaced position, two side branches may be initiated. The polarity in one cell (i) is normal, whilst in the other (ii) it is reversed.

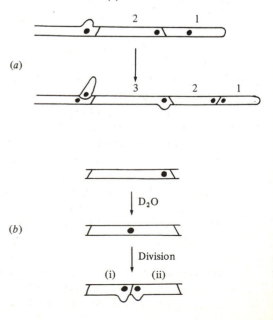

on this sytem have been investigated. Heavy water causes displacement of the nucleus towards the basal portion of the cell. If the treatment is maintained and the nucleus divides, side branches are initiated opposite both of the daughter nuclei following such a division (Fig. 7.9*b*). This means that, in one of the cells resulting from the division, the side branch is in its correct position at the apical end, but in the other, it is at the basal end. Polarity has been reversed in that cell. Centrifugation produced a complex series of reactions depending upon the relative positions of the nucleus when side-branch formation was initiated. If a branch had already been formed prior to the centrifugation, then it would continue to develop as though no treatment had been given. However, new branches were initiated at the correct time with respect to events within the apical cell of the filament, and at positions which reflected the position of the nucleus within the cell initiating the branch.

These experiments show clearly that the position of the nucleus is of great importance in the determination of the site of side branching and, hence, the expression of polarity. They also show that this influence can override any permanent structure at the cell periphery, since centrifugation would not be expected to influence such structure. However, the situation is not as simple as it first appears. All the experimental treatments which have been described only give rise to an expression of reversed polarity if the nucleus happens to be in the wrong place at the time of the initiation of the expression. In other words, whilst the position of the nucleus in a sense can define the polarity of the cell, it is clear that in normal cells the position of the nucleus is itself defined by the organisation of the cytoplasm. This is particularly well illustrated by the recovery from heavy-water treatment or colchicine, where disruption of the distribution of the cytoplasmic contents is fully reversible and the cells assume a normal appearance following the treatment. This implies that some mechanism exists for the determination of the position of the cytoplasmic contents. Thus the hypothesis that the peripheral-cytoplasm/wall interface is the primary site of polarity fixation is not necessarily disproved by these experiments.

The development of polarity in *Fucus*

In the previous sections we have seen some of the effects of polarity in simple systems. One of the most significant theoretical considerations about polarity is that the division of a single cell will give rise to daughter cells with different potentialities. This is precisely the situation in many developmental pathways, even in complex higher organisms. The division of such a cell thus both fixes and accentuates its original polarity; the final result within a tissue, such as a root, is the formation of highly differentiated cell types from divisions occurring in less differentiated cells. The development and accentuation of polarity is difficult to study in multicellular organisms and, for this reason, it is preferable to choose a single cell which is isolated from gradients imposed from surrounding tissues, but which nevertheless develops in a polar fashion to give rise to differentiated daughter cells. Such a system is represented by the fertilised eggs of the marine alga *Fucus*, which has been the object of considerable study over a number of years.

The normal development of the *Fucus* egg results in the formation of a rhizoid at the first division, and the development of the frond from the daughter cell which does not form the rhizoid (Fig. 7.10). This is a highly asymmetric division in terms of the fate of its progeny, just as is the first division of the zygote in the growth of the angiosperm embryo. The division is preceded by development of polarity in the egg; just as in the case of the higher plant embryo, where the division of the zygote is preceded by asymmetry in the disposition of the nucleus and the vacuole. The consequence of developed polarity preceding division is of course that each daughter nucleus is delivered from the division into a different cytoplasmic environment; this in itself will tend to fix and accentuate the polarity.

Within 8–14 h of fertilisation, the egg of *Fucus* forms a localised protuberance which represents an early sign of asymmetric development. The region of the protuberance, which will later become the rhizoid, is characterised by an accumulation of osmiophilic droplets visible in the electron microscope, together with a localised layer of stained material on the outside surface of the cell wall. In the light microscope, the rhizoid region of the cytoplasm can be shown by staining properties to have a different constitution to that of the apical half of the cell. This difference in the cytoplasmic makeup of different parts of the fertilised egg is fixed by the first division of the nucleus and its attendant cytokinesis. The two daughter cells are different, and develop along different pathways. The rhizoid and its derivatives become attached to the substrate and form the holdfast, whilst the derivatives of the thallus cell form the frond. The final polarity of the entire organism can thus be said to rest on the polarity which develops in the fertilised egg prior to its first division.

A variety of gradients can influence the position of the rhizoid site in the developing *Fucus* egg. For example, the rhizoid will form on the shaded side of a light gradient, at the positive pole of an electrical field, at the more concentrated end of a gradient of calcium or potassium ions, and at the low pH end of a pH gradient. Fixation of the polarity does not occur until a few hours before the rhizoid starts to form. Until fixation occurs, this means that if an egg is given a short period of one light gradient followed by a period in a gradient from a different direction, the rhizoid will grow on the shaded side of the second light gradient. In the absence of any artificial gradients, the rhizoid will form at the site of entry of the sperm into the egg. This implies that the fertilised egg has an intrinsic polarity, but that this can be rather easily redirected by environmental conditions, at least for a limited period of time.

The relative timing of fixation of the polar axis and the expression of polarity in the visible formation of a rhizoid is very close. In developing fertilised eggs of *Fucus distichus*, it is found that fixation of the polar axis precedes the emergence of the rhizoid by 2–4 h. Unfortunately, in populations of normally growing eggs there is a period of overlap between these two events; thus fixation can be estimated as occurring between 11–16 h after fertilisation, whereas rhizoid emergence commences in a period amounting to 14–18 h after fertilisation. Observations on normally growing eggs are therefore not sufficient to identify the processes which lead to the fixation of the polar axis.

However, experimental procedures have been developed which appear to allow the separation of these two processes. For example, it was found that the presence of the protein synthesis inhibitor cycloheximide does not affect the fixation process but delays the formation of the rhizoid by several hours. In other words, the presence of cycloheximide during a treatment with a light gradient does not prevent the eventual formation of rhizoids with a fixed orientation to the light source, even though the rhizoids do not appear until after the exposure to cycloheximide had been terminated.

On the other hand, it was found that the agent cytochalasin B does prevent fixation. Thus if cells are exposed to a particular light gradient in the presence of cytochalasin B, during the normal fixation period, and then removed from this environment and exposed to a second light gradient in a different orientation, the formation of rhizoids occurs in a direction

Fig. 7.10. Development of *Fucus*. *a*, The fertilised egg develops a protuberance at the position of the future rhizoid, the first mitosis of the polarised zygote fixes the polarity; *b*, further divisions of the daughter cells produced a polarised plantlet; which *c*, develops a series of fronds and a holdfast.

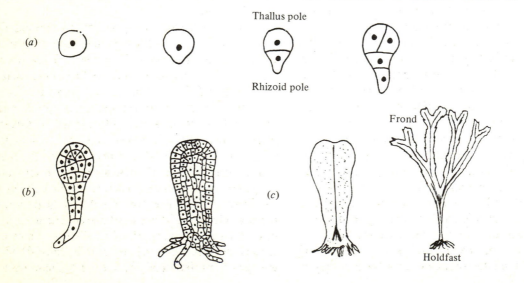

determined by the second light treatment. These results suggest that whilst rhizoid growth requires protein synthesis (and hence is delayed by the presence of an inhibitor of this process for a given period), fixation of the polar axis does not. Conversely, fixation of the polar axis requires ordered movements of cytoplasmic particles, since cytochalasin B is presumed to act as an inhibitor of such movements.

Examination of the developing eggs by electron microscopy showed that rhizoid formation was preceded by a certain degree of detectable polarisation of the cytoplasm, as has already been mentioned. These effects were not striking however, apart from the formation of a localised region of special wall structure at the future rhizoid point. The observation of such a region is of particular significance in view of evidence from other polar systems that the wall/cytoplasm interface is a likely site of the primary determination of polarity. Staining and analytical procedures allowed the demonstration that the future rhizoid point was the site of accumulation of a sulphated polysaccharide, containing fucose.

The appearance of the sulphated polysaccharide in the wall seems to be related to the fixation of the polar axis by several different criteria. First, it is found that when fertilised eggs are grown in media which contained sucrose to increase their osmotic pressure, rhizoid development does not occur. Nonetheless, the sulphated polysaccharide is synthesised and becomes localised in particular regions of the cell wall. Removal of the eggs from the medium with a high osmotic pressure, back into ordinary sea water allowed the development of rhizoids; this development had been fixed during the intervening period, during which time also the polysaccharide had been deposited. In a general way, the polysaccharide appears in the wall over the period during which the polar axis is susceptible to reorientation. Thus, in eggs fertilised 8 h previously, the polysaccharide is absent, but it is present by 16 h in a localised form. However, if the eggs were incubated in the presence of cytochalasin B during the 8–16 h period of susceptibility to external influence, the polysaccharide was found to be present in the wall, but not localised to a particular region. This again suggests that polar axis fixation and the localisation of the polysaccharide are linked, since in the presence of cytochalasin B neither process occurs. Interestingly, it

also seems to show that, whilst the appearance of the material in the wall does not require the presence of microfilaments, its localisation does. The explanation for this observation is unclear.

The role of the plasma membrane

These experiments suggested that the formation of the polar axis might reasonably be assumed to involve localised differentiation of the plasma membrane. This is because it is logical that a localised accumulation of a particular polysaccharide in the wall might correspond to a localised activity of the plasma membrane in secreting that polysaccharide. This proposition has been examined in two ways. By using the freeze–fracture technique, it was possible to show in *Pelvetia* eggs that the site of the future emergence of the rhizoid was marked by the formation of recognisable patches of plasma membrane. These patches were about 0.3 μm in diameter, and could be distinguished from the bulk of the membrane because they were not crossed by traces of the underlying wall fibrils, and the ubiquitous particles within the membrane were present at a lower density in the patches. The concentration of patches of this type at the future rhizoid point was also accompanied by a clear area between the wall and the plasma membrane at this position. These effects could be reoriented by changing the direction of an inducing-light gradient. These results were taken to mean that new membrane was inserted into the plasma membrane at the future rhizoid point and that the site of insertion was also a region at which active secretion was occurring.

The next piece of evidence concerning the plasma membrane relates to its electrical behaviour. By using a very small vibrating probe, it is possible to measure local currents around the circumference of a single egg of *Pelvetia*. When this is done, it is found that prior to axis formation there are many local points at which there is an increased flux of ions across the plasma membrane. However, after fixation of the polar axis, the largest flux is observed at the future rhizoid point. Once again, it was possible to show that the position of this large local flux could be influenced by changes in the direction of light gradients across the egg. The size of the inward current was increased when the concentration of sodium ions was lowered. It was found that calcium ions enter the rhizoid half of the egg cell preferentially by a five-fold factor during axis fixation, and more or less equally into both halves of the cell thereafter.

These results have been incorporated into a scheme for the fixation of the polar axis in the following way. First, the fertilised egg is characterised by the presence in its plasma membrane of local patches able to accumulate calcium ions. Next, these patches move within the membrane under the influence of a gradient until they accumulate in one place. Cell wall materials are then incorporated at the site of the calcium influx; this gives rise to the cortical clearing which is observed at the rhizoid point by electron microscopy. Fixation then occurs by a process which is insensitive to protein synthesis inhibitors but sensitive to cytochalasin B. Finally, localisation of the sulphated polysaccharide takes place.

How are these observations to be explained in terms of cellular events? The current model of the plasma membrane as a fluid mosaic (p. 6) allows for its differentiation into specific regions with different composition and activity, and the migration of such regions within the surface of the membrane. Such movement is known to be stimulated by electrical fields, and electrical fields can cause orientation of the polar axis in *Fucus*. The process of fixation of the polar axis is not explained by this property however, since clearly if specialised areas of membrane can migrate to a site, equally they can migrate away from it. This indeed is precisely what is observed in experiments with *Fucus* where the applied gradient is changed in its orientation before fixation occurs. Fixation is considered likely to involve some cytoskeletal element; sensitivity of the process to cytochalasin B is consistent with the involvement of microfilaments at this stage. Polarisation of the cytoskeleton could also be invoked to explain the ordered deposition of the sulphated polysaccharide which precedes rhizoid emergence.

These results suggest that during its labile phase, polarity is a function of the plasma membrane rather than of the cytoplasm or of the wall. The mediating process between the developing polarity in the membrane and the cytoplasm may be rather direct and mechanical via microfilaments. Alternatively, it might be indirect, involving electrical effects due to localised ions fluxes through the membrane. Once fixation has occurred, usually within 11 h of fertilisation of the egg, it will remain for the entire future life of the organism, perpetuated through cell division and differentiation steps.

Polarity and protoplasts

The fertilised egg of *Fucus* is an example of a single cell which has been produced in a natural sequence of events within the life cycle. It is, however, possible to isolate single cells from tissues of other plants by the use of enzymes. It appears that merely isolating a cell from its neighbours may not be sufficient to disrupt its polarity since, as we have seen (p. 142) in the case of cells isolated from *Zinnia* leaves, more or less normal patterns of xylem differentiation can occur without intervening cell division steps. However, it is possible to go one stage further and remove the wall entirely from an isolated cell, giving rise to a protoplast.

It is clear that from all the studies of cell wall regeneration and on growth of protoplasts isolated from leaves and other tissues of higher plants that there is no evidence for a retention of a polar axis in these circumstances. The cell wall arises at random over the entire surface of the protoplast, and subsequent growth and division of the cells so formed produces formless masses of callus tissue. Such callus tissues may be induced to undergo organogenesis in exactly the same way as callus tissues induced from explants (p. 144). There seems to be no change in behaviour related to the fact that the living material has passed through a protoplast stage. This, of course, reflects the general observation that callus tissues do not have overall polarities, which is precisely why they grow in the formless manner that they do.

The situation is different when protoplasts of a moss, *Physcomitrella patens*, are considered. Protoplasts from this organism can be readily prepared from the filamentous chloronema. When they are cultured, they do not produce a callus mass as do higher plant protoplasts, but instead they quickly regenerate a new filament which grows in an apparently normal manner. The production of the filament is in many ways superficially similar to the germination of a fertilised *Fucus* egg. For example, an initially wall-less cell first produces a wall and, subsequent to this, polarity is developed and expressed in the formation of an elongate outgrowth – a rhizoid in the case of *Fucus*, but a green filament in the case of *Physcomitrella*. Experiments with the moss protoplasts show that the position of the emergence of the filament can be controlled by external conditions; it will emerge on the shaded side of a light gradient, and towards the positive pole of an electrical field

(Fig. 7.11). Here too, there is a resemblance to *Fucus*. Unfortunately the synchrony of growth of the protoplasts is not high, and it is therefore difficult at present to define relative timings of events in the way which is possible with the fertilised eggs. Nonetheless electron microscopy shows that an early manifestation of polarity which definitely precedes the formation of the filament is a localisation of a particularly stained wall component (Fig. 7.12).

The observations confirm the ideas that polarity is made labile by removal of the wall, and becomes fixed again when the wall is reformed. Detailed facts concerning the fixation process in protoplasts are not yet available, but there is no reason to doubt that the process may be similar to that found in naturally occuring single cells.

Polar auxin transport

Studies with simple organisms such as filamentous mosses and ferns indicate something of the general nature of polarity and its importance in key developmental changes. It may be safely inferred from such studies that polarity is equally important in the determination of development and morphogenesis in more complex cell systems although, in the nature of things, analysis of multicellular three dimensional tissues is technically almost impossible at the level of the responses of individual cells. However, one expression of polarity in such tissues has been studied in detail and has important implications for growth and development. This is the phenomenon of polar auxin transport. We have already seen that the polarity of cuttings can be rationalised in terms of the behaviour of growth substances. This will now be discussed in more detail.

The original experiments with, and assay system for, auxin involved the use of the oat (*Avena sativa*) coleoptile (p. 130). It was shown that side illumination of the tip of the coleoptile resulted in curvature of a region distant from the tip. This implied the operation of a transmitted stimulus, and this stimulus could be collected in agar blocks attached to the cut base of the coleoptile. In a refinement of this simple experiment, the coleoptile was cut twice, giving rise to three sections – the tip, a segment from the centre, and the base. It was found that if the coleoptile were 'reassembled' using agar bridges, then curvature was still induced by a light stimulus. However, if the central segments were reversed in orientation during reassembly, then curvature did not take place (Fig. 7.13). This seemed to show that whilst the transmitted stimulus, now thought to be auxin, could travel from the tip to the base of the coleoptile, it could not travel in the reverse direction.

This was confirmed in a further series of experiments. Agar blocks were loaded with auxin by placing them in contact with the cut base of a coleop-

Fig. 7.11. Polar development from protoplasts of the moss *Physcomitrella patens*. The originally spherical protoplasts quickly develop into polar structures by the production of a filamentous outgrowth. The direction of the outgrowth may be controlled by light or electrical fields. In this case a field of 50 V cm^{-1} was applied, with its positive pole to the right of the picture. Picture taken after 4 days growth. Bar = 100 μm.

Fig. 7.12. Electron micrograph of the leading edge of the initial outgrowth of the filament produced by a *Physcomitrella* protoplast. A layer of dark-stained material on the outside of the wall marks the position where the filament will emerge. Bar = 2 μm.

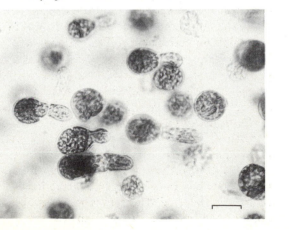

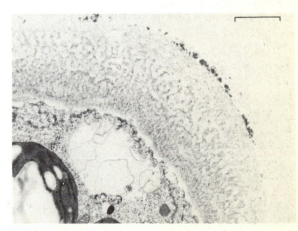

tile tip. Loaded blocks could then be used as donors of auxin into cut segments of coleoptile. It was found that if the donor block were placed at the apical end of a cut segment, with an unloaded receiver block at the cut basal end, then auxin was transported from the donor to the receiver. On the other hand, if the donor block were placed in contact with the cut basal end of the segment, and a receiver block at the cut apical end, no transport occurred (Fig. 7.14).

Apart from its strongly polar nature, the transport process has several interesting properties which can be discovered using this experimental system. The rate of movement of the auxin in the oat coleoptile is 15–18 mm h^{-1}. When segments of different length are used, it is found that the time taken for the first auxin to reach the receiver block from the donor increases with the length of the segment, as would be expected. However, the rate of transport is not dependent upon the length. This suggests immediately that the transport differs from a simple diffusion process. It is also found that a preload of auxin into the receiver block does not prevent the transport of auxin into that block. For example, if a donor block were loaded with say 100 units of auxin activity, after a certain time there might be say 80 units in the receiver block and only 20 units left in the donor block. The characteristics of the system are such that if both donor and receiver blocks were preloaded with 100 units of auxin before the experiment, then afterwards the donor would still only contain 20, whilst the receiver would contain 180. This means that the rate

of movement of the auxin through the segment is independent of the concentration gradient across its ends.

All these early experiments were performed using bioassays to measure the activity of the auxin in the donor and receiver blocks. The availability of radioactive auxin allowed further refinement of the observations. If the distribution of the auxin within a segment which has been loaded from a donor block is examined, it is found that the large peak which occurs close to the donor block at the start of the experiment broadens out as the auxin moves along the segment. This is taken to mean that a proportion of the auxin is immobilised during transport. The moving peak is arrested if the segment is transferred to a nitrogen atmosphere, suggesting that respiration is necessary for the continued transport to occur.

Other factors affecting auxin transport have been identified. Ethylene, for example, is inhibitory, and this inhibition can to some extent be overcome by increased concentrations of auxin itself. It is also found that auxin transport in the coleoptile can be inhibited by light, and that this effect shows the red/far-red reversal which is characteristic of a phytochrome response (p. 218). The inhibition by red light is rather long lived. This is regarded as the explanation for a piece of physiological behaviour. Oat seedlings grown in the dark show a very marked elongation of the mesocotyl, which is the tissue below the coleoptile and separated from it by a node. This elongation is completely inhibited by very low levels of red or green light, and it is reasoned that this is due to an inhibition of transport of auxin from the coleoptile tip into the mesocotyl tissue.

Polar auxin transport in stems

We have already seen that stem cuttings show a strong polarity which is independent of length and

Fig. 7.13. Polar transport of auxin in the coleoptile. *a*, Unilateral lighting causes the intact coleoptile to bend towards the light source; *b*, bending still occurs if the coleoptile is cut into segments and reassembled correctly; *c*, if the central segment is reversed, bending is abolished.

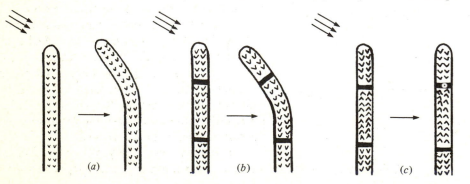

gravitational field. Nonetheless, it is possible in certain cases to produce inverted cuttings where roots have formed at the shoot end; this leads to the development of lateral buds elsewhere on the cutting, giving rise to a plant which appears normal but which has a reversed segment of stem as the leg (p. 187). Prolonged culture of such a plant followed by testing of the leg for its polarity shows that the original polarity is remembered in terms of regeneration behaviour (Fig. 7.2). In a series of experiments with inverted cuttings of *Tagetes* and tomato (*Lycopersicon esculentum*) plants, it was found that even after 4 months' growth, during which time the plants with the inverted stem segment were able to grow considerably in size, flower and set fruit, no change occurred in the direction of polar transport of auxin. This does not

Fig. 7.14. Characteristics of polar auxin transport in stems. *a*, A stem segment can transport auxin applied at the apical end into an agar block placed as received at the basal end; *b*, transport does not take place when the relative positions of donor and receiver blocks are reversed. *c*, Transport can take place against an overall concentration gradient across the segment.

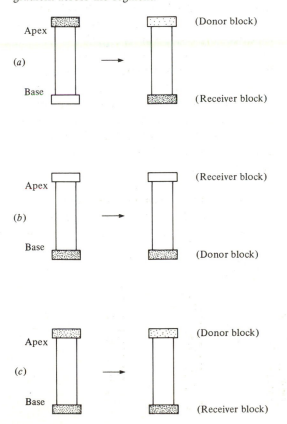

simply mean that auxin transport was not reversed in the tissues. It also implies that cells and vascular tissue which were formed during the course of the experiment were induced to the same polarity as that of the existing stem. This means that an existing polarity is propagated during normal growth processes. That transport of auxin is not only the result of this polarity, but its cause is shown by the fact that polar transport of auxin in stem segments declines with time unless auxin is applied to the apical cut end. In other words the transport of auxin is in some way autocatalytic. This property has implications for pattern formation which will be discussed later (p. 203).

As a result of polar auxin transport away from the shoot tip, the stem contains a gradient of auxin which is actively maintained and so under the control of the living cells within the stem. Dead tissue does not show polarity of auxin movement. This gradient, as we have seen, has possible important functions in the maintenance of apical dominance (p. 172). The transport of auxin is not uniform throughout all the cells of the stem. In a general way, as cells age, their ability to maintain polar auxin transport declines. In a secondarily thickening stem the majority of polar auxin transport takes place in the newly forming vascular tissue. The precise gradient of auxin within a stem therefore is a result of a complex series of variables, together with the additional factor of partial immobilisation in some cells.

The mechanism of polar auxin transport

The mechanism of the polar transport of auxin is clearly of considerable interest and presents a challenge to explain the basis of one of the most important manifestations of polarity at a physiological level. The facts may be summarised quite briefly. The transport is active, requiring metabolic energy for its accomplishment. It can occur against a concentration gradient. Furthermore, it can occur in tissues which have been subjected to a temporary plasmolysis procedure. This last point casts serious doubt on the possibility for a role for plasmodesmata in the polar transport process. It appears rather that the auxin crosses from one cell to the next via the intervening cell wall. In other words, it leaves one cell by the plasma membrane at its base and enters the next via the plasma membrane at its apex.

It requires only a very slight difference in the amounts of auxin arriving and leaving at opposite ends of a cell to account for the polarity which is observed in segments of tissue containing many hundreds or thousands of cells. Each cell, if it works in the same direction as its neighbour in the production of a small concentration gradient across its ends, will contribute to the overall gradient in the tissue. There is no need to postulate the presence of specialised cells which could produce a large gradient across their length. It is safe to assume that the mature cells of the xylem and phloem system are not the major route of the auxin. The need for respiration rules out a role for the dead cells of the xylem, and transport rates for auxin are slower than phloem transport rates. In addition, polar transport has been observed in segments of *Coleus* pith from which the mature vascular tissues had been removed surgically. Dissection experiments with tobacco (*Nicotiana tabacum*) stems suggest strongly that most of the polar auxin transport, in secondary tissues at least, takes place in the newly formed derivatives of the cambium.

The mechanism of polar auxin transport is highly specific to auxin; synthetic analogues do not move as rapidly as IAA, nor is so much flux generated. Polar auxin transport can be inhibited by 2,3,5 triiodobenzoic acid, a compound resembling IAA but with no auxin activity. Polar transport is also very dependent upon pH. Uptake of auxin is much reduced when the pH of the donor block is raised. This is presumably due to ionisation of the IAA. The non-acidic auxin indoleacetonitrile does not show this pH-dependent effect.

We can now consider the model for auxin transport across an individual cell. The pH of the extracellular space is normally around 5. IAA has a pK of 4.7, so that it will be present in the extracellular space in ionised and non-ionised form. The pH of the cytoplasm by contrast is about 7. In this environment, assuming no interactions with other cytoplasmic molecules, IAA will be more ionised than it is in the extracellular space. One assumption now has to be made. IAA passes through the plasma membrane in its non-ionised form. Thus, in an isolated cell bathed in a medium of pH 5 which contains IAA, it would be expected that IAA would accumulate inside the cell, since the concentration of non-ionised IAA is much greater on the outside than the inside due to the difference in pH. It would further be predicted that as the pH of the outside was lowered, then the relative

enrichment in internal concentration of IAA would increase since the acid would become more completely non-ionised on the outside of the cell. This proposition can be tested using tissue culture cells, and it is found to fit the observed facts. This pH-driven accumulation of IAA is not a specific process. It relies for its action only on the fact that IAA is a weak acid, and not on the shape of the molecule. Any other weak acid would be expected to accumulate within cells in exactly the same way. Amongst such weak acids are included several synthetic auxins, but also other substances such as gibberellins and abscisic acid. In order to account for the polar transport of auxin, two other considerations have to be taken into account. First, the specificity of the process, and second, its polarity.

The specificity requirement may be met by postulating a carrier for auxin which is sited in the cell membrane. This carrier is likely to be a protein, since specificity mechanisms are well known in proteins, but not well known amongst lipids for example. The polarity requirement can be met by postulating that the carrier for auxin is located asymmetrically in the cell, with a preferential distribution at the basal end. The mechanism for maintaining the concentration gradient is then explained as follows. At the apical end of the cell, auxin enters the cell by passing through the plasma membrane in its undissociated form, a process which is pH dependent. Once within the cytoplasm, the activity of this non-ionised form is reduced due to ionisation. Distribution of both forms of IAA then occurs through the cell. At the basal end of the cell, proton efflux takes place and, coupled to this, the auxin is exported linked to a specific carrier. This process would be energy dependent and hence inhibited by respiratory inhibitors or a nitrogen atmosphere. Once outside the cell and released from the carrier, the now ionised IAA takes up a proton and can enter the next cell by passing through the plasma membrane.

Considered at the level of a single cell, this model permits of considerable leakage. The non-ionised auxin in the extracellular space can presumably move in any direction and enter any cell. Similarly, there is certainly a slight leakage of ionised IAA out of the cell over its entire surface. However, this outline model can account for all the known facts of auxin transport. More important perhaps is the fact that the model is based on experimentation and deduction, and draws the conclusion that polarity rests at the cytoplasm/wall interface in the localisation of specific carriers in the plasma membrane. This is a conclusion which has been drawn from many studies of polarity in other systems.

The model makes certain predictions, of which the most obvious is that accumulation of IAA within cells should be increased by inhibitors of auxin transport. This is because the transport mechanism depends on a specific efflux of IAA from cells, and therefore if inhibited, this efflux would cease. This prediction is fulfilled in cells of crown gall of Virginia creeper (*Parthenocissus quinquefolia*). The presence of 2,3,5 triiodobenzoic acid stimulates net uptake of IAA. The model also explains how it is that different growth substances may be transported in different directions within the same tissue. All that is required here is a difference in the distribution of the specific carriers for each substance. The model does not, of course, explain how auxin moves about whilst it is inside the cell. It is thought likely that simple diffusion could account for the redistribution which is required by the model. Certainly the supposed inhibitor of cytoplasmic streaming, cytochalasin B, has no effect on polar auxin transport.

This model emphasises strongly the importance of spatial events in development, even at the molecular level. If it is true, then the basis of polar auxin transport is simply an asymmetry in the distribution of a carrier protein within the plasma membrane of cells. The consequences of this simple idea are profound in terms of the physiology of the plant, as we have seen, and as we shall now consider in the context of pattern formation.

Polar auxin transport and pattern formation in the vascular system

The significance of polar auxin transport seems to lie at least in part in the control of vascular differentiation. Even isolated cells can be induced to differentiate into xylem elements (p. 142) which may rightly be taken to show that gradients of hormones are not necessary in order to trigger a subcellular change leading to a particular developmental sequence. Nonetheless, there is convincing evidence that fluxes of a stimulatory substance are required for the continuous development of connected vascular tissue, and that at least one important agent in such flux is auxin.

Many lines of evidence could be presented to support this view, but some of the most direct concern the differentiation of vascular tissue following wounding. When the stem of a plant is partially severed, the formation of a new series of connected vascular elements around the cut is promoted by the presence of leaves above the cut, but not by those below it. The requirement for leaves can be replaced by auxin in stem segments (Fig. 7.15*a*, *b*). The direction of differentiation of the new vascular tissue is downwards from the site of application of the auxin. That such differentiation is controlled by a polar flux of the auxin and not by a sink effect from say the root end of the segment is shown by the type of cut in Fig. 7.15*c*. The differentiation of a continuous vascular system proceeds into that part of the stem which is leading to a dead end. The absolute requirement for the source of auxin is shown by the cuts in Fig. 7.15*d*, *e*, where new differentiation of vascular tissues will only occur if a source of auxin is applied to the surface as shown. The absolute need for a flux of auxin can be demonstrated by isolating a piece of cambium surgically within an intact stem. If the isolated piece is completely encircled by the externally applied auxin source, no continuous differentiation of vascular tissue is found (Fig. 7.15*f*). However, if an outlet for the applied auxin is arranged, then differentiation occurs (Fig. 7.15*g*). In terms of differentiation, an imposed flux of auxin can overcome and even reverse the normal polarity in a tissue (Fig. 7.15*h*).

Experiments with wounding are instructive in more detailed ways. For example, vascular differentiation in response to applied auxin after wounding is not a precise function of the position in the stem at which the wound is made. This means that the differentiation which is seen is not predetermined in some way; it is a consequence of the application of the auxin. Furthermore, were the absolute concentration of auxin the trigger to differentiation, then it would be expected that a cup-shaped zone of differentiated tissue would appear around the point of the application. This does not happen. Instead organised files of vascular elements are formed which connect up to the existing vascular system. This implies not only that it is a flux of auxin which acts as the trigger, but also that once a cell has differentiated it exerts a lateral inhibitory influence, and a stimulatory influence on the next cell in the file.

The inhibition can be seen experimentally. If

auxin is applied to a segment of stem which has been wounded, vascular connections are made between the site of the application of the auxin and the main vascular system (Fig. 7.15*i*). However, if the vascular system is connected to young leaves above it, applied auxin does not elicit this effect (Fig. 7.15*j*). Similarly, if the existing vascular system is preloaded with auxin applied externally to the top of the cut segment, connections are not made between it and a source of auxin applied to the side (Fig. 7.15*k*).

These observations are taken to show that the existing vascular system acts as a sink to produce the flux of auxin which is necessary for connected vascularisation. If the existing system is loaded artificially, or from young leaves, then it cannot act as a sink and the flux is not induced. In the same way it can be envisaged that a newly differentiated vascular element following wounding may act as a sink, withdrawing the stimulus to differentiate from its neighbouring cells and channelling it along the file. It follows from this that whilst auxin can be transported in parenchyma tissue, it would be expected that the formation of new

Fig. 7.15. Patterns of vascular differentiation following wounding instems. *a*, Absence of new vascular differentiation if no leaves or auxin source above the cut. Continuous vascular differentiation occurs if (*b*) an auxin source is applied to the cut segment. This does not (*c*) require a sink, but the source of auxin is essential (*d*), (*e*). If the cambium is exposed (surface views), differentiation only occurs if a flux of auxin can be generated (*f*), (*g*). Polarity reversal can occur if (*h*) no alternative route of differentiation is possible. Newly differentiated vascular tissue can (*i*) connect to existing vascular tissue, provided that the existing tissue is not (*j*) preloaded either from leaves above or (*k*) from externally applied auxin. The solid rectangular blocks are externally applied auxin; the thin wavy lines represent newly formed vascular tissue.

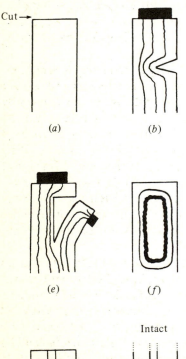

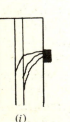

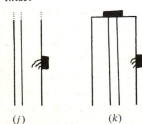

vascular tissue would increase the rate of its transport, since differentiating vascular tissue is the preferred route. This idea has been confirmed in studies of transport rates using radioactive auxin.

Experiments such as these allow some estimate of the time over which the signal must act in order to produce its effect. It is found that applied auxin must be present for at least 1 day for differentiation to be produced, and that the differentiation process from parenchyma cell to death of the xylem element only takes about 2 days. This timing suggests strongly that the stimulus must be present continuously during differentiation and does not simply act as an instantaneous trigger to the process.

The importance of polarity in auxin transport is that all these results are dependent upon the operation of a flux of auxin and not upon a static gradient of concentration. The polar transport of auxin can occur against a concentration gradient, and so it is easy to visualise that a certain inductive flux could occur through a cell which actually only had a very small concentration gradient across it. This is an essential part of the theory, and it can be used to explain why in many circumstances vascular differentiation does not follow the shortest route between the source of auxin and its supposed sink.

From a more general point of view related to pattern formation in normal plants, the essence of the flux hypothesis draws the following conclusions. Cells respond in a particular way to a continuing flux of auxin, and possibly other signals which, in the whole plant, arise in young leaves. The nature of the response – vascular differentiation – is autocatalytic; that is to say, once differentiation begins, it tends to channel the flux along the forming axis of the vascular strands, thus suppressing the differentiation of lateral tissues. Pattern formation in the vascular system according to this hypothesis is therefore due to the continuing and indeed increasing presence of a stimulus, and stresses the importance of intercellular interactions. It does not appear to be due to the operation of switches, either on or off, due to the position of the cells within specific gradients. This aspect of the hypothesis is not necessarily applicable elsewhere. For example, in a stem cutting with its leaves intact, roots are formed at the cut end presumably precisely because the flux is arrested and a concentration of the signal accumulates. The idea of the flux of auxin and other signals therefore should

only be considered relevant in the context of pattern formation in the vascular system and not as a general explanation of morphogenesis.

Polarity and the generation of form

Many examples of developmental events involve prior cell division; sometimes the link between the division and the event itself is a very close one, as for example in the formation of the stomatal complex (p. 118). In such a situation, it is clear that the symmetry and polarity of the division are an integral part of the process and essential for its successful completion. Within meristems, however, it is not at all clear that the generation of form is necessarily linked to a high degree of order in the patterns of formative cell divisions. Indeed, it might be said that the failure to detect such patterning suggests that cell division does not play an important role in the determination of form, although it is as likely that mere technical failures account for much of the lack of understanding of meristem behaviour. Certainly in the *Azolla* root tip, cell divisions do indeed appear to be highly ordered (p. 158).

Nevertheless, the question remains in the context of the present discussion of polarity: what is the relationship between the polarity of a cell, the polarity of any divisions which it may undergo, and the generation of form? This question has been examined in a special way. It is found that if grains of wheat (*Triticum vulgare*) are exposed to very high doses of gamma radiation before being sown, they develop without cell division occurring. Such plants are slower to grow than unirradiated controls, and have only a limited growth potential; they are known as gamma plantlets.

It is found that the gamma-irradiated grain will produce a first leaf which, in 10–12 days, grows to the same size and shape as the first leaf produced within 3 days from an unirradiated control plant. When cell numbers are counted in these two similar leaves it is found, not surprisingly, that the cell numbers in the gamma-plantlet leaf are the same as were present in the embryonic first leaf in the grain, but that the control leaf has many more cells. As a consequence, the cells in the gamma plantlet are much larger than those in the control leaf. In particular, the cells in the gamma-plantlet leaf are much longer than those in the control, although no wider. The gamma plantlet leaf is of course devoid of specialised cells such as guard cells and leaf hairs, since these can only be formed as a result of cell division processes. Nonetheless, the overall form of the leaf is very similar in both types of plant, and this can be interpreted as showing that cell divisions play no essential role in the generation of form in this system.

In a further refinement of these experiments, the production of new leaf primordia was examined. In wheat, this occurs normally within the single outer-most tunica layer, and is accompanied by periclinal cell divisions in this layer to give rise to a small protuberance. In the gamma plantlets, it was found that small protuberances of the expected size could arise in the expected positions from a phyllotactic point of view, without cell division. This again is interpreted as showing that the generation of form may initially at least not involve the reorientation of planes of cell division, although in a normally growing plant, reoriented cell divisions are a consequence of this early development, and not obviously separated from it.

The same sequence of events can of course be inferred from the studies on developing *Fucus* eggs, which have been described. Here the expression of a fixed polar axis is the formation of a protuberance at the rhizoid point, and as we have seen, this precedes by several hours the asymmetric division which gives rise to the partitioning of the egg cytoplasm. These observations all point to a fundamental importance of the concept of polarity.

Finally, it should be stressed that these arguments in no way detract from the significance of cell division and planes of cell division in overall growth and development. Gamma plantlets do not continue to grow beyond the stage of the first leaf, and new leaf primordia which may be initiated at the shoot apex do not develop beyond the stage of a tiny protuberance on the surface of the tissue. Cell division is obviously important for continued growth, as it is for the generation of particular specialised functions within

tissues. However, it is clear that the mechanism for the determination of overall form is, to a great extent, independent of the determination of cell numbers, sizes and shapes.

Summary

Plants grow by a process of extension along growth axes. That these axes are different at either end is an expression of their polarity. Polarity can be manifested at all levels of an organism. In its most easily visible form, it is present in the shoot/root polarity of an entire tree. At a lower level, it is seen as the polar regenerative behaviour of a stem or root cutting. This type of polarity is reflected also in the physiology of the plant, where an important mechanism for its expression is the polar transport of the growth substance auxin. The existence of such polar processes has great importance in the control of the growth of the plant, exemplified by the phenomenon of apical dominance. It seems likely also that polar transport of signals determines the patterning of the vascular system.

At the individual cell level, polarity can be seen in the distribution of cellular organelles and in the localisation of growth processes to particular parts of the cell. When the relative contribution of subcellular components to polar expression is examined, several lines of evidence independently suggest that an early event in the determination of polarity is the formation of polar asymmetries in the structure or chemistry of the plasma membrane. This is exemplified by the development and fixation of the polar axis in a single cell such as the fertilised *Fucus* egg, and the same conclusion may be invoked as an hypothesis to explain polar transport behaviour in higher plants.

In studying polarity therefore we are making a close approach to the molecular basis for some aspects of plant development. In the succeeding chapter this theme will be taken up in more detail.

The regulation of development

Introduction

In the previous chapters we have glimpsed some of the complexities of the development of plants and examined the process from the standpoint of the individual cell, and also in terms of how the reactions of groups of cells determine behaviour. Differentiation has been seen to be expressed at many levels. The organelle complement of a cell may change during its development. The size, shape and wall characteristics of cells are also indicative of developmental status. These structural changes are correlated with changes in the chemical abilities of cells. Differentiation can also be studied in morphogenetic terms; the behaviour of cells or small cell groups in determining the production of new organs from callus tissues or growing meristems.

The regulation of these processes is likewise exerted at a number of different levels. At the present time, biology is still in the immediate aftermath of the discovery of the structure of DNA, and the genetic code by which this structure is translated into metabolically active proteins, the enzymes. Much discussion of the course of development revolves around the central importance of the genes in regulating the process, and this is perfectly sound. In this chapter, it is a theme which we shall consider in some detail. Nevertheless, it is important to grasp the principle that the significance of differentiation and development is to allow the plant to survive in its environment. This is clearly seen when development is viewed in the light of evolutionary change. In broad terms, the progress of evolution has led to the formation of more complex plant bodies, and this has necessitated the development of more complex and specialised structures. With this generalisation in mind it should come as no surprise to realise that many developmental processes are influenced profoundly by environmental circumstances. Naturally, the ability of a plant to react to its environment is a capacity which it inherits in its genetic makeup. The problem of the regulation of development is to attempt to understand the relationship between changes in behaviour and changes in molecular processes, whether or not these be directly concerned with the genetic material.

Throughout the course of the previous chapters the main emphasis has been placed on the cell as the unit of development. In multicellular plants this is a device – an aid to thought – as much as an expression of the reality of the situation. Cells are nearly always in communication with their neighbours and therefore need not act as entirely independent units. Compartmentalisation into cells is not absolutely necessary before differentiation can occur, as is exemplified by such organisms as *Caulerpa* and *Acetabularia*. The capacity to generate overall form is likewise not necessarily linked to the continuous presence of cellular structure. Cells can equally logically be regarded as compartments within the organism, as the organism can be regarded as being 'built up' of cells. The former view is probably the more accurate, as an examination of the growth of the embryo has shown (p. 155).

In considering the individual cell, a further compartmentalisation is encountered. It is almost a matter of definition that a living cell consists of a nucleus and a cytoplasm to sustain that nucleus. This statement admits immediately of the importance of the nucleus with its genetic material, but it is becoming increasingly clear that the contribution of the cytoplasm to developmental processes extends beyond a merely passive role as the translator of the genetic potential of the nucleus. The cytoplasm, after all, is where much of the spatial organisation of the cell must be situated, and such organisation is essential for the development of extended tissues.

Development involves the formation of cells with particular specialised functions. Many of these functions may fairly, if a little simplistically, be regarded as purely biochemical. In order for photosynthesis to occur, it might be argued that the most important event would be the genetic specification of a number of enzymes in the photosynthetic pathway. This is no doubt true, but even with an activity as widespread as photosynthesis the situation is immeasurably more complex than this. Not only do the correct enzymes of the carbon pathway have to be made, but also the pigments and light-harvesting proteins. These components have, furthermore, to be arranged in their correct spatial relationships on the thylakoid membranes. The entire structure of the chloroplast has to be specified during development and, of course, the cells which contain the chloroplasts have to be in a position on the plant where they will receive the light. The fixation of carbon dioxide requires good gas exchange between the green cells and the atmosphere; in a multicellular tissue this implies the formation of air spaces linked to the atmosphere by controllable pores (the stomata). Finally, photosynthesis supplies energy for the rest of the plant; its useful operation therefore requires a further set of enzymes and membrane proteins to export the product, and the development of a vascular system. This illustration serves to show that in seeking an insight into the control of development, there will be no easy answers; the implications of any given situation are far too complex for this to be possible.

The general nature of the development process

Before attempting a more detailed exploration of some of the areas related to developmental regulation, it is as well to grasp a few general aspects of the process. The most obvious is its orderly progress in space and time. This can be illustrated by reference to the vegetative shoot apex. Here leaf primordia are initiated with geometric precision during growth, and each new leaf then undergoes division and differentiation to give rise to the mature and functional organ. The different tissues of the leaf develop along different pathways and cease division at different times. Vascular tissues are formed at a rate which keep pace with the overall growth of the leaf so that its function can be maintained.

This simple description shows immediately the degree of control which is needed in the process of development. First, a small number of cells in the meristem is involved in the formation of a primordium. This small subpopulation of the meristem, at first apparently uniform, is quickly sub-divided into populations of cells with different de-velopmental fates. Both the initial formation of the primordium and the determination of the future of the cells within it are associated with cell division. Divi-sion continues, together with cell enlargement, in the various subpopulations of the primordium. This leads to an increase in cell numbers and growth of the tissues in an integrated manner. The formation of the specialised mature cells of the leaf, such as the guard cells, hairs and glands at the epidermis, involves further cell division associated with renewed potential for growth and differentiation. Thus cell division may be associated with both the proliferation of uniform cell types and also with the initiation of new pathways of development.

At the cellular level, progress towards the attainment of mature function is equally coordinated and precise. Cells of each of the major tissues of the leaf appear to be more or less uniform. Their spatial relationships to one another are determined by a combination of the duration of the period over which cell division continues to occur, and the relative rates and directions of expansion of their surfaces. The organelle populations of each cell change in a way which is fitting to the function it is to serve and, by implication, so does the pattern of enzyme activities which the cell contains. The end result is an organ which is recognisable by its shape, which has a defined position on the plant and which is made up of different cell types, all in their correct relationship to one another and precisely equipped with the correct structural and biochemical attributes.

Determination

It is clear that only certain patterns of develop-mental change can occur together. This is most obviously true in the formation of specialised cells which are peculiar to particular organs. For example, it is a certainty that none of the cells which derive from a root meristem will in the normal course of differen-tiation form stomatal guard cells. Equally, a primordium which arises at the shoot apex will not normally give rise to a root cap or any of the cells characteristic of roots. This limitation on the types of cell differentiation which may occur together is not only seen during the growth of normal plants. The same constraints apply to organs which are formed as a result of the differentiation of callus tissues. If a root develops from a callus tissue, it will be a more or less normal root – it will not form cells which are charac-teristic of leaves or flowers. In other words, although meristems appear to have unlimited potential for development, and although callus tissues are normally described as 'undifferentiated', these states do not exist in the absence of overall control.

This raises the concept of determination. Although all meristematic cells have many features in common, it is clear that they are not all equivalent. The cells in the meristem of a root inevitably give rise to progeny with potentials for development which differ from those which arise in a shoot. Their fate is in some way 'determined' at the outset. We have already seen that this determination is not absolutely fixed. For example, in plants with a pronounced juvenile and adult phase, the products of the shoot apex may differ quite markedly in the two phases. This type of switching is also observed during the change from vegetative to reproductive growth; the same meristem which formerly gave rise to leaves and axillary buds can be switched to produce a fixed sequence of petals, sepals and other floral structures. Again, primordia produced by *Convolvulus* roots can be induced to form buds or roots depending on their nutritional status (p. 145). It is also possible to induce the formation of roots at the shoot tip, exemplified by the production of inverse cuttings (p. 201). However, all these examples serve to emphasise that the pro-ducts of a meristem are under a coarse control which may only be reset in special circumstances.

Coordination of development

The next general consideration is that pathways of development are limited in number and in some way linked together. The complexity of the substruc-ture of cells leads to the expectation that the number of different cell types which could be formed by permutations of different internal and wall structure would be enormous. In fact this is not the case. It is certain that structural analysis underestimates cell diversity, since many changes in cell biochemistry do not lead to visible structural change. Nevertheless, it is clear that the number of cell types within plant bodies is remarkably small. The simplest explanation of this is that controlling elements which determine the presence or absence of a characteristic within a particular cell do not necessarily act independently, but in groups. This can be illustrated in a trivial way by reference to a particular cell type – for example, the sieve element.

In specifying the structure of the sieve element, it is possible to envisage that every characteristic might be specified and controlled independently. Thus the position and development of the sieve pores would involve one or more sets of controls. The synthesis and deployment of P-protein would involve others, and yet more would control the changes seen in the plastids, the endoplasmic reticulum and the nucleus. If complete independence of all these controls were possible it would be predicted that cells might arise with say fully developed sieve pores, but none of the other characteristics of the sieve element or that sieve element plastids might be found in any cell type within the plant. This does not happen. The only way in which partial differentiation of a cell type occurs is during the progress of a cell along a developmental pathway. In such a case it is more logical to suppose that a complete linked set of controls has been activated, and that the observed partial state of differentiation merely expresses the different rates at which the subpathways of development take place.

Of course, linkage of controlling elements can act at different levels, and not all the characteristics of a cell need to be linked together. For example, whilst it is certain that a cell with spirally thickened and lignified walls will not possess green chloroplasts, because it is a xylem element, it could not be inferred that a cell with starch grains in its cytoplasm was necessarily a root cap cell. An analogy may be drawn with the use of language itself. Terms such as 'xylem element', 'sieve element', 'guard cell', 'mesophyll cell', and so on all describe recognisable cell types which have a great deal in common in whichever plant they are studied. One or two words serve adequately to specify structures which require hundreds or thousands of words to describe in detail. In a similar way it is reasonable to assume that the details of their development might be controlled from a small number of switching points.

Environmental controls

Another overall factor in development is easily defined. This is its susceptibility to control by the environment. It is convenient for the purpose of much discussion to regard development as occuring continuously. However, this is not what happens in real plants growing in the natural environment. Many processes are interrupted and others only occur in response to particular circumstances. Periodic dormancy of temperate plants is an example of the former, and the control of flowering by day length or temperature are examples of the latter. Structural and morphological changes brought about by external circumstances have already been encountered in such cases as the development of transfer cell function in response to pathogens (p. 107) and heterophylly in aquatic plants (p. 164).

It is not difficult to appreciate the reasons for this sensitivity to the environment. Continuous growth of all developed structures would be very inefficient in terms of light-harvesting. In temperate regions continuous growth would expose plants during cold or dry seasons to degrees of stress which would be fatal. Being immobile, land plants have to be able to exploit their habitat to its full advantage, whether this be a small local variation in light levels due to shading by other species, or a grosser effect due to a change in latitude or altitude. Thus, in considering the internal controls of development, it is necessary constantly to bear in mind that external controls also operate. In this chapter we shall consider some of the external controls first.

Environmental control and flowering

The best example of the control of a developmental process by an environmental stimulus is the transition to flowering at the vegetative shoot apex. This change signals considerable modification of the structure of the apex and the fate of the cells within it. It is also a profound influence on the life of the entire plant. In many species, which are called monocarpic, flowering is inevitably followed by the death of the plant. This may occur in a single season, as in annual plants, or it may be delayed for many years. The most familiar example of this latter type of behaviour occurs in the so called century plant, *Agave americana*, which flowers once at the age of 10–15 years and then dies. Many perennial plants may of course flower repeatedly, and these are termed polycarpic.

The physiology of flowering is extremely complex and a wide range of factors can affect the induction of reproductive growth. These are all of great interest to physiologists and indeed to growers but, in terms of natural controls of the development of flowers, we can limit ourselves to quite a small proportion of the work which has been carried out. In a general way, flowering is a feature related to the age of the plant. It is obvious that in order successfully to produce seeds or fruits, a certain minimum photosynthetic capacity is necessary. The effect of age is most strikingly seen in forest trees where commonly tens of years of vegetative growth precede the formation of any flowers. At the other extreme, *Chenopodium rubrum* can be induced to flower when it is only a few millimetres high if grown under appropriate conditions of light, nutrition and temperature. Age effects do not represent a very convenient type of control process for the purposes of scientific study, since it is clear that they are closely linked to a variety of other factors.

The second requirement for flowering which is quite widespread is for a period of exposure to low temperatures. This is exemplified by spring-flowering species such as violets (*Viola* spp.) and primroses (*Primula* spp.), and also by biennials such as foxgloves (*Digitalis* spp.) and celery (*Apium graveolens*). It has been particularly well studied in cereal crops, where a marked difference in reproductive behaviour may be found between varieties. For example, a spring wheat planted in the spring will grow and produce a crop following flowering in early summer. A winter wheat sown in autumn will grow slowly through the winter and produce a crop following flowering in early summer. However, a winter wheat sown in late spring will grow but fail to flower. The stimulation of flowering by exposure to a period of low temperatures is called vernalisation.

The cold is detected by the meristematic tissues. It is only necessary to chill the terminal apex of a vegetative shoot in a plant requiring vernalisation for flowering to be induced. Certain plants can be vernalised at the stage of the seed; in this case it has been shown by isolated culture methods that the embryo is the seat of the response. The response which cold treatment elicits can be propagated within the plant. This has been shown by the following type of experiment (Fig. 8.1). A plant of *Chrysanthemum* requiring vernalisation was grown, and its terminal apex was exposed to low temperature. This shoot was then allowed to develop two more leaves before it was removed. One axillary bud from this new leaf pair was allowed to develop and produce two new leaves. The terminal apex of this new branch was removed, with the result that a new branch formed from an axillary bud on the remaining piece of the branch. This process was repeated until seven pairs of leaves had been produced from the original chilled apex. The terminal bud was then allowed to develop, and it produced a flower.

Fig. 8.1. Induction of flowering by chilling in *Chrysanthemum*. If the terminal apex is chilled (far left), the effect is propagated through all the new buds which are formed by growth of the chilled apex.

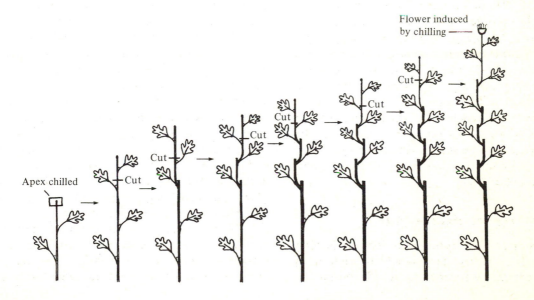

This behaviour strongly suggests that the effect of vernalisation is reproduced in the plant and is not a once and for all phenomenon associated with particular cells of the original chilled apex. This in turn suggests that the change induced by vernalisation may not be the production of a particular chemical substance, but rather a change in the cells which is reproduced during cycles of cell division. The question immediately arises as to whether vernalisation is mediated through the production of a hormone. There is conflicting evidence on this matter. For example, in the case of the chrysanthemum plant just cited, it is found that chilling of the apex results in the flowering of that apex, or any new apex which is formed from it after the chilling has taken place. However, flowering is not induced in non-chilled buds on the same plant. This argues against the idea of a transmissible stimulus, although, of course, it is possible that such a stimulus could not be transported down the plant. With henbane (*Hyoscyamus niger*), it is found that if a leaf from a vernalised plant is grafted on to an unvernalised plant, then flowering is induced in the latter. This would appear to indicate the presence of a transmissible stimulus. Furthermore, grafting of vernalised leaves of henbane is successful with certain other species of plants such as *Petunia* and tobacco (*Nicotiana*). This suggests that any transmissible stimulus is common to more than one species.

Photoperiodic control of flowering

The most intensively studied influence on flowering behaviour is that of daylength. All flowering plants can be separated into one of three classes with regard to their light requirements in the change from vegetative to reproductive growth. The first group – the day neutral plants – will flower equally well in long or short days. Flowering in such plants is subject to environmental controls such as chilling, or to other factors such as age and nutrition. The second group – the short day plants – will only flower if exposed to light periods below a certain maximum length. The precise daylength to stimulate flowering is not fixed, but days of around 6–8 h in length produce the most rapid response. The last group – the long day plants – require light periods above a certain minimum length if they are to flower. Again, the minimum value is not precisely fixed, but is commonly in the region of 12–16 h. Plants may show complicated combined requirements for flowering – for example, short days

following a period of chilling, or short days following long days. The result of such a series of controls is the common observation which everyone perceives – that many plants have characteristic times of the year at which they flower. This may be presumed to have survival value both in terms of the availability of pollinators and the continuation of suitable climatic conditions to allow fruit and seed development. The importance of photoperiodism for the study of plant development is that it represents a system in which the major variable – day length – is easily controlled independently of other conditions.

The light stimulus is perceived by the leaves of the plant. This has been shown by growing plants with the terminal apex in one set of conditions and the leaves in another. It is found that flowering is only induced when the leaves receive the correct daylength conditions. Maintaining a short day plant with its leaves in long days and its terminal apex in short days will not cause it to flower. It is not necessary for all the leaves of the plant to be exposed to the inductive stimulus. If only one leaf is given the inductive lighting, this may be enough to cause the plant to flower. The effect is much increased if all the other leaves are removed. Indeed, if this is done, it may only be necessary to have part of one leaf remaining on the plant for flowering to occur. It has been shown with the short day plant *Xanthium*, for example, that only one eighth of one leaf is needed in order to produce a flowering response to short days.

These results seem to indicate that the flowering response to day length treatments is mediated by a transmissible stimulus. The site of perception is distant from the site of the response, and the antagonism of the stimulus by the presence of other, unstimulated, leaves suggests that there is dilution effect. This has led to the hypothesis that a specific floral stimulus is produced and that it is hormonal in nature. This supposed 'florigen' has not yet been isolated, and of course, it differs from the action of known growth regulators insofar as it is postulated to act specifically in the stimulation of a particular developmental change: the production of flowers.

Several experiments cast doubt on this idea. It is found, for example, that simply connecting a leaf from an induced plant via a gelatin bridge to a leafless plant which has not been induced does not cause that plant to flower. However, if the connection is made by grafting, the floral induction does occur (Fig. 8.2).

This suggests that whatever the nature of the stimulus it is not a small stable molecule which can diffuse out of cells, across a gelatin bridge, and into other cells. A second type of grafting experiment makes a more remarkable point. It is found that if a leaf from an induced plant of *Xanthium* is grafted onto a defoliated uninduced plant, then this plant will flower. This is the original experiment (Fig. 8.2). However, it is found that leaves which form on the plant from buds or the terminal shoot can also act as inducers of flowering, even though they have never been directly exposed to the correct lighting conditions. Were this behaviour to be generally found it would show that the flowering stimulus is in some way propagated and multiplied within the plant. In fact, it is not generally found. In *Perilla*, for example, a defoliated plant can

be caused to flower by an induced-leaf graft, but leaves which subsequently develop under non-inductive conditions do not show inducing ability in further grafts.

The nature of the photoperiodic response

Whilst the end result of the induction of flowering is the same in both long day and short day plants, it is worth asking the question whether the mechanism of induction is the same in both cases. Clearly the environmental requirement is different in the two types of plant, but this does not necessarily imply that different mechanisms exist for the transmission of the stimulus. Various lines of evidence suggest that the two systems have much in common.

The first evidence comes from a consideration

Fig. 8.2. Some photoperiodic effects with a short day plant. *a*, If it is maintained on short days, the plant will flower; *b*, maintained on long days, the plant will not flower; *c*, removal of all the leaves except one does not abolish the flowering response in short days; *d*, grafting a leaf from a plant growing in short days to a defoliated plant maintained in long days causes the receiver plant to flower in long days.

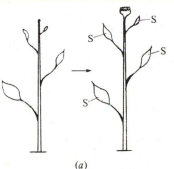
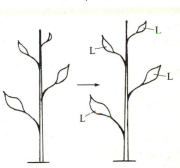
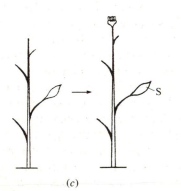
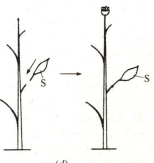

of the effect of light quality on the response. Short day plants require long dark periods for the induction of flowering. If this long dark period is interrupted by a flash of light, the stimulation of flowering may be suppressed. This fact enables the quality of light required to suppress the response to be examined. With the short day plant *Xanthium*, it is found that in order to be most effective, the flash of light – lasting only a minute or so – should be given 8 h from the start of the dark period. If the illumination consists of red light only, then flowering is suppressed. If however, the flash consists of far red light, then flowering is not suppressed. Finally, a flash of red light followed by a flash of far red light does not suppress flowering. In other words, the far red light reversed the effect of the red light. This is a classic symptom of an effect

mediated by the pigment phytochrome (p. 218). The analogous experiments can be performed with long day plants in the following manner. A long day plant held on a suboptimal daylength of say 8 h, will not flower. However, if this short day is supplemented by a further period of red light, flowering may occur. This stimulation of flowering is not produced by supplementary far red lighting, and it can be reversed if far red light is given. In detail, this represents a very complex situation but the general point may be made that both short day and long day plants show symptoms of phytochrome-mediated responses to daylength.

The similarity of the stimulus has been shown by grafting experiments. Two varieties of tobacco were used. *Nicotiana sylvestris* is a long day plant, and

Fig. 8.3. An experiment to show the similarity of the flowering response in long day and short day plants. *a*, A grafted plant consisting of a branch of *Nicotiana tabacum* Mary Mammoth (short day) and *N. sylvestris* (long day), is maintained with a light barrier between the branches. Each branch receives non-inductive lighting. No flowers are induced. *b*, If the light barrier is removed and the plant exposed to long days, both branches are induced to flower.

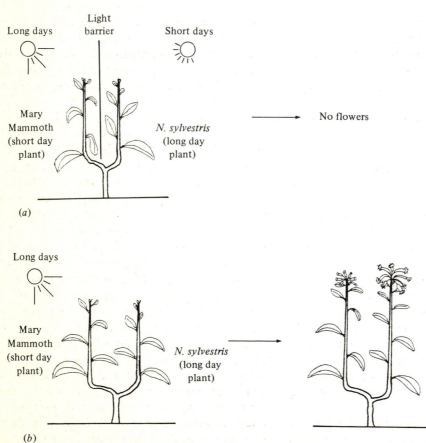

Nicotiana tabacum, variety Mary Mammoth, is a short day plant. These were grafted together to produce a plant with two branches, one of each variety. The two branches were separated by a light barrier, and the *N. sylvestris* branch kept under short days, with the Mary Mammoth branch in long days. Neither branch flowered under these conditions, which is the expected result. However, when the light barrier was removed and the entire plant exposed to long days, it was found that not only did the *N. sylvestris* branch come into flower but that also, after a delay, so did the branch of Mary Mammoth (Fig. 8.3). This is good evidence for the existence of a common flowering stimulus for both long day and short day varieties.

The nature of the changes induced under different photoperiodic conditions is quite unknown. The complexity of the situation is best appreciated by considering the behaviour of plants under interrupted cycles of long or short days. Some plants can be stimulated to flower by a single inductive cycle – one long dark period in the case of a short day plant or one long day in the case of a long day plant. However, the majority of species require several such cycles for the flowering response to be established. It is, of course, possible to arrange matters so that these inductive cycles are not given consecutively in an unbroken sequence. Thus in the case of a plant requiring 5 cycles of inductive conditions, it is easily possible to give 2 cycles, followed by a gap during which non-inductive lighting is given, followed by the remaining 3 inductive cycles. When this type of experiment is carried out a difference emerges in the behaviour of long day and short day plants. It is found that giving the inductive stimulus as an interrupted sequence does not prevent the flowering of long day plants provided that the number of days of the intervening non-inductive period is not too great – up to about 20 days. However, in general, short day plants can be inhibited from flowering by a single long day interposed between sequences of short days. The behaviour of long day plants suggests that the result of the inductive stimulus is quite stable, at least up to 20 days; on the contrary, in the short day plant it appears to be totally destroyed by one long light period. These experiments cannot therefore be interpreted simply by assuming a common basis of response in the long day and short day plants.

The photoperiodic control of flowering is thus extremely complicated and, from the standpoint of explanation, most unsatisfactory. No single hypothesis can yet account for all the known facts and, indeed, there is still considerable controversy over whether flowering is induced by a specific hormone, by a combination of growth substances or by some other and perhaps more direct mechanism. In view of the profound nature of the changes within cells which are the end result of the transition to flowering, this behavioural complexity is perhaps not surprising. What emerges clearly from this work is the importance of the external environment in controlling one of the major developmental transitions in the life of the plant.

Environmental control and dormancy

Aside from flowering, the most obvious examples of periodic change in the growth of a plant are related to the phenomenon of dormancy. Dormancy is most familiar in the growth of temperate plants, where unfavourable environmental conditions occur on an annual basis and elicit the response. Dormancy can be induced as a response to the arrival of short days and low temperatures in winter; equally it may occur in summer during times of drought or extreme high temperatures. Dormancy may occur in various tissue and organ types; in temperate trees axillary buds become dormant, as does the cambium in stems. In other plants, dormant structures include bulbs, tubers and rhizomes. It is also frequently found that the dispersal phase of a plant shows a degree of dormancy. This is true of many seeds of higher plants, but is also shown by the spores of some lower plants.

In physiological terms, dormancy defies an all-embracing definition. When growth is arrested by a direct environmental stress such as drought, we may speak of enforced dormancy or quiescence. On the other hand, many forms of dormancy occur despite the concurrence of apparently favourable growth conditions. This is exemplified by the behaviour of many seeds, which will not germinate immediately after being shed from the plant even though they may be placed in ideal conditions of temperature, humidity, etc., which lead to ready germination at a later stage. Similarly the axillary buds of many trees will not elongate in late summer, even though conditions of light, temperature and so on appear to be favourable. This condition, which may at first be described as an effect of apical dominance (p. 172), is replaced by true dormancy as autumn approaches. During the predormancy period, growth may be restarted by removal of the shoot tip or of the leaves but, in full dormancy, these procedures will not produce any effect. The release from dormancy is likewise a gradual process under natural conditions, with growth restarting only when all the environmental factors are favourable. In some species, apparent absolute dormancy only occurs when they are grown in natural conditions; transfer to different temperature or light regimes may lead to premature restarting of growth.

The dormancy of buds

Bud development may be illustrated by reference to a tree, the sycamore, *Acer pseudoplatanus*. In this species, shoot elongation comes to an end by the formation of a dormant terminal bud. Within the shoot meristem activity continues but, instead of producing new leaves, the outer primordia of the shoot develop into bud scales or cataphylls. Enlargement of the bud may continue over a period of several months, during which time its growth can be restarted with increasing difficulty by such means as the removal of the leaves. In other species extension growth is brought to an end by the abscission of the shoot tip. Under these conditions axillary buds are responsible for continued growth after the period of dormancy.

The onset of dormancy in very many woody species is controlled by daylength. Short days strongly promote the formation of dormant buds, and conversely, maintaining a plant such as birch (*Betula* spp.) under artificial long days will completely remove any tendency for dormant buds to form. This photoperiodic effect is similar in many ways to the photoperiodism which controls flowering. For example, a short burst of light during the long dark period will prevent the dormancy response and, in some species at least, it appears that mature leaves may be the site of perception of the light period. Short daylength is however not a universal inducer of dormancy in buds; certainly it is commonly observed that many trees cease extension growth in June or July in the Northern hemisphere, when days are still long.

The importance to the plant of bud dormancy is that it can pass through the unfavourable conditions of winter in a form which is protected from damage. Growth resumes in the spring, and it is known that the breaking of dormancy in many woody species requires a period of exposure to low temperatures. This period varies from a few hundred to over a thousand hours at a temperature in the range 1–10 °C. In an English winter, it is probable that dormancy-breaking requirements are met by late January or early February, and thereafter growth depends upon a rise in temperature. In regions with mild winters, it is frequently observed that bud break in the spring is irregular and delayed. This represents a problem for fruit growers in California, for example. In some species, dormancy can be broken by exposure to long days without an intervening period of chilling. This behaviour is shown by the common beech (*Fagus sylvatica*), and it suggests that the dormant buds themselves are the site of photoperception.

The dormancy of seeds

A number of different patterns of dormancy can be recognised in seeds. Some, such as those of the Chenopodiaceae, are impermeable to water and so will not germinate until the seed coat has been sufficiently broken down by soil organisms. In such cases, a brief treatment of the seed with concentrated sulphuric acid will serve to overcome dormancy. Some seeds contain immature embryos at shedding, and these will not germinate until the correct stage of maturity has been reached. Grains such as barley (*Hordeum*) and wheat (*Triticum*) will not germinate when freshly shed, but they acquire the ability over a relatively short period without any special stimulus. In these cases, it is difficult to see what advantage dormancy of the seed might confer to the plant.

The advantage of this dormancy is more obvious in those seeds having a chilling requirement. Seeds such as that of the peach (*Prunus persica*) will not germinate if planted immediately in soil at a temperature of say 20 °C. However, if such seeds are left in a moist condition for a period of several weeks at a temperature in the range of 0–5 °C, germination ensues when the temperature is then raised. This type of dormancy in nature will ensure that the seeds of species which show it will not germinate until the spring of the year following that in which they were shed. It is very common behaviour amongst the seeds of woody plants, although it is not confined to this group.

In some cases dormancy is only found in part of the embryo. Acorns, for example will produce a radicle if they are planted directly after shedding, but the epicotyl does not grow until a chilling requirement is met. In a further variation of this pattern, seeds of *Convallaria majalis* (lily of the valley) need one period of chilling for the radicle to emerge, and a second period of chilling to stimulate the growth of the epicotyl. Hence in nature, seedlings will not appear above ground until the second season after planting.

The final group of seeds showing dormancy effects require light for germination; a small related group contains seeds which are inhibited by light. The best known example of a photoblastic seed is that of the lettuce, and this has been studied extensively in the context of phytochrome (p. 218). Phytochrome seems to be the universal photoreceptor for positively photoblastic seeds; that is, those requiring light for germination. The light effect may show interaction with temperature; for example, seeds of Grand Rapids lettuce require light when germination is attempted in the temperature range of 20–30 °C, but not between 10–20 °C. The light requirement also becomes less exacting as the seed ages.

The dormancy of a seed is defined in terms of its ability or inability to produce a radicle as the first step in germination. Since a seed consists not only of an embryo but also of various other tissues including a seed coat, it is clear that the dormancy of the seed may not rest in the embryo itself. For example, removal of the seed coat of seeds of *Acer pseudoplatanus* also removes the chilling requirement for germination by this species. Very few light-requiring seeds have dormant embryos; the dormancy of the whole seed is in fact a function of the presence of the coat. In some cases the dormancy which is expressed by the whole seed may reflect poor gas exchange across the coat, and it may be overcome by slight damage to the coat. In other cases the seed coat may act as a physical barrier to the growth of the radicle, and germination will occur if the coat is removed from the radicle region.

The mechanism of dormancy

Even such a brief survey of dormancy effects as has been made here should be enough to emphasise that there is no single explanation for the effect. Clearly such factors as impermeability to oxygen or to water, and mechanical effects due to the seed coat do not cast much light on the operation of developmental controls, although it has been suggested that one

reason underlying the quiescence of the cells in the quiescent centre at the root tip may be precisely that they are mechanically constrained by the surrounding tissues.

The bud dormancy of woody species which is triggered by short days has been the subject of considerable study, since daylength is an easily controllable experimental variable. It was during work on this problem that the growth inhibitor originally named 'dormin' was first isolated. This substance, now known to be abscisic acid, was found by bioassay to increase in the buds of *Betula pubescens* and *Acer pseudoplatanus* in response to short day conditions. It is now known that externally applied ABA can act to induce dormant bud formation in a variety of species although, as has been discussed previously (Chapter 6), an effect shown by an applied hormone does not necessarily mean that the same substance controls the same process in the normal growth of the plant. There is some evidence that endogenous gibberellin levels increase during the chilling period which releases buds from dormancy and this, together with the ability of applied gibberellin to overcome the chilling requirement, has led to the suggestion that dormancy in buds may result from a change in the balance between growth inhibitors and growth promoters.

The original research using bioassays would of course have detected just such a change in the balance of inhibition to promotion. More precise later work with buds has failed to demonstrate a close correlation between increasing degrees of dormancy and increased levels of (chemically identified) ABA. Similar comments may be made of the situation in dormant seeds. Apart from physical effects, there is evidence for the presence of growth inhibitors within the seed coat. The release from dormancy can thus be regarded as reflecting a gradual decline in the level of inhibitory substances. In some seeds it appears that chilling increases the amounts of growth promoters in the seed, so that here too, a balance may be postulated between growth inhibition and promotion in the control of dormancy expression.

The metabolic changes which accompany release from the resting state include increased respiration, increased rates of RNA and protein synthesis, increased uptake of potassium ions, and the initiation of DNA synthesis. In seeking a unifying hypothesis to explain all these changes, one suggestion which has been made is that resting cells show low ion selectivity and have 'leaky' membranes. In consequence of this, such cells would be expected to have a reduced ability to synthesise ATP, which would mean a general reduction in the rate of all reactions requiring ATP. The key to recovery of the normally active state is proposed to be calcium ions; an increase in calcium ion concentration causes membrane potentials to rise and prevents the relative loss of potassium ions. Recovery of membrane function would then result in recovery of ATP production and hence of synthetic capability. Of course, such an idea merely takes the problem on to a different level – why do calcium levels vary in any case? This is clearly an unsatisfactory state of affairs. Similarly, where it has been shown that chromatin extracted from dormant buds has less template activity than comparable preparations from actively growing buds, it remains unclear whether such a correlation is a cause of dormancy or merely an expression of its effect

Phytochrome

We have now seen that several developmental processes in plants are sensitive not merely to light quantity, but also to the quality of light. The transition to two dimensional growth in fern gametophytes is one example (p. 189), and the same can be said of aspects of flower induction in higher plants (p. 214), the promotion of dormancy in the buds of woody trees (p. 216) and its breaking in the germination of seeds (p. 217). All these phenomena have one thing in common. This is that an effect is produced by red light which can be reversed by a subsequent exposure to far red light.

The nature of this effect has been studied in detail using the germination of lettuce seeds as the experimental system. It is found that germination was promoted by light of a wavelength around 600 nm, but that this promotion can be abolished by a subsequent exposure to far red light of around 730 nm wavelength. The promotion/inhibition cycle can be repeated many times, and the effect on germination rate is always dependent on the last light exposure; when it is red, germination takes place, but when it is far red, it does not (Fig. 8.4).

This behaviour is now known to be due to the presence of a single pigment which can exist in two separate states within the cell. The pigment was named phytochrome, and the two states in which it

can exist are referred to as P_r and P_{fr}. Each form is capable of being converted into the other by the presence of light of the appropriate wavelength, and this equilibrium can be expressed in the manner of an equation:

$$P_r \underset{\text{Far red}}{\overset{\text{Red}}{\rightleftharpoons}} P_{fr}.$$

The pigment has been isolated and consists of a protein of molecular weight 120 000 and a chromophore in the form of a tetrapyrrole compound. It is probable that the chromophore undergoes change in the transition between the two states, and that this change is accompanied by a conformational rearrangement in the protein. Since the discovery of phytochrome, many processes in plants have been shown to be mediated by light in a red/far-red reversible manner which is characteristic of its involvement. Apart from those which we have already seen, may be mentioned spore germination in lower plants, chloroplast movements in the alga *Mougeotia*, internode extension in higher plants, the initiation of root primordia and membrane permeability. This represents a very wide range of activities, some of which are major developmental switches, whereas others are growth processes and yet others subcellular movements.

At the present time it is quite unclear what the cellular location of the phytochrome is in most of the situations where its action may be inferred. In the case of the orientation of chloroplasts in *Mougeotia*, it has been shown by microbeam irradiation that the phytochrome is located at the cell periphery; it is therefore a reasonable assumption that it is located within the plasma membrane. It has been further shown in this situation that the two forms of phytochrome are present in different orientations. When polarised light is used to irradiate the cells, red light is only effective when it is supplied when the plane of the electric vector is parallel to the cell surface, and far red light only has its effect when the plane of the electric vector is at right angles to the cell surface (Fig. 8.5).

The active form of the pigment is P_{fr}. It appears that in exerting its action it is destroyed. The observations behind this assumption are as follows. If red light is given, its effect may be reversed by an immediate exposure to far red light. However, if the exposure to far red light is delayed, the reversibility declines with time. The time taken for the reversibility to disappear entirely is known as the escape time, and it varies from a few minutes to several hours. In the

Fig. 8.4. Effects of light on germination of seeds of lettuce (*Lactuca sativa*) 'Grand Rapids'. Germination is poor in darkness or in far-red light. Red light stimulates germination and can overcome the inhibitory effect of far-red light. Far-red light can reverse the stimulatory effect of red light. The effects of red and far-red light are reversible over many cycles of application.

In darkness or far-red light

Red light

Red, then far-red light

Far-red, then red light

dark, following an exposure to red light, it is found that the amount of P_{fr} declines with time, and so does the total amount of phytochrome. This suggests that P_{fr} is relatively labile. Continued periods in the dark following a red-light exposure eventually lead to the restoration of the levels of phytochrome through net synthesis. This suggests that in the plant phytochrome may be in a continuous state of turnover, with synthesis supplying new material in the form of P_r, and P_{fr} being lost both through general degradation and as a result of biological activity of this form of the pigment (Fig. 8.6). Thus the relative levels of P_r and P_{fr} within a tissue may depend not only on the wavelength characteristics of the light to which it is exposed, but also on the relative rates of synthesis and destruction, since phytochrome is synthesised as P_r and destroyed in the form of P_{fr}.

The mechanism of action of phytochrome is not known in detail. It is very probable that it has effects on the permeability of the plasma membrane as one of its more general properties. Some quite direct evidence for this is available. For example, the sleep movements of leaves of *Mimosa pudica* (the sensitive plant) can be inhibited by exposure to far red light at the end of the day, and this inhibition can be reversed by red light. These movements are believed to be related to the control of the turgor of cells by the movement of potassium ions, and hence it is reasonable to conclude that membrane permeability may be regulated by phytochrome. Some membrane-related responses to red and far red light are extremely rapid. The exposure of etiolated coleoptiles to red light causes a redistribution of electrical charge within seconds to produce a positive potential at the tip with respect to the base. In such conditions it seems very probable that the photoreceptor (phytochrome) is very close to the site of production of the response, since there is insufficient time between the stimulus and the response for secondary triggers to be involved.

On the other hand, it is unlikely (to say the very least) that major developmental switches such as the transition to flowering at the shoot apex could be accounted for simply in terms of a change in the permeability of the plasma membrane. It has been suggested that cells may use other molecules (cytokinins and gibberellins have been postulated) as 'second messengers'. This means that the primary role of phytochrome in such cells would be to increase or decrease the level of a growth regulator, and this change would then cause the response. This idea is built on the analogy with animal cells in which there is good evidence that cyclic AMP acts as a second messenger for some hormones. As an explanation of behaviour in plants it suffers from a lack of convincing evidence of a direct link between a change in the state of phytochrome and a change in levels of growth regulators, together with the all-embracing doubts concerning endogeneous levels of growth regulators which have already been expressed (Chapter 6).

Environmental and internal control: a summary

In the previous sections we have considered some effects of the external environment on whole plants. It is very difficult to suggest ways in which such effects may be linked to the behaviour of individual cells. By and large no attempt is made to explain the effect of say daylength on the behaviour of individual cells in a shoot tip. The best that can be done is to propose that behavioural changes brought about by the external environment might result from changes

Fig. 8.5 The role of phytochrome in the movement of the chloroplast in *Mougeotia*. The diagram represents a view along the length of the cell, with light presented in the plane of the page. *a*, Red light causes the chloroplast to rotate; this effect is reversed by subsequent far red illumination. *b*, Polarised red light causes the chloroplast to rotate when the plane of the electric vector is parallel to the upper cell surface. Far red light polarised in the same plane does not reverse the effect. *c*, Far red light does reverse the effect when the plane of the electric vector is perpendicular to the upper cell surface.

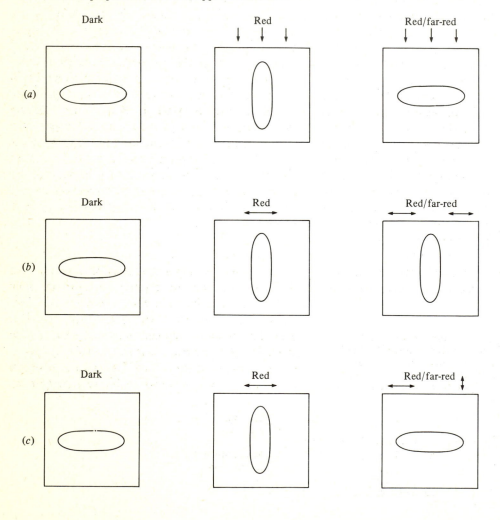

in hormone levels, or general metabolic activities such as respiration or ion transport. However, even when correlations of this type can be demonstrated, it is quite uncertain that they are linked in the manner of cause and effect.

However, the fact of control of development in whole plants by the external environment does emphasise that development does not occur continuously simply on the basis of internal programming of cells. An important logical conclusion follows from this. Just as the plant finds itself within the external world as its 'environment', so the cell must develop in an environment which consists of other cells, at least in multicellular plants. The direct environment of a cell is its cell wall. The state of this wall, its chemical composition, shape, the ions and small molecules which it contains all reflect the past history of the cell, and the nature of its neighbours. 'Environmental' control viewed in this way is simply another name for the coordination of growth and development within a tissue.

The concept need not stop at the cell periphery. As we saw in Chapter 1, a cell can in its most simple form be regarded as a nucleus together with the cytoplasm to sustain it. If we regard the nucleus as predominant in this partnership, which is reasonable since it contains most of the genetic material of the cell, then the idea of environmental control takes on a further meaning. The environment of the nucleus is the cytoplasm in which it exists. Just as in the case of the cell and its wall and neighbours, so too with the nucleus and its cytoplasm, there is a strong degree of interaction between the two partners in the relationship. The nucleus cannot exist independently of its cytoplasm. On the other hand, the cytoplasm is dependent upon the nucleus. The importance of this

interdependence has already been glimpsed in the case of highly asymmetric divisions such as occur in the developing *Fucus* egg. Here the nucleus divides, and identical genetic units are delivered into different cytoplasmic environments. The result of this difference is that the two daughter cells embark upon completely different developmental pathways – one forming the holdfast, the other forming the frond. This simple example clearly shows that within the partnership of nucleus and cytoplasm, both members are of equal importance.

In the following sections we shall consider some more detailed evidence concerning the nature of the interaction between the nucleus and the cytoplasm.

Nucleocytoplasmic interaction

The relationship between the nucleus and the cytoplasm has been extensively studied using the marine alga *Acetabularia*. *Acetabularia* is a single-celled organism with a characteristic morphology, and it has a remarkable ability to withstand surgical operations which would be fatal to most other plant cells. In particular it is possible to isolate the large single vegetative nucleus, wash it clean of its adhering cytoplasm, and implant it into the cytoplasm of a different cell without loss of function. This ability has enabled a wide range of experiments to be carried out which relate not only to the specificity of the genetic material in controlling cell morphology, but also to the regulation of the expression of such specificity.

Nuclear control

Normal development of *Acetabularia* consists of the formation from the zygote of an elongated stalk with a rhizoid at its base. The stalk eventually gives rise to a series of hairs in whorls, and then a cap (Fig. 8.7). The shape of the cap is species-specific. For example, *Acetabularia mediterranea* (MED) has a circular cap with many radial ridges and a more or less smooth edge; *A. crenulata* (CREN) has a circular cap

Fig. 8.6. Phytochrome interrelationships. The pigment is synthesised in the P_r form. Red light converts this to P_{fr} which may then exert an action, be destroyed, revert to P_r in the dark, or be converted to P_r by irradiation with far red light.

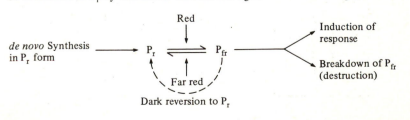

de novo Synthesis in P_r form $\longrightarrow$ P_r $\rightleftharpoons$ P_{fr} $\longrightarrow$ Induction of response / Breakdown of P_{fr} (destruction)

Red / Far red / Dark reversion to P_r

with a few ridges and a lobed edge (Fig. 8.8). During vegetative development, the nucleus of the plant remains at the base of the stalk within one of the branches of the rhizoid. If the rhizoid is removed, development of the cap can still proceed and its shape will be correct for the species (Fig. 8.8). However, if a nucleus from a different species is introduced into the enucleate stalk, then the morphology of the cap which subsequently develops will be altered. Thus a MED nucleus implanted into an enucleate CRED stalk will cause that stalk to produce a cap with MED characteristics, and vice versa. It is possible to implant more than one nucleus into an enucleate fragment of stalk. When such an experiment is done the cap which develops reflects the nuclear complement of the stalk in its shape. Thus if say three MED nuclei and one CREN nucleus were implanted into an enucleate fragment, the cap which results would be mainly MED in character. These results show clearly that the nucleus controls the shape of the cap. In more general terms they also show that it is possible to control form by genetic factors in the absence of complications due to cell division.

However, whilst it is reassuring to see that genetic control of form is possible even at the level of a single cell, these experiments cast no light on the mechanism of that control. Work with *Acetabularia* has approached this aspect in a number of ways. First, it is not necessary for the nucleus to be continuously present in order to achieve normal cap morphology. Enucleate fragments of stalk are quite capable of generating a cap of the correct shape. It was also noticed that, in implantation experiments involving the introduction of a single nucleus into an enucleate

fragment of a different species, the effect of the implanted nucleus was not always absolute. In other words, a MED nucleus implanted into a CREN stalk would not always result in the formation of a completely normal MED cap. It was further noticed that if the first formed cap were removed, regeneration of a second cap could occur, and if it did, this second cap had more MED character.

This was explained in terms of 'morphogenetic substances'. It was proposed that the nucleus produces species-specific morphogenetic substances which it releases into the cytoplasm. Thus a CREN enucleate stalk fragment would inevitably contain CREN morphogenetic substances, even after the removal of that part of the cell which contained the nucleus. An introduced MED nucleus would release MED morphogenetic substances into the same cytoplasm, so that any cap which was subsequently formed by the activity of the cytoplasm would be likely to have both MED and CREN character.

The morphogenetic substances were shown to be sensitive to UV irradiation. If an enucleate CREN stalk were irradiated before implantation of a MED

Fig. 8.8. Control of cap morphology in *Acetabularia*. *a, Acetabularia mediterranea* (solid black nucleus). The typical cap has many ridges and a smooth outline. It can form on a stalk without the presence of a nucleus. *b, Acetabularia crenulata* (nucleus an open circle) has a cap with fewer ridges and a lobed edge. *c,* If a CREN nucleus is introduced into an enucleate MED stalk, the resultant cap will have some CREN characteristics.

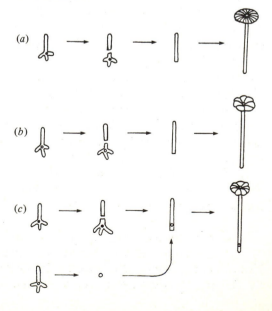

Fig. 8.7. The life cycle of *Acetabularia*. The zygote develops a stalk and a branched rhizoid. As the stalk grows, it differentiates first a whorl of hairs and then a cap. The single vegetative nucleus remains in one of the branches of the rhizoid until the cap is formed. It then gives rise to thousands of secondary nuclei which migrate up the stalk and into the cap to form cysts.

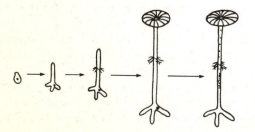

nucleus, the cap which resulted would have a more purely MED character. These experiments elegantly show that the ultimate form of the cap is determined by the nucleus, but that nuclear control is exerted by a product which is released into the cytoplasm. It is very tempting to conclude that the morphogenetic substances are messenger RNAs. If this is the case, then they are unusually stable and long lived in *Acetabularia*.

The presence of long-lived messenger RNA (mRNA) in the cytoplasm of *Acetabularia* has been demonstrated by biochemical techniques. For example, it is found that the activity of the enzyme malic dehydrogenase increases in enucleate fragments for several weeks after the removal of the nucleus. This was shown to be due to new enzyme synthesis, by growing enucleate fragments in the presence of heavy water in the medium. The enzyme was isolated after different times of culture, and it was found that its density increased as the period of growth in heavy water increased. Net new synthesis of an enzyme in the continued absence of a nucleus suggests strongly that long lived mRNAs are present in the cytoplasm.

Environmental control in *Acetabularia*

The morphogenesis of the cap is not solely dependent upon the presence of morphogenetic substances in the cytoplasm. The overall morphology of *Acetabularia* is also controlled by the light conditions under which it is grown. The cap formed in strong light is on a shorter stalk than that formed in dim light, despite the fact that the stalk elongates at a faster rate in strong light. Under conditions of very dim lighting, *A. crenulata* does not form a cap at all, and the stalk continues to elongate. In one experiment, cells were grown for 5 years without the formation of any caps. However, if the tip of a stalk which has been grown in dim light is cut off and placed in bright light, cap formation will ensue, in the absence of a nucleus. This experiment shows that the failure to produce a cap in dim light was not due to the absence of morphogenetic substances, or to a failure of gene expression. It was due to an absence of light. This is a very simple and clear example of the control exerted on developmental processes by external factors.

Cytoplasmic control in *Acetabularia*

The results discussed so far has emphasised the idea of the dominance of the nucleus in controlling development. However, there is also evidence from *Acetabularia* of an effect of the cytoplasm on the nucleus. One experiment is as follows. If a rhizoid which contains a vegetative nucleus is grafted onto an enucleate stalk, then a normal cap is formed. Subsequent to this, the vegetative nucleus gives rise to thousands of secondary nuclei which migrate up the stalk and into the cap, where they form cysts. The interval between the grafting operation and the formation of cysts has been shown to depend upon the age of the cytoplasm in the stalk. Thus if a rhizoid is grafted onto a young stalk, development takes longer than if it is grafted onto an old stalk.

During ageing, the vegetative nucleus can be observed to undergo certain ultrastructural changes in addition to its enormous increase in volume. The young nucleus is characterised by a nucleolus which has many fibrillar finger-like projections at its surface, and the perincluear cytoplasm exists only as a thin layer. By contrast, a nucleus from an old cell (one which has formed a cap) has a spherical nucleolus which is vacuolate, and the zone of cytoplasm around the nuclear periphery is thickened. It is found that if a young nucleus is implanted into the cytoplasm of an 'old' enucleate stalk, it assumes the appearance of an old nucleus over a period of 5–10 days, although its normal development would have taken several weeks. Conversely, an old nucleus implanted into the cytoplasm of an enucleate young stalk assumes a juvenile appearance within 10 days. If such ultrastructural variations do indeed reflect the state of activity of the nucleus, then this experiment shows that such activity may be controlled or at least influenced by the cytoplasm.

Finally, a biochemical analogue of these processes has been found. Immediately prior to the maxium development of the cap, the normal cell of *Acetabularia* shows a marked increase in the levels of thymidine kinase activity. This increase coincides with the start of the formation of secondary nuclei, and it is therefore not unreasonable to suppose that they are linked together, since the enzyme is involved in the metabolism of a DNA precursor. It was found that the increase in thymidine kinase activity occurred even in enucleate fragments of stalk which developed a cap. This increase could be inhibited by puromycin, and was shown to be due to net new enzyme synthesis, rather than say the formation of an activator. Puromycin inhibits protein synthesis both on cytoplasmic ribosomes and on the smaller ribosomes of mitochondria and plastids. However, by using other inhibitors it was shown that the synthesis of thymidine kinase was occurring within cytoplasmic organelles, and it was also shown that the protein was encoded in the genome of the organelle, and not by the nucleus. The final site of action of the enzyme is on the plasma membrane.

This represents an extremely complicated biochemical situation in which a protein is produced in one cytoplasmic compartment, migrates through the cytoplasm to the plasma membrane, and there exerts an action to produce a substrate for DNA synthesis (thymidine triphosphate), which then migrates to the nucleus to be incorporated into DNA. It is a clear demonstration of the dependence of the nucleus upon the cytoplasm. More remarkable still is the fact that the production of the enzyme is synchronised to the state of development of the cap, and the presence of the nucleus is not required for this synchronisation to occur. This gives a small insight into one aspect of the comparative autonomy of the cytoplasm and its organelles.

It has to be emphasised finally that in many ways *Acetabularia* is an extremely unusual and atypical plant cell. Nevertheless, within it we are shown a microcosm of many of the problems related to development in other cells. The nucleus carries within its genes most of the information necessary for the correct development of the cell. It is not totally dominant, however; it requires the cytoplasm for its sustenance and its own differentiation, and the entire cell requires the correct external conditions for its successful growth.

Genes and development

We come now to consider the central theme of the control of development, namely the role of the genes. Genetic control of development can be inferred from countless observations of inheritable differences in developmental abilities. Some of these, such as leaf form in the pea (*Pisum sativum*) have been shown to be due to single genes (Fig. 8.9). It must be realised at the outset that such a statement describes nothing at all of the way in which a developmental process is regulated at the molecular or cellular level. It is simply a description of the result of breeding experiments.

As we have seen, there is considerable evidence to support the view that differentiation of plant cells does not involve overall loss of genes. Cells can be isolated from a variety of mature tissues and, under the right conditions, can be induced to grow back into perfectly normal plants. Therefore the behavioural changes in cells which are associated with differentiation presumably arise as a result of selective expression of parts of the total genome. In order to understand the possible basis for this selective expression it is first necessary to consider what is known about the organisation of the genetic material itself.

Organisation of the genetic material

The first thing to consider concerning the genetic material is its quantity. When absolute amounts of DNA are measured in a variety of cells it is found, not surprisingly, that simple prokaryotic cells such as bacteria have considerably less DNA within them than do cells of animals or higher plants. Roughly speaking the DNA of a bacterium consists of between 10^5 and 10^6 base pairs, whereas a haploid set of chromosomes from an angiosperm might contain as many as 10^9 base pairs. This three orders of magnitude difference between the amount of DNA per cell in prokaryotes and eukaryotes should

Fig. 8.9. Genetic control of leaf form in the garden pea (*Pisum sativum*). The plant on the right is the normal form, with leaflets and tendrils. The plant on the left has no tendrils, but correspondingly more leaflets. This change is controlled by a single gene.

immediately alert us to the possibility that theories of genetic control which have been developed by studying bacteria may not be directly applicable to higher organisms.

However, even within eukaryotes there are wide and at first sight inexplicable differences between the amounts of DNA per haploid chromosome set. It might be expected that a simple alga would require less DNA to code for its range of structures and activities than a higher plant and, in general, this is found to be the case. Usually, algal cells contain about one tenth the amount of DNA found in a higher plant cell. However, it is quite unexpected that a species such as *Lilium longiflorum* appears to require some 20 times as much DNA for the completion of its life cycle as does a human being. Even within a single genus, for example *Vicia*, the amounts of DNA per haploid chromosome set may vary between species by a factor of 5–10.

There are several possible explanations for this phenomenon. The first is that the extra amounts of DNA seen in closely related species may simply be nonfunctional. This is difficult to accept as a concept since generally speaking, useless structures should be eliminated through processes of evolutionary change. In the case of DNA sequences, it is entirely possible however that non-functional sequences could be linked to essential sequences and therefore retained. The second explanation lies in the function of the genes themselves. The most direct function of genes is to code for protein structure by the intermediate formation of RNA. Such protein-coding genes are known as structural genes. However, structural genes are not expressed at random or continuously, and work with bacterial cells has shown that there are other classes of genes (called regulator genes) which control the activities of the structural genes. In view of this, the enormous disparity between the amounts of DNA in prokaryotic and eukaryotic cells could result from the greater degree of regulation required by the more complex organism. In other words, eukaryotic cells contain enormously greater numbers of regulator genes. In fact, it is estimated that only about 1% of the genome of a higher plant is made up of structural genes. This description however clearly does not account for large differences in amounts of DNA per haploid chromosome set in closely related species, where the need for regulatory mechanisms would be expected to be very similar.

When the nature of the sequences within DNA is examined, it is found that so far from being unique throughout the length of the molecule, many sequences are repeated. This can be shown in experiments involving the renaturation of the separated DNA strands. By such means it has been shown that in plants characterised by large amounts of DNA per haploid chromosome set, over 80% of the DNA may be in the form of repeated sequences. In onion (*Allium cepa*) the figure reaches 95%. When plants with smaller absolute amounts of DNA are examined, it is found that smaller proportions of the DNA are present as repeated sequences, often of the order of 60% (Table 8.1). Thus there is a statistical correla-

Table 8.1. *The variation in amounts of DNA and the proportion of repeated sequences. In general terms there is an increase in DNA content and the percentage of repeated sequences from yeasts to monocotyledons, but many anomalies exist in the correlation, even between closely related species.*

Species	Amount of DNA (2C)	Repetitive DNA (%)
Yeasts		
Candida macedoniensis	8.31×10^9 d	7.1
Saccharomyces exiguus	1.13×10^{10} d	10.8
Debaryomyces hansenii	6.05×10^9 d	13.8
Fungi		
Neurospora crassa	3.2×10^7 d	13.5
Coprinus lagopus	3.6×10^7 d	15.0
Phycomyces blakesleeanus	1.9×10^{10} d	30.0
Dicotyledons		
Raphanus sativus	3.1 pg	18
Anemone coronaria	19.9 pg	53
Anemone blanda	32.0 pg	57
Anemone pavoniana	29.3 pg	62
Anemone cylindrica	21.9 pg	65
Anemone riparia	21.0 pg	67
Pisum sativum	9.9 pg	75
Tropaeolum majus	7.3 pg	82
Vicia faba	29.3 pg	85
Monocotyledons		
Zea mays	11.0 pg	78
Poa trivialis	6.9 pg	82
Secale cereale	18.9 pg	92
Allium cepa	35.5 pg	95

tion between the amount of DNA present and the proportion of it which exists as repeated sequences. Whether this finding is the entire explanation of the differences in absolute amounts of DNA in different species is still a matter of opinion, but it is certainly the best so far put forward.

The most widespread examples of repeated sequences are found to be genes coding for ribosomal RNAs – up to 30 000 copies may be present in a single cell.

Organisation of chromatin

As was seen in Chapter 1 (p. 11), the DNA of a eukaryotic nucleus does not exist in the form of naked molecules simply lying free in the nucleoplasm. The DNA is always associated with a range of proteins, and this complex is the chromatin of the nucleus. The proteins fall into two groups. The histones are basic proteins, whilst the non-histone proteins are more diverse.

The histones may be classified into five major fractions. These are called H1, H2A, H2B, H3 and H4. All are basic proteins in which more than 25% of the amino acids are positively charged. H1 is the largest with 219 residues; the others are smaller, with 129 (H2A), 125 (H2B), 135 (H3) and 102 (H4) residues. They all have highly charged terminal regions with a less charged central section. Sequence determinations of histones from different sources have shown that they are characterised by great stability in terms of their chemical composition throughout the plant and animal kingdoms. This is particularly true of the histones other than H1, which does show some variability. Thus H4 from peas and from calf differ by only two amino acids. Even these differences are small – a valine replaced by an isoleucine (both neutral amino acids) and a lysine in place of an arginine (both basic amino acids). The high degree of conservation of chemical structure suggests that the histones play a fundamental role in the organisation of the DNA within all eukaryotic cells.

The present model of chromatin structure reflects this expectation. It is based upon a unit called a nucleosome, which consists of a complex of eight histone molecules (two of each, except H1) acting as a core for the winding of $1\frac{3}{4}$ turns of DNA (Fig. 8.10). Each nucleosome therefore consists of a nucleoprotein particle joined to its neighbours by a short link. The DNA associated with the particle is about 146 base pairs long, and the linker region is variable in length, up to 100 or so base pairs. Histone H1 is thought to be associated with the linker region.

The evidence for this model consists of the finding that partial digestion of chromatin with nucleases gives rise to the release of resistant cores of DNA and histone. In the electron microscope chromatin can be made to appear as a series of spherical particles linked together by fine fibrils, rather in the manner of beads on a string. When the stoichiometry of the histones in the resistant cores is examined, it is found that they are present in the ratio $1:2:2:2:2$. That the DNA is on the outside of the particle is inferred from its sensitivity to enzyme attack. These facts are reconciled by the model, which also emphasises the universal role of the histones in the formation of chromatin. It can be calculated that a protein of 400 amino acids in length would be coded for a length of DNA associated with six nucleosomes,

Fig. 8.10. Nucleosomes in chromatin. *a*, The DNA (solid black line) forms a beads on a string structure (four nucleosomes are depicted, separated by linker regions); *b*, each nucleosome contains two molecules of each histone except H1.

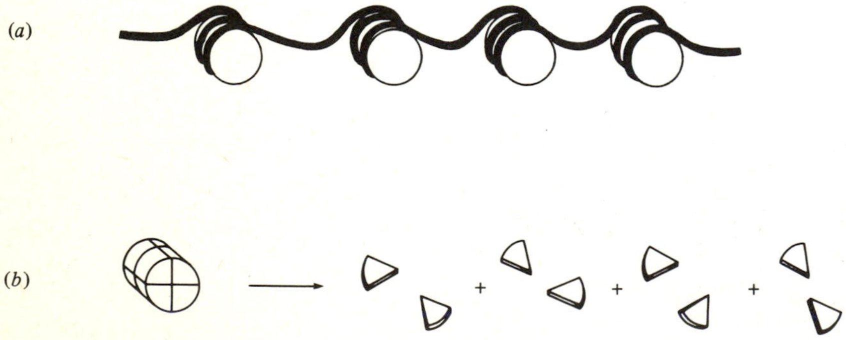

(*a*)

(*b*)

since each nucleosome with its link is (on average) associated with 200 base pairs.

The organisation of the nucleosome itself is as yet poorly understood. If histones and purified DNA are mixed together *in vitro*, a precipitate forms. This is not surprising in view of the highly charged nature of both types of molecule. It is becoming clear that, within the nucleus, other types of protein may be involved to prevent such non-specific clumping reactions. One of these, nucleoplasmin, represents up to 10% of the nuclear protein present in oocytes of *Xenopus*. Other features of the organisation of

chromatin are not fully understood. For example, the length of the linker region between the nucleosome particles is variable, but the basis for this variation is not clear. It is also uncertain exactly how nucleosomes behave during DNA replication, although current studies suggest that newly synthesised DNA rapidly becomes associated with histones, and that old and new histones are not mixed together in a single nucleosome. It is not clear however whether new DNA is associated with new histone or old histone.

Higher degrees of organisation of chromatin are necessary to account for the ability of the DNA, up to 10 m in length, to be accommodated in a functional state within the space of the nucleoplasm. The fundamental 10 nm fibril (the beads on a string structure of nucleosomes and linker regions) is thought to be wound into a 'solenoid' structure with a diameter of about 30 nm. These single coils are then themselves wound into supercoils (Fig. 8.11). This double level of coiling of the basic fibril is thought to

Fig. 8.11. Different levels of folding of DNA. *a*, The DNA helix associates with nucleosomes to give *b*, the basic beads on a string structure. This is then *c*, wound on itself like a solenoid; a further folding of this solenoid structure corresponds to *d*, the state of chromatin in an extended section of a chromosome. When the chromosome condenses *e*, further coiling takes place.

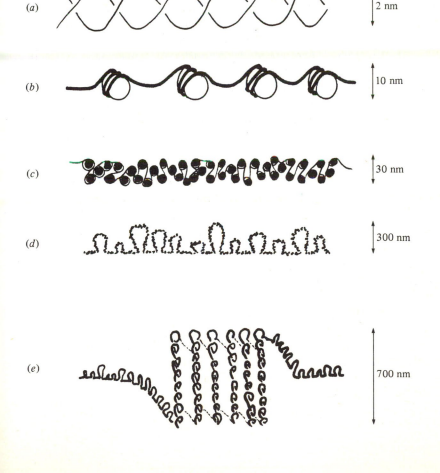

(*a*)   2 nm

(*b*)   10 nm

(*c*)   30 nm

(*d*)   300 nm

(*e*)   700 nm

be sufficient to account for the observed contraction of the length of the DNA in chromatin.

The control points of gene expression

Differentiation involves the sequential expression of different parts of the genome. Clearly this implies regulation of gene expression since in any given cell many of the possible products of the genome may not be required. Some of the possible control points will now be summarised.

Consider the activity of a structural gene. It is first transcribed to give rise to a corresponding RNA molecule within the nucleus. It used to be thought that this molecule corresponded to messenger RNA, but it is now realised that considerable processing of this first-formed 'heterogenous nuclear RNA' occurs before it leaves the nucleus as mRNA. The RNA is translated in the cytoplasm in association with ribosomes and transfer RNAs (tRNAs) to give rise to a protein. The protein itself may subsequently be modified before its final active state is achieved.

The possibilities of control in this sequence of events are clearly very diverse. The gene itself may be subject to control (transcriptional control), and this is logically the most efficient and economical way of controlling the activity of the protein to which it corresponds. Nevertheless, control can also be envisaged as occurring at the level of RNA processing within the nucleus, or at the translation step in the cytoplasm, or even after translation. Furthermore, when it is realised that all the enzymes required for protein synthesis, and all the ribosomal and transfer RNAs are themselves encoded in gene structure, it is clear that a simple explanation of the control of gene expression is not likely to be found.

Control in prokaryotes

The most clear model of the control of gene action comes from work with bacteria. Here it is found that genes are linked in functional groups called operons. An operon consists of a structural gene or a number of structural genes which are all controlled by a contiguous gene called the operator. The operator gene is sensitive to substances produced by another type of gene called the regulator. The regulator gene may be sited remote from the operon which it controls. The regulator can act as a switch and thus switch on or off a sequence of genes which may produce a sequence of products. This system was discovered as a result of studies on inducible and repressible enzymes in the bacterium *Escherichia coli*.

This organism is capable of controlling the enzymes which it makes in response to the presence of small organic molecules in the medium in which it is grown. For example, if a galactoside is added to the medium, the bacterium is able to respond to its presence by synthesising β-galactosidase, which hydrolyses the galactoside. Thus in the presence of lactose, an enzyme is produced which can hydrolyse the disaccharide to a mixture of galactose and glucose. The mechanism of this induction of the enzyme is as follows. The regulator gene for the galactosidase produces a protein, called a repressor. This repressor interacts with the operator of the galactosidase operon and prevents the formation of the enzyme. In the presence of lactose, the repressor protein interacts with a derivative of the lactose called allolactose. This interaction causes a change in the shape of the repressor protein molecule which means that it no longer interacts with the operator. Thus transcription of the operon proceeds, and the gene is said to be de-repressed (Fig. 8.12).

In a similar way it is possible to explain the repression of enzyme synthesis by small molecules. For example, if the bacterium is grown in the presence of the amino acid histidine, then the enzymes required for its synthesis are not produced by the cell. In this case, it is proposed that histidine binds to the repressor protein for the histidine operon, this time resulting in a change in the protein which allows it to interact with the operator. Therefore in the presence of the histidine, the operon is switched off, whilst if histidine is absent, the operon is switched on. In other words, in the case of inducible enzymes, the repressor protein alone is the form which is able to interact with the operator; whereas for repressible enzymes, the repressor protein has to bind to the small effector molecule before its active state is reached.

This model has two very important implications. The first is that gene action can be regulated by small organic molecules, by the intermediary binding of such molecules to proteins (the repressors). This idea is very attractive since it might offer an explanation of one way in which plant growth substances could affect gene expression. The second implication is that genes are linked together in groups. The linkage of controlled characteristics is something which we have already identified as a feature of the developmental process in plants (p. 209). A further refinement of this concept is that complex control is possible. For example, the product of one group of genes may affect the expression of another group, either because it is itself a repressor for the second group, or because it is an enzyme which catalyses the formation or destruction of a small effector molecule. In this way it is easily possible to visualise ways in which complex trains of genetic groups could be switched on in regulated sequences. The analogy with computers is very tempting, although perhaps scarcely flattering.

It is essential to realise however that as attractive as an hypothesis may be, it is only defensible if experimental evidence supports it. The general principle of this work with *E. coli*, that the expression of one gene may be controlled by another, is probably of wide application; indeed it was proposed to occur in maize (*Zea mays*) long before the work was done with bacteria. However, the details of the process are unlikely to be the same in prokaryotes and eukaryotes. Two examples may be cited. First, in any higher plant

cell it is likely that over 90% of the DNA is not being transcribed. It is therefore more probable that controls which directly affect gene action are exerted in the form of specific activations, rather than repressions. It is as though the genome were generally repressed and specifically activated by gene activator proteins, rather than specifically repressed as in the examples given above. Second, the linkage of characteristics during development has been cited. It is tempting to equate this with the existence of operons in higher plants. In fact it is known that many linked characters occur in different parts of the genome, even on different chromosomes. The situation relating to the control of gene expression in higher organisms is immeasurably more complex than it is in the bacteria. •

Levels of control in higher organisms

An important and indeed fundamental difference between prokaryotic and eukaryotic genetic systems is the presence in the latter of the proteins which are associated with the DNA to form

Fig. 8.12. Model of the regulation of gene action in prokaryotes. The diagram shows the situation in the case of an inducible enzyme. *a*, In the absence of the inducing molecule (the effector), the regulator gene produces a repressor protein which binds to the operator gene and prevents transcription of the structural gene(s); *b*, in the presence of the effector, the repressor protein is inactivated and the structural gene is depressed.

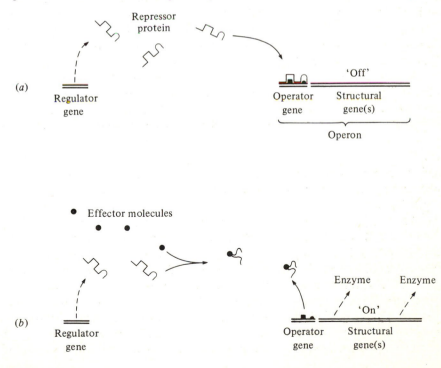

chromatin. As we have seen (p. 226) the histones can be assigned a more or less structural role in the condensation of the DNA and its folding. In this context they may have a general genetic action of repressing gene expression. It is unlikely, however, that a small group of highly conserved proteins like the histones would be used for gene regulation in a specific developmental context. Evidence is beginning to emerge that actively transcribed portions of the chromatin (which at any one time represent only a small proportion of its length) may be associated with some of the other non-histone proteins of the nucleus. These too are fairly highly conserved, but they form a larger and more diverse group than the histones.

These ideas can be demonstrated experimentally by isolating chromatin and removing its associated proteins. The naked DNA thus prepared shows a higher level of transcriptional activity in a test system *in vitro*; this activity is greatly reduced by adding back the histone protein fraction. The loss of transcriptional activity is not so great, however, when both the histones and the acidic non-histone proteins are added back to the DNA. This, of course, is a crude analogy with any probable control *in vivo*. However, there is evidence that modification of histones by acetylation may modulate the activity of the nucleosome in transcription, and also that actively transcribed genes are able to bind selectively two abundant non-histone proteins in the so called high-mobility group (HMG). Such modulation as is achieved by these means should probably be regarded as increasing accessibility to the genes by decondensing the chromatin, rather than as representing specific activation of single genes or gene groups.

Chromatin is described as existing in two forms (p. 11). Heterochromatin is so called because of its dense-staining properties, and is thought to consist of highly condensed chromatin which is not available for transcription. Euchromatin, on the other hand, is less condensed and accessible for transcription. It should be realised, however, that some of the heterochromatin within an organism is always condensed and never transcribed. This is known as constitutive heterochromatin, to distinguish it from facultative heterochromatin which may be present in some cells and not others. The change from euchromatin to heterochromatin is thus another example at a crude level (visible staining properties) of the control of gene transcription. The formation of heterochromatin

signals the end of transcriptional activity of that part of the genome.

Evidence for transcriptional control at a specific level (with no implication as to its exact mechanism) has come from the analysis of mRNAs in plants of tobacco (*Nicotiana tabacum*). Here it is found that when different tissues are analysed for their mRNA populations, each has about 25 000–30 000 different types. Of these, some 8000 are common to all the tissues, whereas the majority are specific to particular tissue types. This is quite good evidence for the proposition that transcriptional control is exercised during tissue differentiation. It is of course not entirely convincing, since mRNA is already one stage removed from the DNA of the genes – it is a product of the processing of the first-formed heterogenous nuclear RNA. It is entirely possible therefore that at least some of the specificity in messenger RNA arises during the processing stage. Some evidence for this idea is available. In polysomes isolated from the stem of the tobacco plant, it is found that only 75% of the mRNA sequences are common to those found in polysome preparations of leaves. However, when nuclear RNA is examined in the same tissues, it is found that all the sequences of the leaf mRNA are present in the stem nuclear RNA. This is exactly what would be expected if control were being exercised at the level of nuclear RNA processing.

Translational control has already been implied in the case of *Acetabularia* (p. 221). Here developmental changes may occur in the absence of the nucleus, so that clearly control must be exerted at a later stage than anything which occurs within the nucleus. It is probable also that some of the rapid effects mediated by phytochrome may represent control at the stage of translation or later. For example, in cotyledons of *Sinapis alba*, an increase in the level of the activity of the enzyme nitrate reductase is observable within 5 min of a short burst of red light. This response is not inhibited by cycloheximide; this fact, coupled with the rapidity of the response, suggests that control is being exerted post-translation.

It therefore appears that some evidence can be obtained to support the view that control of gene expression takes place during differentiation of cells, and that this control can be exercised at a variety of levels. It has to be admitted, however, that in no case within higher plants is a molecular description of any of these events possible in detail. Gene regulation presents a striking example of a field of biology in which ideas and possibilities at present far outstrip experimental verification. The best example of this is perhaps the mechanism of action of plant growth substances. Here it is easy to propose that plant growth substances, possibly in cooperation with receptors in the cell membrane, could act as effector molecules to control gene expression. It is only a small conceptual step further to suggest that the receptors could vary both qualitatively and quantitatively within cells of different tissues, to account for the known facts of differential responses and sensitivity to a given stimulus. As we have seen, once the idea of effectors of gene action is accepted, it is another small step in concept to propose a multitude of linked events by which sequences of genes might be activated in precise order, or switched on or off permanently. Nonetheless, such a construction is built on sand: receptors for plant hormones have as yet evaded identification and study; unless and until this situation is altered, the rest of the foregoing description must be totally speculative.

The course of differentiation

Despite the obscurity of its details, there is sufficient evidence already available to conclude that differentiation is indeed accompanied by the sequential appearance of different gene products, controlled at various levels. How much of the general nature of the differentiation process might this explain? Clearly it can account for the temporal sequence of events quite well. It does not matter that developmental steps involve a multiplicity of processes since models of gene activation can accommodate this in terms of the switching on or off of many genes simultaneously. Orderly progress in time can be explained in terms of the products of one sequence of events acting as a stimulus to the next.

However, other aspects of development remain to be considered: in particular, the phenomena of the stability of the differentiated state, and that of determination. As we have seen, the cells within a meristem are in some sense already determined in their eventual behaviour; root meristems do not normally give rise to cells characteristic of leaves or flowers, for example. On the other hand, determination at this level may change during the life of the meristem; for example during the juvenile/adult transition (phase change) in ivy (*Hedera helix*) and woody plants, and indeed during the transition to flowering. These facts imply that overall patterns of genetic activity may be being controlled within different regions of the organism. They also require that differentiated states remain stable through the process of cell division. Many developing cells or cell types undergo division, and this does not lead to a return to some pre-existing, less differentiated state. The development of the stomatal complex may be cited as an illustration of this. It proceeds by division of epidermal cells, themselves differentiated. Three different types of division occur in sequence in cereal stomatal development; none of them has the effect of interrupting the process of development by returning the progeny to a less differentiated state. In considering the effects of cell division in this context, it is as well to remember however that cell division often signals the start of a change in a developmental pathway.

Clearly if differentiation involves the sequential activation or repression of parts of the genome of individual cells, then the stability of the differentiated state through cell division requires that whatever is responsible for activation or repression of genes should itself be replicated along with the DNA and other components of the cell. Ideas concerning the details of this process are once again more freely available than experimental evidence, particularly in higher plants.

It has been proposed that regulation of gene action may be achieved by the cooperative binding of regulatory proteins to the DNA. This is known to occur in the case of the histone H1 – the position at which one molecule of the histone is bound to the DNA becomes a favoured site for other H1 histone molecules to bind. If this were to happen with a regulatory protein, then at DNA replication it can be envisaged that some of the regulatory proteins could be 'inherited' by each of the daughter DNA helices. These would then favour the binding of new proteins at the same site, and thus propagate the regulation through the cycle of DNA replication. Again, there is evidence from mammalian systems that methylation of DNA may correlate with gene inactivation. Methylation of bases within DNA is known to be propagated through DNA replication because of the properties of the enzymes involved; this, too, represents a possible mechanism for the maintenance of a particular state of regulation through cycles of cell division. Finally, in some bacteria, it is known that whole sequences of DNA may undergo re-arrangements which cause the inactivation of genes within such sequences. Again, since subsequent DNA replication copies the rearranged sequence, regulatory states are maintained through cycles of cell division.

It is possible however that in multicellular systems at least, such processes are not the only way in which the differentiated state could be maintained. Within limited regions states of differentiation in dividing cells could be maintained by an effect of adjacent non-dividing cells, mediated by intercellular communication. This is consistent with much of the observed behaviour of meristems, and could even be invoked in a negative way to explain the failure to maintain states of differentiation in isolated cells. This, of course, returns us to the idea of environmental control, since the environment of a cell in a multicellular tissue is precisely its neighbours. It can be justifiably argued that the obvious correlation which exists between the state of differentiation of a cell and its position in a tissue rests equally on this type of environmental modulation as it does on the 'distance' from a supposed starting point of genetic change. Clearly, in practice, these two factors are not exclusive and during normal growth no doubt act in a cooperative manner.

Summary

Development is primarily controlled by the genetic makeup of an organism. In plants the differentiation of specialised cells is a result of the selective expression of parts of the total genome. The factors that decide which genes are active in any particular cell can be identified at many levels. In terms of the whole plant, gross variations in the natural environment may, by some as yet unknown mechanism, cause whole new patterns of behaviour and gene expression to be activated. Within particular tissues or organs, cells exhibit behaviour which depends closely on their position within the tissue and their relationships to other cells. Within the cell itself, gene expression may be influenced by the particular cytoplasmic environment in which the nucleus finds itself.

At the present time there is little detailed knowledge available concerning the mechanisms of any of these effects in higher plants. Evidence suggests that the control of expression of genes may take place at the level of transcription, RNA processing, translation and post-translation. It is quite uncertain as yet how the stability of the differentiated state is maintained, and how whole populations of cells in particular parts of a plant may show determination in their potential behaviour.

Finally, one important point remains to be made. Discussion of genetic regulation attacks only the question of the temporal features of development. It leaves entirely alone the enormous problem of the spatial component of all cellular processes. Growth does not occur in all directions simultaneously at the cell surface; plant growth substances do not move at random throughout the plant body: products of genes do not drift aimlessly about in the cytoplasm. We have no insight at all into why an RNA molecule produced in a particular place in the nucleus should move after processing to the cytoplasm, there to encounter a particle made of more RNA and protein. We have no idea why the product of that encounter should then itself move to a specific site in the cell in order to carry out its function. What is certain is that movements as precise as these occur continuously in all living cells, and not one of them is understood. The present work has been an attempt to point out the complexities of the results of these spatial events. It remains for the future to describe their cause.

FURTHER READING

Chapter 1

Beevers, H. Microbodies in higher plants. *Ann. Rev. Plant Physiol.*, **30**, 159–93 (1979).

Clowes, F. A. L. & Juniper, B. E. *Plant Cells*. Blackwell Scientific Publications (1968).

Cutter, E. *Plant Anatomy*, Part I. *Cells and Tissues*. Edward Arnold (1978).

Esau, K. *Anatomy of Seed Plants*, 2nd edn. John Wiley & Sons (1977).

Franke, W. W., Scheer, U., Krohne, G. & Jarasch, E. The nuclear envelope and the architecture of the nuclear periphery. *J. Cell Biol.*, **91**, 39s–50s (1981).

Gunning, B. E. S. & Hardham, A. R. Microtubules. *Ann. Rev. Plant Physiol.*, **33**, 651–98 (1982).

Gunning, B. E. S. & Steer, M. W. *Ultrastructure and the Biology of Plant Cells*. Edward Arnold (1975).

Hepler, P. K. & Palevitz, B. A. Microtubules and microfilaments. *Ann. Rev. Plant Physiol.*, **25**, 309–62 (1974).

Juniper, B. E. Junctions between plant cells. In: Graham, C. F. & Wareing, P. F. (eds), *The Developmental Biology of Plants and Animals*, pp. 111–26. Blackwell Scientific Publications (1976).

Ledbetter, M. C. & Porter, K. R. *Introduction to the Fine Structure of Plant Cells*. Springer-Verlag (1970).

Lloyd, C. W. (ed.) *The Cytoskeleton in Plant Growth and Development*. Academic Press (1982).

Matile, P. Biochemistry and function of vacuoles. *Ann. Rev. Plant Physiol.*, **29**, 193–213 (1978).

Reinert, J. & Ursprung, H. (eds) *Origin and Continuity of Cell Organelles*. Springer-Verlag. (1971).

Robards, A. W. *Electron Microscopy and Plant Ultrastructure*. McGraw-Hill (1970).

Roberts, K. & Hyams, J. S. *Microtubules*. Academic Press (1979).

Robinson, D. G. & Kristen, U. Membrane flow via the Golgi apparatus of higher plant cells. *Int. Rev. Cytol.*, **77**, 89–127 (1982).

Tolbert, N. E. (ed.) *The Plant Cell*. Vol. I of *The Biochemistry of Plants*, edited by P. K. Stumpf & E. E. Conn. Academic Press (1980).

Whaley, W. G. & Dauwalder, M. The Golgi apparatus, the plasma membrane and functional integration. *Int. Rev. Cytol.*, **58**, 199–245 (1979).

Chapter 2

Bedbrook, K. J. R. & Kolodner, R. The structure of chloroplast DNA. *Ann. Rev. Plant Physiol.*, **30**, 593–620 (1979).

Boardman, N. K. Development of chloroplast structure and function. In: Trebst, A. & Avron, M. (eds), *Encyclopaedia of Plant Physiology*, new series, vol. 5, part I, pp. 583–600. Springer-Verlag (1977).

Bogorad, L. Chloroplasts. *J. Cell Biol.*, **91**, 256s–70s (1981).

Bradbeer, J. W. Chloroplasts – structure and development. In: Smith, H. (ed.), *The Molecular Biology of Plant Cells*, pp. 64–84. Blackwell Scientific Publications (1977).

Craig, I. W. & Gunning, B. E. S. Organelle development. In: Graham, C. F. & Wareing, P. F. (eds), *The Developmental Biology of Plants and Animals*, pp. 270–301. Blackwell Scientific Publications (1976).

Gibbs, M. *Structure and Function of Chloroplasts*. Springer-Verlag (1971).

Jenner, C. F. Storage of starch. In: Loewus, F. A. & Tanner, W. (eds), *Encyclopaedia of Plant Physiology*, new series, vol. 13A, pp. 700–47 Springer-Verlag (1982).

Kirk, J. T. O. & Tilney-Bassett, R. A. E. *The Plastids*, 2nd edn. Freeman & Co. (1975).

Possingham, J. V. Plastid replication and development in the life cycle of higher plants. *Ann. Rev. Plant Physiol.*, **31**, 113–29 (1980).

Reinert, J. (ed.) *Chloroplasts: Results and Problems in Cell Differentiation*, vol. 10. Springer-Verlag (1980).

Sane, P. V. The topography of the thylakoid membrane of the chloroplast. In: Trebst, A. & Avron, M. (eds), *Encyclopaedia of Plant Physiology*, new series. vol. 5, part I, pp. 522–42. Springer-Verlag (1977).

Thomson, W. W. Ultrastructure of mature chloroplasts. In: Robards, A. W. (ed.), *Dynamic Aspects of Plant Ultrastructure*, pp. 138–77. McGraw-Hill (1974).

Thomson, W. W. & Whatley, J. M. Development of nongreen plastids. *Ann. Rev. Plant Physiol.*, **31**, 375–394 (1980).

Chapter 3

Albersheim, P., McNeil, M. & Labavitch, J. M. The molecular structure of the primary cell wall and elongation growth. In: Pilet, P. E. (ed.), *Plant Growth Regulation*, pp. 1–12. Springer-Verlag (1977).

Bauer, W. D. Plant cell walls. In: Smith, H. (ed.), *The Molecular Biology of Plant Cells*, pp. 6–23. Blackwell Scientific Publications (1977).

Brown, R. M. (ed.) *Cellulose and Other Natural Polymer Systems*. Plenum Publishing Corp. (1982).

Chrispeels, M. J. Biosynthesis, intracellular transport and secretion of extracellular macromolecules. *Ann. Rev. Plant Physiol.*, **27**, 19–38 (1976).

Colvin, J. R. Ultrastructure of the plant cell wall; biophysical viewpoint. In: Tanner, W. & Loewus, F. A. (eds), *Encyclopaedia of Plant Physiology*, new series, vol. 13B, pp. 9–24. Springer-Verlag (1981).

Fincher, G. B. & Stone, B. A. Metabolism of non-cellulosic polysaccharides. In: Tanner, W. & Loewus, F. A. (eds). *Encyclopaedia of Plant Physiology*, new series, vol. 13B, pp. 68–132. Springer-Verlag (1981).

Kolattukudy, P. E. Structure, biosynthesis and biodegradation of cutin and suberin. *Ann. Rev. Plant Physiol.*, **32**, 539–67 (1981).

Labavitch, J. M. Cell wall turnover in plant development. *Ann. Rev. Plant Physiol.*, **32**, 385–406 (1981).

Loewus, F. *Biogenesis of Plant Cell Wall Polysaccharides*. Academic Press (1973).

Maclachlan, G. A. Cellulose metabolism and cell growth. In: Pilet, P. E. (ed.), *Plant Growth Regulation*, pp. 13–20. Springer-Verlag (1977).

Masuda, Y. Wall extensibility in relation to auxin effects. In: Pilet, P. E. (ed.), *Plant Growth Regulation*, pp. 21–6. Springer-Verlag (1977).

Muller, S. C., Brown, R. M. & Scott, T. K. Cellulose microfibrils: nascent stages of synthesis in a higher plant cell. *Science*, **194**, 949–51 (1976).

Preston, R. D. *The Physical Biology of Plant Cell Walls*. Chapman & Hall (1974).

Roland, J.-C. & Vian, B. The wall of the growing plant cell: its three-dimensional organisation. *Int. Rev. Cytol.*, **61**, 129–66 (1979).

Willison, J. H. M. & Brown, R. M. A model for the pattern of deposition of microfibrils in the cell wall of *Glaucocystis*. *Planta*, **141**, 51–8 (1978).

Wilson, B. F. & Archer, R. R. Reaction wood: induction and mechanical action. *Ann. Rev. Plant Physiol.*, **28**, 23–43 (1977).

Chapter 4

Buchen, B. & Sievers, A. Sporogenesis and pollen grain formation. In: Kiermayer, O. (ed.), *Cytomorphogenesis in Plants*, pp. 349–76. Springer-Verlag (1981).

Catesson, A. M. Cambial cells. In: Robards, A. W. (ed.), *Dynamic Aspects of Plant Ultrastructure*, pp. 358–90. McGraw-Hill (1974).

Cronshaw, J. Phloem structure and function. *Ann. Rev. Plant Physiol.*, **32**, 365–484 (1981).

Evert, R. F. Phloem structure and histochemistry. *Ann. Rev. Plant Physiol.*, **28**, 199–222 (1977).

Fineran, B. A. Distribution and organisation of non-articulated laticifers in mature tissues of poinsettia (*Euphorbia pulcherrima* Willd.). *Ann. Bot.*, **50**, 207–20 (1982).

Galatis, B. The organisation of microtubules in guard cell mother cells of *Zea mays*. *Can. J. Bot.*, **60**, 1148–66 (1982).

Gunning, B. E. S. & Pate, J. S. Transfer cells. In: Robards, A. W. (ed.) *Dynamic Aspects of Plant Ultrastructure*, pp. 441–80. McGraw-Hill (1974).

Hepler, P. K. Morphogenesis of tracheary elements and guard cells. In: Kiermayer, O. (ed.), *Cytomorphogenesis in Plants*, pp. 327–47. Springer-Verlag (1981).

Roberts, L. W. *Cytodifferentiation in Plants*. Cambridge University Press (1976).

Sexton, R. & Roberts, J. A. Cell biology of abscission. *Ann. Rev. Plant Physiol.*, **33**, 133–162 (1982).

Schnepf, E. Gland cells. In: Robards, A. W. (ed.) *Dynamic Aspects of Plant Ultrastructure*, pp. 331–57. McGraw-Hill (1974).

Shininger, T. L. The control of vascular development. *Ann. Rev. Plant Physiol.*, **30**, 313–37 (1979).

Chapter 5

Fukuda, H. & Komamine, A. Establishment of an experimental system for the study of tracheary element differentiation from single cells isolated from the mesophyll of *Zinnia elegans*. *Plant Physiol.*, **65**, 57–60 (1980).

Letham, D. S., Goodwin, P. B. & Higgins, T. J. V. (eds) *Phytohormones and Related Compounds, A Comprehensive Treatise* (two volumes). Elsevier-North Holland (1978).

Liebermann, M. Biosynthesis and action of ethylene. *Ann. Rev. Plant Physiol.* **30**, 533–91 (1979).

Phillips, R. Cytodifferentiation. *Int. Rev. Cytol., Supp.* 11A 55–70 (1980).

Raven, J. A. & Rubery, P. H. Coordination of development: hormone receptors, hormone action and hormone transport. In: Smith, H. & Grierson, D. (eds), *The Molecular Biology of Plant Development*, pp. 28–48. Blackwell Scientific Publications (1982).

Rubery, P. H. Auxin receptors. *Ann. Rev. Plant Physiol.* **32**, 569–96 (1981).

Sharp, W. R., Larsen, P. O., Paddock, E. F. & Raghavan, V. (eds) *Plant Cell and Tissue Culture: Principles and Applications*. Ohio State University Press.

Sunderland, N. Induction of growth in the culture of pollen. In: Yeoman, M. M. & Truman, D. E. S. (eds), *Differentiation in vitro: British Society for Cell Biology Symposium*, no. 4, pp. 1–24. Cambridge University Press (1982).

Thorpe, T. A. Organogenesis *in vitro*: structural, physiological and biochemical aspects. *Int. Rev. Cytol., Suppl.* 11A 71–111 (1980).

Tran Thanh Van, K. M. Control of morphogenesis in *in vitro* cultures. *Ann. Rev. Plant Physiol.*, **32**, 291–311 (1981).

Walton, D. C. Biochemistry and physiology of abscisic acid. *Ann. Rev. Plant Physiol.*, **31**, 453–89 (1980).

Yeoman, M. M. & Forche, E. Cell proliferation and growth in callus cultures. *Int. Rev. Cytol. Suppl.*, 11A 1–24 (1980).

Chapter 6

Barlow, P. W. Root development. In: Smith, H. & Grierson, D. (eds), *The Molecular Biology of Plant Development*, pp. 185–222. Blackwell Scientific Publications (1982).

Behrens, H. M., Weisenseel, M. H. & Sievers, A. Rapid changes in the pattern of electric current around the root tip of *Lepidium Sativum* L. following gravistimulation. *Plant Physiol*, **70**, 1079–1083 (1982).

Cutter, E. *Plant Anatomy*, part 2, *Organs*. Edward Arnold (1978).

Green, P. B. Organogenesis – a biophysical view. *Ann. Rev. Plant Physiol*, **31**, 51–82 (1980).

Gunning, B. E. S. Microtubules and cytomorphogenesis in a developing organ: the root primordium of *Azolla pinnata*. In: Kiermayer, O. (ed), *Cytomorphogenesis in Plants*, pp. 301–25. Springer-Verlag (1981).

Halperin, W. Organogenesis at the shoot apex. *Ann. Rev. Plant Physiol*, **29**, 239–62 (1978).

Jaffe, L. F. Electrical controls of development. *Ann. Rev. Biophys. Bioeng*, **6**, 445–76 (1977).

Juniper, B. Geotropism. *Ann. Rev. Plant Physiol.*, **27**, 385–406 (1976).

Leopold, A. C. & Kriedemann, P. E. *Plant Growth and Development*, 2nd edn. McGraw-Hill (1975).

Pilet, P. E. (ed). *Plant Growth Regulation*. Springer-Verlag (1977).

Steeves, T. A. & Sussex, I. M. *Patterns in Plant Development*. Prentice-Hall Inc. (1972).

Steward, F. C. *Growth and Organisation in Plants*. Addison-Wesley (1967).

Thimann, K. V. *Hormone Action in the Whole Life of Plants*. University of Massachusetts Press (1977).

Trewavas, A. J. Growth substance sensitivity: the limiting factor in plant development. *Phys. Plant.*, **55**, 60–72 (1982).

Wareing, P. F. & Phillips, I. D. J. *Growth and Differentiation in Plants*, 3rd edn. Pergamon Press (1981).

Weisenseel, M. H. & Kicherer, R. M. Ionic currents as control mechanism in cytomorphogenesis. In: Kiermayer, O. (ed.), *Cytomorphogenesis in Plants*, pp. 379–99. Springer-Verlag (1981).

Wilkins, M. B. (ed.) *Physiology of Plant Growth and Development*. McGraw-Hill (1969).

Chapter 7

Bloch, R. Polarity and gradients in plants: a survey. In: Ruhland, W. (ed.), *Handbuch der Pflanzenphysiologie*, XV/1, pp. 146–88. Springer-Verlag (1965).

Dyer, A. F. *The Experimental Biology of Ferns*. Academic Press (1979).

Goldsmith, M. H. M. The polar transport of auxin. *Ann. Rev. Plant Physiol*, **28**, 439–78 (1977).

Jacobs, W. P. & Olson, J. Developmental changes in the algal coenocyte *Caulerpa prolifera* (Siphonales) after inversion with respect to gravity. *Amer. J. Bot.*, **67**, 141–6 (1980).

Jaffe, L. F. Localization in the developing *Fucus* egg and the general role of localizing currents. *Adv. Morph.*, **7**, 295–328 (1978).

Novotny, A. M. & Forman, M. The relationship between changes in cell wall composition and the establishment of polarity in *Fucus* embryos. *Dev. Biol.*, **40**, 162–73 (1974).

Quatrano, R. S. Development of cell polarity. *Ann. Rev. Plant Physiol.*, **29**, 487–510 (1978).

Rubery, P. H. & Sheldrake, A. R. Effect of pH and surface charge on cell uptake of auxin. *Nat. New Biol.* **244**, 285–8 (1973).

Sachs, T. The control of the patterned differentiation of vascular tissues. *Adv. Bot. Res.*, **9**, 151–62 (1981).

Schmiedel, G. & Schnepf, E. Side branch formation and orientation in the caulonema of the moss, *Funaria hygrometrica*:

experiments with inhibitors and with centrifugation. *Protoplasma*, **101**, 47–59 (1979).

Schmiedel, G. & Schnepf, E. Polarity and growth of caulonema tip cells of the moss, *Funaria hygrometrica. Planta*, **147**, 405–13 (1980).

Sheldrake, A. R. The polarity of auxin transport in inverted cuttings. *New Phytol.*, **73**, 637–42 (1974).

Sinnott, E. W. *Plant Morphogenesis*. McGraw-Hill (1960).

Torrey, J. G. & Galun, E. Apolar embryos of *Fucus* resulting from osmotic and chemical treatment. *Amer. J. Bot.*, **57**, 111–19 (1970).

Chapter 8

Alberts, B., Bray, D., Lewis, J., Raff, M., Roberts, K. & Watson, J. D. *Molecular Biology of the Cell*. Garland Publishing (1982).

Davidson, R. J. & Britten, E. H. Regulation of gene expression: possible role of repetitive sequences. *Science*, **204**, 1052–9 (1979).

Ford, P. J. Control of gene expression during differentiation and development. In: Graham, C. F. & Wareing, P. F. (eds), *The Developmental Biology of Plants and Animals*. Blackwell Scientific Publications (1976).

Gall, J. G. Chromosome structure and the C-value paradox. *J. Cell Biol.*, **91**, 3s–14s (1981).

Grierson, D. The nucleus and the organisation and transcription of nuclear DNA. In: Smith, H. (ed.), The Molecular Biology of Plant Cells. pp. 213–55. Blackwell Scientific Publications (1977).

Hancock, R. & Boulikas, T. Functional Organisation in the Nucleus. *Int. Rev. Cytol.*, **79**, 165–213 (1979).

Heslop-Harrison, J. Differentiation. *Ann. Rev. Plant Physiol.*, **18**, 325–48 (1967).

Johnson, C. B. Photomorphogenesis. In: Smith, H. & Grierson, D. (eds), *The Molecular Biology of Plant Development*, pp. 365–404. Blackwell Scientific Publications (1982).

Kloppsteck, K. *Acetabularia*. In: Smith, H. & Grierson, D., *The Molecular Biology of Plant Development*, pp. 136–58. Blackwell Scientific Publications (1982).

Nagl, W. Nuclear Organisation. *Ann. Rev. Plant Physiol.*, **27**, 39–69 (1976).

Schweiger, H. G. & Berger, S. Pattern Formation in *Acetabularia*. In: Kiermayer, O. (ed.), *Cytomorphogenesis in Plants*, pp. 119–45. Springer-Verlag (1981).

Taylorson, R. B. & Hendricks, S. B. Dormancy in seeds. *Ann. Rev. Plant Phsyiol.*, **28**, 331–54 (1977).

Trewavas, A. J. Possible control points in plant development. In: Smith, H. & Grierson, D. (eds), *The Molecular Biology of Plant Development*, pp. 7–27. Blackwell Scientific Publications (1982).

Verbelen, J.-P., Pratt, L. H., Butler, W. L. & Tokuyasu, K. Localisation of phytochrome in oats by electron microscopy. *Plant Physiol.*, **70**, S4–S8 (1982).

Vince-Prue, D. *Photoperiodism in Plants*. McGraw-Hill (1975).

Wareing, P. F. Determination and related aspects of plant development. In: Smith, H. & Grierson, D. (eds), *The Molecular Biology of Plant Development*, pp. 517–41. Blackwell Scientific Publications (1982).

GLOSSARY

acropetal
> Occurring towards the apex of an organ. Thus, acropetal transport refers to the movement of a substance from the base of an organ to its tip.

angiosperm
> A plant whose seed is carried inside a matured ovary (a fruit).

anticlinal
> Perpendicular to the nearest surface; commonly used to describe the plane of cell division or the orientation of a cell wall.

apoplasm
> The continuous space within a plant body which is outside each living protoplast.

autoradiography
> A microscopical technique whereby the position of radioactive tracers within a specimen is determined by placing a layer of photographic emulsion over a histological section of the specimen and allowing it to become exposed by the radioactivity. Subsequent development of the emulsion leaves a pattern of silver grains corresponding to the position of the tracer in the section. May be used both at the light and electron microscope level.

basipetal
> Occurring towards the base of an organ (the opposite of acropetal).

bioassay
> The detection of the presence of a chemical activity by use of a living test system. For example, the detection of auxin activity by the curvature induced in an oat coleoptile. Bioassays may be very sensitive but do not discriminate between effects caused by different molecular species or by mixtures of promoting and inhibitory substances.

bryophyte
> A member of the phylum comprising the mosses, liverworts and hornworts.

calyptrogen

A meristem in the root apex which gives rise to the root cap independently of the rest of the root.

cambium

Commonly used unqualified to mean the vascular cambium. This is the lateral meristem which forms the secondary xylem and secondary phloem in a root or a stem. The cork cambium (phellogen) is a meristem which produces the periderm, a protective layer in many stems and roots.

caulonema

The secondary form of the protonema of a moss. Consists of filaments of cells characterised by oblique cross walls, few chloroplasts and the ability to produce buds.

chimaera

A plant which consists of a combination of tissues of different genetic composition (applied equally to differences in nuclear or plastid genomes).

chloronema

The juvenile form of the protonema of a moss. Consists of filaments of cells characterised by perpendicular cross walls and many chloroplasts.

coenocyte

A cell which contains many nuclei which are not separated by partitioning walls. It may exist as a transient phase in development (embryogenesis in gymnosperms) or throughout a stage in the life cycle (vegetative phase of *Caulerpa*).

collenchyma

A supportive tissue whose cells are characterised by the presence of unevenly thickened non-lignified primary walls.

companion cell

A cell in the phloem of angiosperms which is associated with a sieve element. The companion cell and the sieve element develop from the division products of the same mother cell.

determinate

Growth which is characterised by the formation of a restricted number of lateral organs (from an apical meristem). Thus the growth of a flower is determinate, whereas the growth of a bud is not, since it may produce an indefinite number of leaves and axillary buds.

dicotyledon

Any plant of the class Magnoliopsida (all having two cotyledons).

distal

Farthest from the point of attachment or origin (opposite of proximal).

epicotyl

The shoot part of the embryo which is above cotyledon(s) and consists of an axis with leaf primordia.

eukaryote

An organism which has a membrane-bound nucleus, genetic organisation into chromosomes and membrane-bound cytoplasmic organelles.

freeze etch technique

An electron microscope technique in which the specimen is frozen and shattered *in vacuo*. The frozen surfaces thus exposed are coated with a layer of metal; this metal replica is used as the specimen in the electron microscope.

frond

The leaf of a palm or of a fern; otherwise a leaf-like thalloid shoot.

gametophyte

A plant of the haploid generation which produces gametes.

guard cell

One of a pair of cells which flank the opening of a stomatal pore and control the size of the pore. Formed by division of the guard cell mother cell.

gymnosperm

A plant which has mature ovules at the time of pollination and seeds which are not enclosed in an ovary.

indeterminate

Growth of an unrestricted nature, as in the growth of a vegetative bud (the opposite of determinate).

meristem

A tissue primarily concerned with the formation of new cells by division.

mesophyll

The photosynthetic parenchyma of the leaf blade which is situated between the two epidermal layers. Spongy mesophyll has large intercellular air spaces; Palisade mesophyll consists of elongated cells arranged with their long axis perpendicular to the leaf surface.

middle lamella

The layer which marks the position of the first formed cell wall between contiguous cells.

monocotyledon

Any plant of the class Liliopsida (all having a single cotyledon).

parenchyma

A tissue consisting of parenchyma cells. These are characterised as not distinctly specialised, nucleate, often vacuolate. May be qualified by location (cortical parenchyma, phloem parenchyma) or function (photosynthetic parenchyma, storage parenchyma).

pedicel

The stalk of an individual flower; as opposed to peduncle, the stem of an inflorescence of many flowers.

periclinal

Parallel to the nearest surface – commonly used to denote planes of cell division or orientation of cell walls (the opposite of anticlinal).

pericycle

Part of the ground tissue of the stele situated between

the phloem and the endodermis. In roots, the originating tissue for lateral roots.

ploidy
The number of complete chromosome sets in a nucleus.

polarity
Applied to an axis, polarity is the property of having unequal ends to an axis.

procambium
Primary meristem which differentiates into primary vascular tissue. 'Cambium' (properly, vascular cambium) differentiates into secondary vascular tissue.

prokaryote
An organism which does not have its genetic material organised into chromosomes within a membrane-bound nucleus, and which does not have membrane-bound organelles.

prothallus
The stage of growth on the gametophyte of a Pteridophyte which takes the form of a flat green thallus with rhizoids.

protoderm
Primary meristem which gives rise to the epidermis.

protoplast
The entire living contents enclosed by (but not including) the cell wall. Also used unqualified to mean the isolated unit released by dissolving away the cell wall — more properly called an isolated protoplast.

radicle
The embryonic root.

rhizoid
A root-like structure which anchors a plant to its substrate (found in fungi, mosses, liverworts and ferns).

rhizome
A thick horizontal stem, which often forms buds on its upper surface and roots on its lower surface.

schlerenchyma
A tissue composed of schlerenchyma cells, which are characteristically lignified, and act as supporting elements. They may or may not be devoid of a protoplast at maturity.

subsidiary cell
An epidermal cell which is associated with a stoma, and is morphologically distinguishable from an epidermal cell.

suspensor
The extension at the base of the embryo which anchors the embryo within the embryo sac.

symplasm
The continuous space within the plant body which is inside the plasma membrane.

tracheary element
A general term for a water conducting cell, whether a tracheid or vessel member.

trichome
An outgrowth from the epidermis (e.g. leaf hairs and surface glands).

Index